LikeWar

Named a Best Book of 2018 by *Foreign Affairs*
An Amazon Best Book of 2018

"*LikeWar* should be required reading for everyone living in a democracy and all who aspire to." — *Booklist* (starred review)

"Backed by over 100 pages of notes, *LikeWar* is sober, deeply researched, and still compulsively readable. Comparisons to *On War* and *The Art of War* are apt." — Amazon Best Book of the Month selection

"Fascinating." — CBS News

"*LikeWar* is an engaging and startling work." — *Vice*

"Excellent . . . a deeply researched page-turner." — *Foreign Policy*

"It's extremely timely and fascinating — reading it in bed with my husband, I kept pulling him away from his book to show him passages."
— Michelle Goldberg, *New York Times*

"*LikeWar* is a magical combination of history, technology, and early warning wrapped in a compelling narrative of how today's information space can threaten the truth, our polity, and our security. It's a page-turner, too, chock full of deep insights and fascinating detail. Sun Tzu tells us to know ourselves, our enemy, a̶n̶ ̶b̶a̶t̶t̶l̶e space, and *LikeWar* delivers on all three."
— G̶e̶n̶e̶r̶a̶l̶ ̶M̶i̶c̶h̶a̶e̶l̶ ̶H̶ayden, former director of the CIA and NSA, author of *The Assault on Intelligence*

". . . o̶u̶r social intuitions about its power. In . . . i̶n̶g offer insight into the ways that social media . . . m̶a̶n̶i̶pulate beliefs and attitudes for self-serving purposes."
— Vint Cerf, coinventor of the internet, recipient of the Presidential Medal of Freedom

"Much as Clausewitz did for conventional war, *LikeWar* lays out the new twenty-first-century principles of war. Mixing fascinating stories and the front edge of research, it explains the twilight battlegrounds of politics and war on social media — a frightening future where truth is the first casualty and our fundamental values are deeply in danger. I loved it."

— Admiral James Stavridis, US Navy (ret.),
former supreme allied commander, NATO

"Reading *LikeWar* will help you to avoid being part of this Internet of Idiots . . . While students of history, strategic studies, political science, and international relations will all find *LikeWar* on their required reading list, anyone else who wishes to understand the world we live in must add *LikeWar* to the top of the pile on their nightstand."   — *Forbes*

"The picture Singer and Booking paint of how social media is being weaponized is compelling, and one that ought to give pause to any practitioner in the field of national security. I am reluctant to be so effusive in my praise, but this is truly a must-read book."   — *Lawfare*

"Essential reading for anyone interested in how social media has become an important new battlefield in a diverse set of domains and settings."

— O'Reilly Media

"A compelling read . . . *LikeWar* . . . is not a warning about tomorrow's war — it's a map for those who don't understand how the battlefield has already changed."   — *Washington Post*

"*LikeWar: The Weaponization of Social Media* is a book that anyone active in social media should read. That is, everyone."   — *Irish Tech News*

"The stories Singer and Brooking tell and the lessons they teach are fundamental for military leaders at the tactical, operational, and strategic levels of war . . . A blueprint on how to think, operate and survive in this operational environment."   — *Army Magazine*

"Consider *LikeWar* your must-read book of 2019. Informative, insightful, yet also cautionary."   — *Chicago Now*

"Essential reading if today's leaders (both in and out of uniform) are to understand, defend against, and ultimately wield the nonkinetic yet violently manipulative effects of social media."

— US Army Training and Doctrine Command

"A great read . . . A must-read for the intelligence community . . . A compelling story of the opportunities and risks of social media that every intelligence analyst must appreciate, as publicly available information will increasingly dominate how we read or misread the operational and strategic environment."

— Lieutenant General Robert Ashley, US Army,
director of the Defense Intelligence Agency

"*LikeWar* is an eye-opening literary experience. Most of us access social media in some form on a near-daily basis, but do we really understand the phenomenon?"

— Modern War Institute at West Point

"Although the book is titled *LikeWar*, it isn't so much about warfare as about how social media is affecting society broadly: how we consume information, why social media is so addictive, how it has been capitalized on by social movements, celebrities, politicians, terrorists, and states. It's worth reading for the history of the internet alone, which bounces along as vignettes about individuals that personalize the story (they clearly apply the elements of effective social media they identify: narrative, emotion, authenticity, community, and inundation) . . . A valuable primer on where social media came from and how it's currently being used. It also has some useful suggestions for taming its effects."

— *War on the Rocks*

"My films have specialized in realistic horror. *LikeWar* is scary as hell, as it shows how people can be manipulated online to make our worst fears come true."

— Jason Blum, producer of *The Purge* and *Get Out*

"Through a series of vivid vignettes, *LikeWar* shows how the internet has become a new battlefield in the twenty-first century, in ways that blur the line between war and peace and make each of us a potential target of postmodern conflict."

— Francis Fukuyama, author of *The End of History*,
director of the Center on Democracy, Development, and
the Rule of Law at Stanford University

LikeWar

LikeWar

The Weaponization
of Social Media

P. W. Singer and
Emerson T. Brooking

Mariner Books
An Imprint of HarperCollins*Publishers*
Boston   New York

First Mariner Books edition 2019
Copyright © 2018 by P. W. Singer and Emerson T. Brooking

Mariner Books
An Imprint of HarperCollins Publishers, registered in the United States of America and/or other jurisdictions.

www.marinerbooks.com

Library of Congress Cataloging-in-Publication Data is available.
Names: Singer, P. W. (Peter Warren), author. | Brooking, Emerson T., author.
Title: Likewar : the weaponization of social media /
P.W. Singer and Emerson T. Brooking.
Other titles: Like war
Description: Boston : Eamon Dolan/Houghton Mifflin Harcourt, 2018. |
Includes bibliographical references and index.
Identifiers: LCCN 2018017519 (print) | LCCN 2018035802 (ebook) |
ISBN 9781328695758 (ebook) | ISBN 9781328695741 (hardback) |
ISBN 9780358108474 (pbk.)
Subjects: LCSH: Social media—Political aspects. | Internet—Political aspects. |
Mass media and propaganda. | Cyberterrorism. | Hacking—Political aspects. |
BISAC: POLITICAL SCIENCE / Government / International. | SOCIAL SCIENCE /
Future Studies. | TECHNOLOGY & ENGINEERING / General.
Classification: LCC HM742 (ebook) | LCC HM742 .S5745 2018 (print) |
DDC 302.23/1—dc23
LC record available at https://lccn.loc.gov/2018017519

Book design by Kelly Dubeau Smydra
Printed in the United States of America
ScoutAutomatedPrintCode

Chapter 8 epigraph quote is from the film *Sneakers,* directed by Phil Alden Robinson. Universal Pictures, 1992. Used by permission of Walter F. Parkes.

Title page illustration copyright © by Doan Trang

22 23 24 25 26 LBC 10 9
8 7 6

It sounds a dreadful thing to say, but these are things that don't necessarily need to be true as long as they're believed.

— ALEXANDER NIX

Contents

LikeWar

1

The War Begins

An Introduction to LikeWar

> It was an extraordinary life that we were living —
> an extraordinary way to be at war, if you could call it war.
>
> — GEORGE ORWELL, *Homage to Catalonia*

THE OPENING SHOT OF THE WAR was fired on May 4, 2009. By all appearances, it had nothing to do with war.

"Be sure to tune in and watch Donald Trump on Late Night with David Letterman as he presents the Top Ten List tonight!"

When @realDonaldTrump blasted his first bland tweet into the ether, there was little to distinguish the account from the horde of other brands, corporations, and celebrities who had also joined "social media." This constellation of emerging internet services, where users could create and share their own content across a network of self-selected contacts, was a place for lighthearted banter and personal connections, for oversharing and pontification, for humblebrags and advertising. That the inveterate salesman Donald John Trump would turn to it was not surprising.

Yet beneath the inanity, Twitter, Facebook, YouTube, and the like were hurtling toward a crossroads — one that would soon see them thrust into the center of civic life and global politics. Just a few years earlier, Twitter had begun as a way for groups of friends to share their "status" via text message updates. Now with 18 million users spread around the world, the startup was on the brink of a revolutionary success. But it would be driven by a different celebrity. A few weeks after Trump's first tweet, superstar entertainer Michael Jackson died. His passing convulsed the internet in grief. Pop music's irreplaceable loss, however, proved Twitter's gain. Millions of people turned to the social network to mourn, reflect, and speculate. The platform's traffic surged to a record 100,000 tweets per hour before its servers crashed. People were using social media for something new, to *experience* the news together online.

Trump was also at a crossroads. The 63-year-old real estate magnate had just suffered his fourth bankruptcy when Trump Entertainment Resorts (the holding company for his casinos, hotels, and Trump Marina) collapsed under a $1.2 billion debt and banished him from the executive board. Although he had successfully rebranded himself as a reality-television host, that shine was starting to wear off. *The Apprentice* had fallen from its early prime-time heights to the 75th most watched show before being put on hiatus. The celebrity spin-off that Trump was promoting was still on the air, but its ratings were plummeting. His appearance on *Letterman* was an attempt to stanch the bleeding. It wouldn't work. In *The Celebrity Apprentice*'s season finale, just six days after Trump's first tweet, more Americans would elect to watch *Desperate Housewives* and *Cold Case.*

But beneath his shock-blond dome, Trump's showman's brain was already moving on to the Next Big Thing.

The transformation played out slowly, at least for the internet. Trump's initial online messaging was sporadic, coming once every few days. In the first years of life, @realDonaldTrump was obviously penned by Trump's staff, much of it written in the third person. The feed was mostly announcements of upcoming TV appearances, marketing pitches for Trump-branded products like vitamins and key chains, and uninspired inspiring quotes ("Don't be afraid of being unique — it's like being afraid of your best self").

But in 2011, something changed. The volume of Trump's Twitter messages quintupled; the next year, it quintupled again. More were written in the first person, and, most important, their tone shifted. This @realDonaldTrump was *real*. The account was also real combative, picking online fights regularly — comedian Rosie O'Donnell was a favorite punching bag — and sharpening the language that would become Trump's mainstay. His use of "Sad!," "Loser!," "Weak!," and "Dumb!" soon reached into the hundreds of occurrences. Back then, it still seemed novel and a little unseemly for a prominent businessman to barrel into online feuds like an angst-ridden teenager. But Trump's "flame wars" succeeded at what mattered most: drawing attention.

As the feed became more personal, it became more political. Trump issued screeds about trade, China, Iran, and even Kwanzaa. And he turned President Barack Obama, whom he'd praised as a "champion" just a few years earlier, into the most prominent of his celebrity targets, launching hundreds of bombastic attacks. Soon the real estate developer turned playboy turned reality show entertainer transformed again, this time into a right-wing political power. Here was a voice with the audacity to say what needed to be said, all the better if it was "politically incorrect." Not coincidentally, Trump began to use the feed to flirt with running for office, directing his Twitter followers to a new website (created by his lawyer Michael Cohen). ShouldTrumpRun.com, it asked.

The technology gave Trump immediate feedback, both validation that he was onto something and a kind of instant focus-testing that helped him hone and double down on any particularly resonant messages. Resurrecting an old internet conspiracy, Trump attacked not just Obama's policies but his very eligibility to serve. ("Let's take a closer look at that birth certificate.") The online reaction spiked. Together, Trump and Twitter were steering politics into uncharted territory.

Through social media, Trump was both learning how the game was played online and creating new rules for politics beyond it. All those over-the-top tweets didn't just win fans. They also stoked an endless cycle of attention and outrage that both kept Trump in the spotlight and literally made him crave more and more.

The engineers behind social media had specifically designed their platforms to be addictive. The brain fires off tiny bursts of dopamine as a user posts a message and it receives reactions from others, trapping the brain in a cycle of posts, "likes," retweets, and "shares." Like so many of us, Donald Trump became hooked on social media. In the three years that followed, he would personally author some 15,000 tweets, famously at all hours of the day and night.

Exactly 2,819 days after his first tweet, @realDonaldTrump would broadcast a vastly different announcement to an incomprehensibly different world. It was a world in which nine-tenths of Americans now had social media accounts and Twitter alone boasted 300 million active users. It was a world shaped by online virality and "alternative facts." And it was one in which the same account that had once informed hundreds of readers that "everybody is raving about Trump Home Mattress" now proclaimed to hundreds of millions, "I am honered [sic] to serve you, the great American people, as your 45th President of the United States!"

Although our story begins here, this is not a book about the Trump presidency. Instead, this is a book about how a new kind of communications became a new kind of war. Trump's quest to rebrand himself and then win the White House wasn't just a marketing or political campaign; it was also a globe-spanning information conflict, fought by hundreds of millions of people across dozens of social media platforms, none of which had existed just a generation earlier. Not just the battlespace was novel, but the weapons and tactics were, too. When Trump leveled his first digital barbs at Rosie O'Donnell, he was pioneering the same tools of influence that he would use to win the presidency — and to reshape geopolitics soon thereafter.

Nor was Trump alone. As his battle for attention and then election was taking place, thousands of others were launching their own battles on social media. The participants ranged from politicians and celebrities to soldiers, criminals, and terrorists. Conflicts of popularity and perception began to merge with conflicts of flesh and blood. As the stakes of these online struggles increased, they began to look and feel like war. Soon enough, they would *become* war.

WAR GOES VIRAL

The invasion was launched with a hashtag.

In the summer of 2014, fighters of the self-declared Islamic State (also known as ISIS or *Daesh* in Arabic) roared into northern Iraq, armed with AK-47 rifles, grenades, and even swords. Their dusty pickup trucks advanced quickly across the desert. Far from keeping their operation a secret, though, these fighters made sure everyone knew about it. There was a choreographed social media campaign to promote it, organized by die-hard fans and amplified by an army of Twitter bots. They posted selfies of black-clad militants and Instagram images of convoys that looked like *Mad Max* come to life. There was even a smartphone app,

created so that jihadi fans following along at home could link their social media accounts in solidarity, boosting the invaders' messages even
further. To maximize the chances that the internet's own algorithms
would propel it to virality, the effort was organized under one telling
hashtag: #AllEyesOnISIS.

Soon #AllEyesOnISIS had achieved its online goal. It became the
top-trending hashtag on Arabic Twitter, filling the screens of millions
of users — including the defenders and residents of cities in the Islamic
State's sights. The militants' demands for swift surrender thus spread
both regionally and personally, playing on the phones in their targets'
hands. ISIS videos also showed the gruesome torture and execution of
those who dared resist. And then it achieved its real-world goal: #All
EyesOnISIS took on the power of an invisible artillery bombardment,
its thousands of messages spiraling out in front of the advancing force.
Their detonation would sow terror, disunion, and defection.

In some ways, Iraq had changed dramatically in the years since the
2003 U.S.-led invasion. Where dictator Saddam Hussein had once
banned mobile phones because ease of communication threatened his
grip on power, three-quarters of Iraqis now owned one. The 150,000
Iraqis online in 2003 had grown to nearly 4 million. Phone-obsessed
and internet-savvy, Iraqi teenagers weren't all that different from their
American counterparts.

But in other ways, Iraq hadn't changed enough. The bloody sectarian war between the Shia majority and the dispossessed Sunni minority — a conflict that had claimed the lives of over 200,000 Iraqi civilians and 4,500 U.S. soldiers — still simmered. Especially in the west
and north, where most Sunnis lived, the army was undertrained and
often unpaid. Soldiers and police barely trusted each other. Sunni civilians trusted both groups even less. As it laid the groundwork for invasion, ISIS didn't have to look far for willing spies and insurgents, recruited via online forum boards and coordinated via the messaging
service WhatsApp.

The prized target for ISIS was Mosul, a 3,000-year-old multicultural
metropolis of 1.8 million. As the ISIS vanguard approached and #All
EyesOnISIS went viral, the city was consumed with fear. Sunni, Shia,
and Kurdish neighbors eyed each other with suspicion. Were these
high-definition beheadings and executions real? Would the same things

happen here? Then young Sunni men, inspired by the images of the indomitable black horde, threw themselves into acts of terror, doing the invaders' work for them.

The Iraqi army stood ready to protect the city from this tiny but fearsome horde — in theory, at least. In reality, most of Mosul's 25,000-strong garrison existed only on paper, either having long since deserted or been invented by corrupt officers eager to fatten their paychecks. Worse, the roughly 10,000 who actually did exist were able to track the invading army's highly publicized advance and atrocities on their smartphones. With #AllEyesOnISIS, soldiers began to ask each other whether they should fight or flee. The enemy hadn't even arrived, but fear already ruled the ranks.

Defenders began to slip away, and then the trickle became a flood. Thousands of soldiers streamed from the city, many leaving their weapons and vehicles behind. Most of the city's police followed. Among Mosul's citizens, the same swirling rumors drove mass panic. Nearly half a million civilians fled. When the invading force of 1,500 ISIS fighters finally reached the city's outskirts, they were astounded by their good fortune. Only a handful of brave (or confused) soldiers and police remained behind. They were easily overwhelmed. It wasn't a battle but a massacre, dutifully filmed and edited for the next cycle of easy online distribution.

ISIS militants gleefully posted pictures of the arsenal they had captured, mountains of guns and ammunition, and thousands of American-made, state-of-the-art vehicles that ranged from Humvees to M1A1 Abrams battle tanks to a half dozen Black Hawk helicopters. They staged gaudy parades to celebrate their unlikely triumph. Those so inclined could follow these events in real time, flipping between the posts of ISIS fighters marching in the streets and those watching them march. Each point of view was different, but all promised the same: more — much more — to come.

How had it gone so wrong? This was the question that haunted Iraqi officials ensconced in the capital, U.S. military officers now working marathon shifts in the Pentagon, and the hundreds of thousands of refugees forced to abandon their homes. It wasn't just that entire cities had been lost to a ragtag army of millennials, but that four whole Iraqi army divisions — trained and armed by the most powerful nation in the world — had essentially evaporated into thin air.

In the surprising loss of Mosul and collapse of the defending Iraqi forces, though, a student of history could detect echoes of another strange defeat. In 1940, amid the opening stages of World War II, France had seemed unassailable. The nation boasted an army of 5 million soldiers, equipped with modern tanks and artillery. Its Maginot Line, 60 massive fortresses stretched over 900 miles, loomed as the mightiest defensive fortification in the world. French generals had spent twenty years studying the last war with Germany, drawing up precise new battle plans. As 2.5 million Nazi soldiers amassed at the border, French commanders thought they were ready.

They weren't.

France would fall in less than two months. German tanks tore through forests the French had thought impassable, making the vaunted Maginot Line irrelevant. The German forces then moved faster than the French generals could think. Commanders received belated orders to halt German units that had already blown past them, gone around them, or simply weren't there. When French armies retreated, they had no time to establish a new defensive line before the Germans were already upon them, forcing further retreat.

The true power of the German blitzkrieg was speed: a pace of advance so relentless that French defenders were consumed with an unease that turned swiftly to panic. The weapon that made all this possible was the humble radio. Radio allowed armored formations to move in swift harmony. Radio spread reports of their attacks — sometimes real, sometimes not — which spread confusion across the entire French army. Radio also let the Germans bombard the French civilian leaders and populace with an endless stream of propaganda, sowing fear and doubt among what soon became a captive audience.

Marc Bloch, a French historian and soldier who would ultimately meet his death at the hands of a Nazi firing squad, recorded his memories of the French rout almost as soon as it happened. His recollections survive in a book aptly titled *Strange Defeat*. Bloch described the fear that swept through the French ranks. Soldiers were given continuous orders to fall back, while French fire brigades clogged the roads as they preemptively abandoned their towns to burn. "Many instructions to evacuate were issued before they need have been," he recalled. "A sort of frenzy of flight swept over the whole country."

Where the Germans had harnessed radio and armored vehicles, ISIS pioneered a different sort of blitzkrieg, one that used the internet itself as a weapon. The same Toyota pickup trucks and secondhand weapons of countless guerrilla groups past had taken on a new power when combined with the right Instagram filter, especially when shared hundreds of thousands of times by adoring fans and automated accounts that mimicked them. With careful editing, an indecisive firefight could be recast as a heroic battlefield victory. A few countering voices might claim otherwise, but how could they prove it? These videos and images moved faster than the truth. Their mix of religiosity and ultraviolence was horrifying to many; to some, however, it was intoxicating.

Of course, Iraqis weren't the only ones who watched the Islamic State's relentless advance. Anyone anywhere in the world with an internet connection could track each agonizing twist and turn of the conflict, using Google Translate to fill in the gaps. Observers could swoop from official Iraqi news sources to the (usually more interesting) social media feeds of the jihadists themselves. You could check the war like you checked the @ESPN Twitter feed. If you were so inclined, you could message with the people fighting it. Sometimes, they'd talk back. Even ISIS militants were addicted to the feedback loop that social media provided.

It was a cruel, surreal spectacle. To us, two internet junkies and defense analysts, it also sounded an alarm bell. Many articles and books had been written on "cybersecurity" and "cyberwar" (including by one of us) — raising the specter of hackers breaking into computers and implanting malicious lines of software code. When the next war came, we'd often been told, it would be a techno-nightmare marked by crashing networks, the disruption of financial markets, and electrical outages. It would show the "true" power of the internet in action.

But the abrupt fall of Mosul showed that there was another side to computerized war. The Islamic State, which had no real cyberwar capabilities to speak of, had just run a military offensive like a viral marketing campaign and won a victory that shouldn't have been possible. It hadn't hacked the network; it had hacked the information on it.

In the months that followed, ISIS's improbable momentum contin-ued. The group recruited over 30,000 foreigners from nearly a hundred countries to join the fight in its self-declared "caliphate." The export of its message proved equally successful. Like a demonic McDonald's, ISIS opened more than a dozen new franchises, everywhere from Libya and Afghanistan to Nigeria and Bangladesh. Where franchises were not possible, ISIS propaganda spurred "lone wolves" to strike, inspir-ing scores of terrorist attacks from Paris and Sydney to Orlando and San Bernardino. And that same contagion of fear spread wider than ever be-fore. Polling showed Americans were suddenly more frightened of ter-rorism than they'd been in the immediate aftermath of 9/11. All thanks, essentially, to the fact that ISIS was very good at social media.

ISIS was just the leading edge of a broader, globe-spanning phenom-enon. The technology it was using — rather than any unique genius on the part of the jihadists — lay at the heart of its disruptive power and outsize success. And it was a technology available to everyone. Others could do the same thing. Indeed, they already were.

In the Syrian civil war where ISIS first roared to prominence, nearly every rebel group used YouTube to recruit, fundraise, and train. In turn, the regime of Syrian president Bashar al-Assad used Instagram to project a friendly face to the world, while it gassed its own citizens. When Russian forces annexed Crimea and chomped away at east-ern Ukraine, the Russians made their initial forays online, foment-ing unrest. During the battles that followed, opposing soldiers trolled each other's social media pages. So, too, the Israeli Defense Forces and Hamas militants fought multiple "Twitter wars" before a global audi-ence. The IDF took this fight, and how it influenced world opinion, so seriously that the volume of "likes" and retweets influenced the tar-gets it chose and its pace of operations on the ground. In Afghanistan, NATO and the Taliban had taken to sniping at each other's Twitter feeds, mixing mockery with battle footage. Everywhere, armed groups and governments had begun generating information operations and war propaganda that lived alongside the internet's infinite supply of silly memes and cat videos.

It all represented a momentous development in the history of con-flict. Just as the modern internet had "disrupted" the worlds of enter-tainment, business, and dating, it was now disrupting war and politics.

It was a revolution that no leader, group, army, or nation could afford to ignore.

How much the novel had become normal was evidenced when a reconstituted Iraqi army swept back into Mosul in 2016, two years after #AllEyesOnISIS had chased it away. This time it came equipped for the new battlefield that extended far beyond Mosul's battered streets. Eighteen-wheel trucks lumbered after tanks and armored personnel carriers, dragging portable cellphone towers to ensure bandwidth for its own messaging. The Iraqi military issued a rapid-fire stream of Facebook, YouTube, and Twitter updates both practical (the status of the operation) and bizarre (grinning selfies of Iraqi soldiers as they detonated leftover ISIS suicide-bomb trucks). Naturally, the operation had its own hashtag: #FreeMosul.

The Iraqis' U.S. military allies also threw themselves into this new fight. Just as U.S. forces coordinated air strikes and targeting data for the Iraqi army, they also sought to shape the flow of online conversation in Iraq and beyond. For months, U.S. special operators and information warfare officers had trained for the assault by practicing "cognitive maneuvering" against pretend ISIS propagandists. Now they pushed out message after message that reflected what they had learned. Meanwhile, hundreds of contractors in the employ of the U.S. State Department stalked the conversations of potential ISIS recruits, reminding them of ISIS's barbarity and its impending defeat.

Because the Islamic State was also online, the result could be surreal, almost circular moments. At one point, the Iraqi army proudly announced on Facebook that it had shot down a drone used by ISIS to film battles to put on Facebook. It also meant that combat could be followed live, now from both sides of the front lines. You could "like" whichever version you preferred, your clicks enlisted in the fight to determine whose version got more views.

The physical and digital battlefields could drift eerily close together. The Kurdish news network Rudaw didn't just dispatch camera crews to embed with soldiers on the front lines; it also livestreamed the whole thing, promising "instant access" to the carnage across Facebook, Twitter, and YouTube. When an ISIS car bomb hurtled toward the screen and exploded, friends, family, and tens of thousands of strangers watched together as a Rudaw reporter struggled to his feet before screaming the name of his cameraman into the billowing smoke. Because the livestream included emojis — smiling and frowning faces, hearts, and the universal "like" symbol — the scene unleashed a cascade of cartoon emotions. Most viewers were fearful for the crew's safety, so their yellow emoji faces registered shock. When the cameraman's friend emerged safely, the emojis changed to a wave of online smiles. Scattered among them, however, were a few frowning faces. These were the ISIS sympathizers and fighters who had wanted the journalists to die.

The online crowd didn't just watch and cheer; it even got involved in other, more positive ways. In a reversal of how ISIS had first exploited the technology in taking Mosul, a global network of online volunteers formed, dedicated to using social media to save lives there. They scanned online networks for any snippet of information about where civilians were caught in the crossfire, steering rescuers from the local hospital to their location. A hub for this effort was @MosulEye, run by an Iraqi man working behind ISIS lines as a new kind of online fifth column for peace. He described this effort as "a huge change . . . To be able to reach out to those who were rescued and hear their voices, knowing that I helped rescue them and spare their lives is priceless."

Social media had changed not just the message, but the dynamics of conflict. How information was being accessed, manipulated, and spread had taken on new power. Who was involved in the fight, where they were located, and even how they achieved victory had been twisted and transformed. Indeed, if what was online could swing the course of a battle — or eliminate the need for battle entirely — what, exactly, could be considered "war" at all?

The very same questions were being asked 6,000 miles from Mosul and, for most readers, quite a bit closer to home.

THE INTERNET WORLD COLLIDER

Like so many young men, Shaquon Thomas lived his life online. For "Young Pappy," as Thomas was known, it was a life of crime, his online brand extolling murders and drug deals.

Thomas had grown up with a loving family and a talent for music. At the age of 4, he started rapping, taught by his brother. But his future would be shaped by the intersection of old geography and new technology. The family lived in a neighborhood in Chicago that was caught between three street gangs: the Conservative Vice Lords, the Gangster Disciples, and the Black P. Stones.

Thomas was a Gangster Disciple and wanted everyone to know it. So he trumpeted the fact online, using it to build up that essential new currency: his personal brand. But there were real consequences of revealing your gang status online. On two separate occasions, he was shot at in broad daylight. Several bystanders were killed, including a young man who was waiting at the same bus stop to go to his first day of a new job. But each time, Thomas got away.

Although Thomas had survived the attacks, his online Young Pappy persona had to respond. So he did the only logical thing after two near-death incidents: he dropped another video on YouTube. "You don't even know how to shoot," he taunted his would-be killers. It was a hit, receiving over 2 million views. The 20-year-old was now a star, both in social media and gangland.

They killed Thomas a week later and just one block away from where he had recorded the video. Four days after that, a high school sophomore shot another rival gang member. The reason? He'd made disparaging posts about the deceased Young Pappy.

Shaquon Thomas's fate has befallen thousands of other young men across the United States. His hometown of Chicago has famously become the epicenter of a new kind of battle that we would call "war" in any other nation. Indeed, more people were killed by gang violence in 2017 in Chicago than in all U.S. special operations forces across a decade's worth of fighting in Iraq and then Syria. At the center of the strife is social media.

"Most of the gang disputes have nothing to do with drug sales, or gang territory, and everything to do with settling personal scores," explains Chicago alderman Joe Moore, who witnessed one of the shootings of Young Pappy. "Insults that are hurled on the social media."

Much of this violence starts with gangs' use of social media to "cybertag" and "cyberbang." Tagging is an update of the old-school practice of spray-painting graffiti to mark territory or insult a rival. The "cyber" version is used to promote your gang or to start a flame war by including another gang's name in a post or mentioning a street within a rival gang's territory. These online skirmishes escalate quickly. Anyone who posts about a person or a street belonging to a rival gang is making an online show of disrespect. Such a post is viewed as an invitation to "post up," or retaliate.

Digital sociologists describe how social media creates a new reality "no longer limited to the perceptual horizon," in which an online feud can seem just as real as a face-to-face argument. The difference in being online, however, is that now seemingly the whole world is witnessing whether you accept the challenge or not. This phenomenon plays out at every level, and not just in killings; 80 percent of the fights that break out in Chicago schools are now instigated online.

In time, these online skirmishes move to the "bang," sometimes called "Facebook drilling." (There are regional variants of this term. In Los Angeles, for instance, they use the descriptor "wallbanging.") This is when a threat is made via social media. It might be as direct as one gang member posting to a rival's Facebook wall, "I'm going to catch you. I'm going to shoot you." Or it might be symbolic, like posting photos of rival gang members turned upside down.

As with the distant lone wolf attacks of ISIS, cybertagging vastly increases the reach of potential violence. In the past, gangs battled with their neighbors; the "turf war" literally was about the border of their neighborhoods. Now, as journalist Ben Austen explained in an exhaustive investigation of the lives of young Chicago gangbangers, "the quarrel might be with not just the Facebook driller a few blocks away but also the haters ten miles north or west." You can be anywhere in the city and never have met the shooter, but "what started as a provocation online winds up with someone getting drilled in real life."

The decentralized technology thus allows any individual to ignite this cycle of violence. Yet by throwing down the gauntlet in such a public way, online threats have to be backed up not just by the individual, but by the group as well. If someone is fronted and doesn't reply, it's not just the gang member but the gang as a whole that loses status. The outcome is that anyone can start a feud online, but everyone has a collective responsibility to make sure it gets consummated in the real world.

One death can quickly beget another. Sometimes, the online memorialization of the victim inspires vows of revenge; other times, the killers use it to taunt and troll to further bloodshed. One teen explained, "Well, if you're a rival gang, I'll probably send you a picture of your dead boy's candles (at a public memorial) ... or I'll take a picture by your block with a gun and say, 'Where you at?' It depends on how you take it ... and things will go from there."

What has taken place in Chicago has been replicated across the nation. In Los Angeles, for instance, gangs use social media not just to beef on a new scale, but also to organize their chapters, recruit across the country, and even negotiate drug and arms deals with gangsters in other nations. Robert Rubin is a former gang member who now runs a gang intervention group called Advocates for Peace and Urban Unity. With mournful eyes and a goatee flecked with grey, he looks like a poet. He summarized the problem best. Social media is "the faceless enemy ... ," he said. "The old adage 'sticks and stones may break my bones but words do not hurt me,' I believe is no longer true. I believe words are causing people to die."

This shift ranges well beyond U.S. shores. Wherever young men gather and clash, social media now alters the calculus of violence. It is no longer enough for Mexican drug cartel members to kill rivals and seize turf. They must also *show* their success. They edit graphic executions into shareable music videos and battle in dueling Instagram posts (gold-plated AK-47s are a favorite). In turn, El Salvadoran drug gangs — notably the Mara Salvatrucha (MS-13) — have embraced the same franchise model as ISIS, rising in global prominence and power as groups in other countries link and then claim affiliation, in order to raise their own social media cachet. The result is a cycle of confrontation in which the distinction between online and offline criminal worlds has essentially become blurred.

Although these information conflicts all obey the same basic principles, their level of physical violence can vary widely. One example can be seen in the evolution of the Revolutionary Armed Forces of Colombia (FARC), whose fifty-four-year war against the Colombian government ended with a fragile 2016 peace. As FARC transitions to domestic politics, its struggle has shifted from the physical to the digital front. At its camps, former guerrilla fighters now trade in their rifles for smartphones. These are the "weapons" of a new kind of war, a retired FARC explosives instructor explained. "Just like we used to provide all our fighters with fatigues and boots, we're seeing the need to start providing them with data plans."

On the other hand, ostensibly nonviolent movements can launch information offensives to support shockingly violent acts. When Rodrigo Duterte was elected president of the Philippines in 2016, he was hailed as the country's first "social media president." His upstart election campaign had triumphed by matching his penchant for bombastic statements with an innovative online outreach effort that drew attention away from his rivals. Twitter even rewarded him with a custom emoji. But Duterte was also a demagogue. Dismissive of human rights, he had swept into office by promising a brutal crackdown on not just criminals but also his political opponents, two categories he rapidly began to blend. Backed by an army of feverishly supportive Facebook groups and Twitter bots, his administration set about discrediting journalists and rights activists, and bullying them into silence. At the same time, his followers sowed false stories and rumors that provided justification for Duterte's increasingly authoritarian actions. What started as an online deception campaign took a terrible toll. In its first two years, Duterte's "drug war" killed more than 12,000 people — not just dealers, but addicts, children, and anyone the police didn't like.

The very same changes can also affect the interplay of nations — and, by extension, the entire global system. As diplomats and heads of state have embraced the social media revolution, they've left behind the slow-moving, ritualistic system that governed international relations for centuries. In a few seconds of Twitter typing, President Trump threatened nations with nuclear war, dismissed cabinet secretaries, and issued bold policy proclamations that skipped a dozen layers of U.S. bureaucracy, sometimes running counter to the stated policies of his own administration. With each Trump tweet, U.S. diplomats and foreign embassies alike scrambled to figure out whether they should treat these online messages seriously. Meanwhile, official social media accounts of governments ranging from those of Russia and Ukraine to Israel and the Palestinian Authority have fallen into disputes whose only objective seems to be finding the wittier comeback. Diplomacy has become less private and policy-oriented and more public and performative. Its impact is not just entertainment, however. As with the gangs, each jab and riposte is both personal and witnessed by the entire world, poisoning relations and making it harder for leaders to find common ground.

And it's not just diplomats. For the first time, entire *populations* have been thrown into direct and often volatile contact with each other. Indians and Pakistanis have formed dueling "Facebook militias" to incite violence and stoke national pride. In times of elevated tensions between the nuclear-armed powers, these voices only grow louder, clamoring for violence and putting new pressure on leaders to take action. In turn, Chinese internet users have made a habit of launching online "expeditions" against any foreign neighbors who seem insufficiently awed by China's power. Notably, these netizens also rally against any perceived weakness by their own governments, constantly pushing their leaders to use military force. Attending a U.S. military–sponsored war game of a potential U.S.-China naval confrontation in the contested Senkaku Islands, we learned that it wasn't enough to know what actions the Chinese admirals were planning; we also had to track the online sentiment of China's 600 million social media users. If mishandled in a crisis, their angry reactions could bubble up into a potent political force, thus limiting leaders' options. Even in authoritarian states, war has never been so democratic.

Running through all these permutations of online conflict is one more troubling, inescapable theme. Sometimes, the terrible consequences of these internet battles may be the only "real" thing about them.

Even as we watched the Islamic State rampage through Iraq, there was another conflict taking place in the United States — in plain sight, but unrealized by too many at the time. Agents of the Russian Federation were organizing an online offensive that would dwarf all others before it. Throughout the 2016 U.S. presidential election, thousands of human trolls, backed by tens of thousands of automated accounts, infiltrated every part of the U.S. political dialogue. They steered discussion, sowed doubt, and obfuscated truth, launching the most politically consequential information attack in history. And that operation continues to this day.

WAR BY OTHER MEMES

Carl von Clausewitz was born a couple of centuries before the internet, but he would have implicitly understood almost everything it is doing to conflict today.

Raised in Enlightenment Europe, Clausewitz enlisted in the army of the Prussian kaiser at the age of twelve. When Napoleon plunged Europe into a decade of conflict and unleashed a new age of nationalism, Clausewitz dedicated the rest of his life to studying war. For decades, he wrote essay after essay on the topic, exchanging letters with all the leading thinkers of the day and rising to become head of the Prussian military academy. His extensive writing was dense, complicated, and often confusing. But after Clausewitz died in 1831, his wife, Marie, edited his sprawling library of thoughts into a ten-volume treatise modestly titled *On War*.

Clausewitz's (and his wife's) theories of warfare have since become required reading for militaries around the world and have shaped the planning of every war fought over the past two centuries. Concepts like the "fog of war," the inherent confusion of conflict, and "friction," the way plans never work out exactly as expected when facing a thinking foe, all draw on his monumental work.

His most famous observation, though, regards the nature of war itself. In his view, war is politics by other means, or, as he put it in more opaque terms, "the continuation of political intercourse with the addition of other means." The two are intertwined, he explained. "War in itself does not suspend political intercourse or change it into something entirely different. In essentials that intercourse continues, irrespective of the means it employs." War is political. And politics will always lie at the heart of violent conflict, the two inherently linked. "The main lines along which military events progress, and to which they are restricted, are political lines that continue throughout the war into the subsequent peace."

In other words, Clausewitz thought, war is part of the same continuum that includes trade, diplomacy, and all the other interactions that take place between peoples and governments. This theory flew in the face of the beliefs of older generations of soldiers and military theorists, who viewed war as a sort of "on/off" switch that pulled combatants into an alternate reality governed by a different set of rules. To Clausewitz, war is simply another way to get something you want: an act of force to compel an enemy to your will.

Winning is a matter of finding and neutralizing an adversary's "center of gravity." This is often a rival's army, whose destruction usually ends its ability to fight. But routing an army is not always the most effective path. "The moral elements are among the most important in war," Clausewitz wrote. "They constitute the spirit that permeates war as a whole ... They establish a close affinity with the will that moves and leads the whole mass of force." Figure out how to shatter a rival's spirit, and you might win the war, while avoiding the enemy army entirely.

Easier said than done. Modern warfare has seen numerous efforts to target and drain an enemy's spirit, almost never with success. In World War II, Britain bore the Blitz, years of indiscriminate bombing by German planes and rockets that sought to force the nation to capitulate. The British instead turned what Winston Churchill called their "Darkest Hour" into triumph. Similar logic drove the United States' Rolling Thunder bombing campaign against the cities and industry of North Vietnam during the late 1960s. U.S. warplanes dropped more than 6.5 million tons of bombs and killed tens of thousands of people. But the North Vietnamese never seriously contemplated surrender.

Similar logic has applied to war propaganda, another attempt by combatants to target the enemy spirit, which has proven historically ineffective. During the Blitz, the most popular radio station in Britain was an English-language propaganda station produced by the Nazis — because the British loved to laugh at it. So, too, the United States air-dropped tens of millions of leaflets across North Vietnam — which were promptly used as toilet paper.

But in the space of a decade, social media has changed all that. Attacking an adversary's most important center of gravity — the spirit of its people — no longer requires massive bombing runs or reams of propaganda. All it takes is a smartphone and a few idle seconds. And anyone can do it.

Today, it is possible to communicate directly with people you're ostensibly at war with — to send them "friend" requests, to persuade or debate with them, or to silently stalk their digital lives. Opposing soldiers on a battlefield might find each other online and then "like" or troll their foes. Based on what they share on social media, a few dozen sympathizers might be identified out of a population of millions and then groomed to commit acts of violence against their fellow citizens. Eager volunteers might join nationalist brigades to stir the pot of hatred and resentment between rival peoples, sparking a war or genocide. They might even divide and conquer a nation's politics from afar.

None of these scenarios is hypothetical. Each of them has already happened. Each will happen many more times in the years to come.

From the world's most powerful nations to the pettiest flame war combatants, all of today's fighters have turned social media into a weapon in their own national and personal wars, which often overlap. They all fight to bend the global information environment to their will. The internet, once a light and airy place of personal connection, has since morphed into the nervous system of modern commerce. It has also become a battlefield where information itself is weaponized.

For the internet's optimistic inventors and fiercest advocates, so sure of its capacity to bring peace and understanding, this is a bitter pill to swallow. "I thought once everybody could speak freely and exchange information and ideas, the world [was] automatically going to be a better place," confessed Twitter cofounder Evan Williams. "I was wrong about that."

Yet this is where we stand. Just as the internet has reshaped war, war is now radically reshaping the internet.

This book is an attempt to make sense of this seismic shift — to chart its history, identify its rules, and understand its effects. Our own backgrounds aided in this effort. We're a pairing of digital immigrant and digital native. One of us grew up in the dark ages before the internet, forced to learn its arcane ways, systems, and procedures; the other was born into them, seeing what was once impossible as utterly normal.

Over five years, we studied the history of communications and propaganda; the evolution of journalism and open-source intelligence; the bases of internet psychology; social network dynamics and virality; the evolution of Silicon Valley corporate responsibility; and the applications of machine intelligence. At the same time, we tracked dozens of conflicts and quasi conflicts in every corner of the globe, all playing out simultaneously online. We cast our net wide, scooping up everything from the spread of YouTube battle clips to a plague of Nazi-sympathizing cartoon frogs. We interviewed experts ranging from legendary internet pioneers to infamous "reality stars," weaving their insights together with those of viral marketers and political hacks, terrorist propagandists and preteen reporters, soldiers and generals (including one who may have committed some light treason).

We visited the offices and bases of the U.S. defense, diplomatic, and intelligence communities; traveled overseas to meet with foreign government operatives; and made trips to both brightly colored offices of social media companies and dark labs that study the tech of war. Meanwhile, we treated the internet as a laboratory itself. We leapt into online battles just to experience the fight and see where it would lead. We downloaded apps and joined distant nations' digital armies. We set traps for trolls, both to learn from them and have some fun at their expense. Then we found ourselves being enlisted into the fight in new ways, asked to advise the investigations trying to figure out how other nations had attacked the United States with these new weapons, as well as aid the U.S. military information operations tasked to fight back. At the end of our adventure, we even found ourselves the targets of "friend" requests from U.S. government officials who didn't exist — and whose puppet masters worked out of St. Petersburg instead of Washington, DC.

The following chapters lay out the lessons from this journey. We begin with the history of the communications technology that has so swiftly remade the world, a history that set the patterns for all the changes that would follow in war and politics.

We then trace the ways that social media has created a new environment for conflict. It has transformed the speed, spread, and accessibility of information, changing the very nature of secrecy. Yet, while the truth is more widely available than ever before, it can be buried in a sea of "likes" and lies. We explore how authoritarians have thus managed to co-opt the once-liberating force of the social media revolution and twist it back to their own advantage. It has granted them not just new ways to control their own populations, but also a new global reach through the power of disinformation.

From there, we sketch the terrain of this new battlefield. Social networks reward not veracity but virality. Online battles and their real-world results are therefore propelled by the financial and psychological drivers of the "attention economy," as well as by the arbitrary, but firmly determinative, power of social media algorithms.

Next we review the basic concepts of what it takes to win. Narrative, emotion, authenticity, community, and inundation are the most effective tools of online battles, and their mastery guides the efforts of most successful information warriors. These new wars are not won by missiles and bombs, but by those able to shape the story lines that frame our understanding, to provoke the responses that impel us to action, to connect with us at the most personal level, to build a sense of fellowship, and to organize to do it all on a global scale, again and again.

We then explore what happens when this all comes together, with a focus on the theory and conduct of information warfare, the memes that drive ideas on the web, and the differences between public and secret campaigns that win online battles.

Finally, we examine the last unprecedented change that the internet has brought to war and politics, the one part that would have stumped even Clausewitz. While social media has become a battlefield for us all, its creators set its rules. On a network of billions, a tiny number of individuals can instantly turn the tide of an information war one way or another, often unintentionally. We examine the role of social media companies as they reckon with a growing political power that they have too often proven ill-equipped to wield. We identify the challenges that lie ahead, especially as these engineers of communication yet again try to solve human problems with new technology, in this case artificial intelligence — machines that mimic and may well surpass humans.

We end by considering the far-reaching implications of a world in which every digital skirmish is a "war" and every observer a potential combatant. These implications inform a series of actions that governments, companies, and all of us as individuals can undertake in response.

Our research took us around the world and into the infinite reaches of the internet. Yet we continually found ourselves circling back to five core principles, which form the foundations of this book.

First, the internet has left adolescence. Over decades of growth, the internet has become the preeminent medium of global communication, commerce, and politics. It has empowered not just new leaders and groups, but a new corporate order that works constantly to expand it. This pattern resembles the trajectory of the telegraph, telephone, radio, and television before it. But the rise of social media has allowed the internet to surpass those revolutions. It is now *truly* global and instantaneous — the ultimate combination of individual connection and mass transmission. Yet tumultuous as the past few years have been, social media — and the revolution it represents — is only now starting to flex its muscles. Half the world has yet to come online and join the fray.

Second, the internet has become a battlefield. As integral as the internet has become to business and social life, it is now equally indispensable to militaries and governments, authoritarians and activists, and spies and soldiers. They all use it to wage wars that observe no clear borders. The result is that every battle seems personal, but every conflict is global.

Third, this battlefield changes how conflicts are fought. Social media has rendered secrets of any consequence essentially impossible to keep. Yet because virality can overwhelm truth, what is known can be reshaped. "Power" on this battlefield is thus measured not by physical strength or high-tech hardware, but by the command of attention. The result is a contest of psychological and algorithmic manipulation, fought through an endless churn of competing viral events.

Fourth, this battle changes what "war" means. Winning these online battles doesn't just win the web, but wins the world. Each ephemeral victory drives events in the physical realm, from seemingly inconsequential celebrity feuds to history-changing elections. These outcomes become the basis of the next inevitable battle for online truth, further blurring the distinction between actions in the physical and digital realms. The result is that on the internet, "war" and "politics" have begun to fuse, obeying the same rules and inhabiting the same spectrum; their tactics and even players are increasingly indistinguishable. Yet it is not the politicians, generals, lawyers, or diplomats who are defining the laws of this new fight. Rather, it's a handful of Silicon Valley engineers.

Fifth, and finally, we're all part of this war. If you are online, your attention is like a piece of contested territory, being fought over in conflicts that you may or may not realize are unfolding around you. Everything you watch, like, or share represents a tiny ripple on the information battlefield, privileging one side at the expense of others. Your online attention and actions are thus both targets and ammunition in an unending series of skirmishes. Whether you have an interest in the conflicts of *LikeWar* or not, they have an interest in you.

The modern internet is not just a network, but an ecosystem of nearly 4 billion souls, each with their own thoughts and aspirations, each capable of imprinting a tiny piece of themselves on the vast digital commons. They are the targets not of a single information war but of thousands and potentially millions of them. Those who can manipulate this swirling tide, to steer its direction and flow, can accomplish incredible good. They can free people, expose crimes, save lives, and seed far-reaching reforms. But they can also accomplish astonishing evil. They can foment violence, stoke hate, sow falsehoods, incite wars, and even erode the pillars of democracy itself.

Which side succeeds depends, in large part, on how much the rest of us learn to recognize this new warfare for what it is. Our goal in *Like-War* is to explain exactly what's going on and to prepare us all for what comes next.

2

Every Wire a Nerve

How the Internet Changed
the World

> You ask what I am? . . . I am all the people who thought of me
> and planned me and built me and set me running . . . I am all the
> things they wanted to be and perhaps could not be, so they built
> a great child, a wondrous toy.
>
> — RAY BRADBURY, *I Sing the Body Electric!*

"WHAT *IS* INTERNET, ANYWAY?"

It was 1994, and *Today* show cohost Bryant Gumbel was struggling on live TV to read the "internet address" for NBC's newly unveiled email hotline. In confusion, the anchorman turned from the teleprompter. Was "Internet" something you could mail letters to? he asked. Cohost Katie Couric didn't know. Eventually, an offscreen producer bailed them out. "Internet is the massive computer network," he shouted, "the one that's becoming really big now."

Today, their exchange seems quaint. Roughly half the world's population is now linked by this computer network. It is not just "really big"; it is the beating heart of international communication and commerce. It supports and spreads global news, information, innovation, and discovery of every kind and in every place. Indeed, it has become woven into almost everything we do, at home, at work, and, as we shall see, at war. In the United States, not only is internet usage near-universal, but one-fifth of Americans now admit that they essentially never stop being online. The only peoples who have remained truly ignorant of the internet's reach are a few un-networked tribes of the Amazon and New Guinea. And for them, it is just a question of time.

But *using* the internet isn't really the same as *understanding* it. The "internet" isn't just a series of apps and websites. Nor is it merely a creature of the fiber-optic cable and boxy servers that provide its backbone. It is also a galaxy of billions of ideas, spreading through vast social media platforms that each pulse with their own entropic rhythm. At the same time, it is a globe-spanning community vaster and more diverse than anything before it, yet governed by a handful of Silicon Valley oligarchs.

As revolutionary as the internet may seem, it is also bound by history. Its development has followed the familiar patterns etched by the printing press, telegraph, television, and other communications mediums before it. To truly understand the internet — the most consequential battlefield of the twenty-first century — one must understand how it works, why it was made, and whom it has empowered.

In other words, "What *is* Internet, anyway?"

The answer starts with a memo that few read at the time, for the very reason that there was no internet when it was written.

LO AND BEHOLD

"In a few years, men will be able to communicate more effectively through a machine than face to face. That is a rather startling thing to say, but it is our conclusion."

That was the prediction of J.C.R. Licklider and Robert W. Taylor, two psychologists who had seen their careers pulled into the relatively new field of computer science that had sprung up during the desperate days of World War II. The "computers" of their day were essentially giant calculators, multistory behemoths of punch cards and then electric switches and vacuum tubes, applied to the hard math of code breaking, nuclear bomb yields, and rocket trajectories.

That all changed in 1968, the year Licklider and Taylor wrote a paper titled "The Computer as a Communication Device." It posited a future in which computers could be used to capture and share information instead of just calculating equations. They envisioned not just one or two computers linked together, but a vast constellation of them, spread around the globe. They called it the Intergalactic Computer Network.

Reflecting their past study of the human mind, Licklider and Taylor went even further. They prophesied how this network would affect the people who used it. It would create new kinds of jobs, build new "interactive communities," and even give people a new sense of place, what Licklider and Taylor called "to be on line." So long as this technology was made available to the masses, they wrote, "surely the boon to humankind would be beyond measure."

The information that could be transmitted through this prospective network would also be fundamentally different from any communication before. It would be the most essential form of information itself. The "binary information digit," or "bit," had first been proposed in 1948 by Claude Shannon of Bell Labs. The bit was the smallest possible unit of data, existing in either an "on" or "off" state. By stringing bits together, complex instructions could be sent through computers, with perfect accuracy. As acclaimed physicist John Archibald Wheeler observed, breaking information into bits made it possible to convey *anything*. "Every particle, every field of force, even the space-time continuum itself," Wheeler wrote, "derives its function, its meaning, its very existence, from answers to yes-no questions, binary choices, bits."

By arranging bits into "packets" of information and establishing a system for sending and receiving them (packet switching), it was theoretically possible for computers to relay instructions instantly, over any conceivable distance. With the right software, the two men foretold, these bits could be used to query a database, type a word, even (in theory) generate an image or display video. Years before the internet would see its first proof of concept, its theoretical foundation had been laid. Not everyone recognized the use for such a system. When a team of researchers proposed a similar idea to AT&T in 1965, they were bluntly rejected. "Damn it!" exclaimed one executive. "We're not going to set up a competitor to ourselves!"

Luckily, Licklider and Taylor were in a position to help make their vision of computers exchanging bits of information a reality. The two were employees of the Pentagon's Advanced Research Projects Agency (ARPA). After the surprise of the Sputnik space satellite launch in 1957,

ARPA was established in 1958 to maintain U.S. parity with Soviet science and technology research. For the U.S. military, the potential of an interconnected (then called "internetted") communications system was that it could banish its greatest nightmare: the prospect of the Soviet Union being able to decapitate U.S. command and control with a single nuclear strike. But the real selling point for the other scientists working for ARPA was something else. Linking up computers would be a useful way to share what was then an incredibly rare and costly commodity: computer time. A network could spread the load and make it easier on everyone. So a project was funded to transform the Intergalactic Computer Network into reality. It was called ARPANET.

On October 29, 1969, ARPANET came to life when a computer at UCLA was linked to a computer at Stanford University. The bits of information on one machine were broken down into packets and then sent to another machine, traveling across 350 miles of leased telephone wire. There, the packets were re-formed into a message. As the scientists excitedly watched, one word slowly appeared on the screen: "LO."

Far from being the start of a profound statement such as "Lo and behold," the message was supposed to read "LOGIN." But the system crashed before the transmission could be completed. Fittingly, the first message in internet history was a miscommunication.

THE RACE TO COMMUNICATE

Despite the crash, the ARPANET team had accomplished something truly historic. They hadn't just linked two computers together. They had finished a race that had spanned 5,000 years and continually reshaped war and politics along the way.

The use of technology to communicate started in ancient Mesopotamia sometime around 3100 BCE, when the first written words were pressed into clay tablets. Soon information would be captured and communicated via both the most permanent forms of marble and metal and in the more fleeting forms of papyrus and paper.

But all this information couldn't be transferred easily. Any copy had to be painstakingly done by hand, usually with errors creeping in along the way. A scribe working at his maximum speed, for instance, could only produce two Bibles a year. This limitation made information, in any form, the scarcest of commodities.

The status quo stood for almost 4,000 years, until the advent of the printing press. Although movable type was first invented in China, traditional Mandarin Chinese — with its 80,000 symbols — was too cumbersome to reproduce at scale. Instead, the printing press revolution began in Europe in around 1438, thanks to a former goldsmith, Johannes Gutenberg, who began experimenting with movable type. By 1450, he was peddling his mass-produced Bibles across Germany and France. Predictably, the powers of the day tried to control this disruptive new technology. The monks and scribes, who had spent decades honing their hand-copying techniques, called for rulers to ban it, arguing that mass production would strangle the "spirituality" of the copying process. Their campaign failed, however, when, short on time and money, they resorted to printing their own pamphlet on one of Gutenberg's new inventions. Within a century, the press became both accessible and indispensable. Once a rare commodity, books — some 200 million of them — were now circulating widely in Europe.

In what would become a familiar pattern, the new technology transformed not just communications but war, politics, and the world. In 1517, a geopolitical match was struck when a German monk named Martin Luther wrote a letter laying out 95 problems in the Catholic Church. Where Luther's arguments might once have been ignored, the printing press allowed his ideas to reach well beyond the bishop to whom he penned the letter. By the time the pope heard about the troublesome monk and sought to excommunicate him, Luther had reproduced his 95 complaints in 30 different pamphlets and sold 300,000 copies. The result was the Protestant Reformation, which would fuel two centuries of war and reshape the map of Europe.

The technology would also create new powers in society and place the old ones under unwelcome scrutiny. In 1605, a German printer named Johann Carolus found a way to use his press's downtime by publishing a weekly collection of "news advice." In publishing the first newspaper, Carolus created a new profession. The "press" sold information itself to customers, creating a popular market model that had never before existed. But in the search to profit from news, truth sometimes fell by the wayside. For instance, the *New-England Courant*, among the first American newspapers, published a series of especially witty letters by a "Mrs. Silence Dogood" in 1722. The actual writer was a 16-year-old apprentice named Benjamin Franklin, making him — among many other things — the founding father of fake news in America.

Yet the spread of information, true or false, was limited by the prevailing transportation of the day. In ancient Greece, the warrior Pheidippides famously ran 25 miles from Marathon to Athens to deliver word of the Greeks' victory over the Persian army. (The 26.2-mile distance of the modern "marathon" dates from the 1908 Olympics, where the British royal family insisted on extending the route to meet their viewing stands.) It was a race to share the news that would literally kill him. When Pheidippides finally sprinted up the steps of Athens's grand Acropolis, he cried out to the anxious city leaders gathered there, "Nikomen! We are victorious!" And then, as the poet Robert Browning would write 2,400 years later, he collapsed and died, "joy in his blood bursting his heart."

From the dawn of history, any such messages, important or not, could only be delivered by hand or word of mouth (save the occasional adventure with carrier pigeons). This placed a sharp upper bound on the speed of communication. The Roman postal service (*cursus publicus,* "the public way"), established at the beginning of the first millennium CE, set a record of roughly fifty miles per day that would stand essentially unchallenged until the advent of the railroad. News of world-changing events — the death of an emperor or start of a war — could only travel as fast as a horse could gallop or a ship could sail. As late as 1815, thousands of British soldiers would be mowed down at the Battle of New Orleans simply because tidings of a peace treaty ending the War of 1812 — signed two weeks earlier — had yet to traverse the Atlantic.

The world changed decisively in 1844, the year Samuel Morse successfully tested his telegraph (from the Latin words meaning "far writer"). By harnessing the emerging science of electricity, the telegraph ended the tyranny of distance. It also showed the important role of government in any communication technology now able to extend across political boundaries. Morse spent years lobbying the U.S. Congress for the $30,000 needed to lay the thirty-eight miles of wire between Washington, DC, and Baltimore for his first public test. Critics suggested the money might be better spent testing hypnotism as a means of long-distance communication. Fortunately, the telegraph won — by just six votes.

This was the start of a telecommunications revolution. By 1850, there were 12,000 miles of telegraph wire and some 20 telegraph companies in the United States alone. By 1880, there would be 650,000 miles of wire worldwide — 30,000 miles under the ocean — that stretched from San Francisco to Bombay. This was the world that Morse's brother had prophesied in a letter written while the telegraph was still under development: "The surface of the earth will be networked with wire, and every wire will be a nerve. The earth will become a huge animal with ten million hands, and in every hand a pen to record whatever the directing soul may dictate!"

Morse would be applauded as "the peacemaker of the age," the inventor of "the greatest instrument of power over earth which the ages of human history have revealed." As observers pondered the prospect of a more interconnected world, they assumed it would be a gentler one. President James Buchanan captured the feeling best when marking the laying of the first transatlantic cable between the United States and Britain, in 1858. He expressed the belief that the telegraph would "prove to be a bond of perpetual peace and friendship between the kindred nations, and an instrument designed . . . to diffuse religion, liberty, and law throughout the world." Within days, that transatlantic cable of perpetual peace was instead being used to send military orders.

Like the printing press before it, the telegraph quickly became an important new tool of conflict, which would also transform it. Beginning in the Crimean War (1853–1856), broad instructions, traveling weeks by sea, were replaced — to the lament of officers in the field — by micromanaging battle orders sent by cable from the tearooms of London to the battlefields of Russia. Some militaries proved more effective in exploiting the new technology than others. In the Wars of German Unification (1864–1871), Prussian generals masterfully coordinated far-flung forces to the bafflement of their foes, using real-time communications by telegraph wire to replace horseback couriers. As a result, the telegraph also spurred huge growth in war's reach and scale. In the American Civil War (1861–1865), Confederate and Union soldiers, each seeking an edge over the other, laid some 15,000 miles of telegraph wire.

The telegraph also reshaped the public experience and perception of conflict. One journalist marveled, "It gives you the news before the circumstances have had time to alter ... A battle is fought three thousand miles away, and we have the particulars while they are taking the wounded to the hospital."

This intimacy could be manipulated, however. A new generation of newspaper tycoons arose, who turned sensationalism into an art form, led by Harvard dropout turned newspaper baron William Randolph Hearst. His "yellow journalism" (named for the tint of the comics in two competing New York dailies, Hearst's *New York Journal* and Joseph Pulitzer's *New York World*) was the kind of wild rumormongering American readers couldn't get enough of — and that helped spark the Spanish-American War of 1898. When one of his photographers begged to return home from Spanish-controlled Cuba because nothing was happening, Hearst cabled back: "Please remain. You furnish the pictures and I'll furnish the war." Concern over the issue of telegraphed "fake news" grew so great that the *St. Paul Globe* even changed its motto that year: "Live News, Latest News, Reliable News — No Fake War News."

The electric wire of the telegraph, though, could only speak in dots and dashes. To use them required not just the infrastructure of a telegraph office, but a trained expert to operate the machine and translate its coded messages for you. Alexander Graham Bell, an amateur tinkerer whose day job was teaching the deaf, changed this with the telephone in 1876. Sending sound by wire meant users could communicate with each other, even in their offices and homes. Within a year of its invention, the first phone was put in the White House. The number to call President Rutherford B. Hayes was "1," as the only other phone line linked to it was at the Treasury Department. The telephone also empowered a new class of oligarchs. Bell's invention was patented and soon monopolized by Bell Telephone, later renamed the American Telephone and Telegraph Company (AT&T). Nearly all phone conversations in the United States would be routed through this one company for the next century.

Telegraphs and phones had a crucial flaw, though. They shrank the time and simplified the means by which a message could travel a great distance, but they did so only between two points, linked by wire. Guglielmo Marconi, a 20-year-old Irish-Italian tinkering in a secret lab in his parents' attic, would be the first to build a working "wireless telegraphy" system, in 1894.

Marconi's radio made him a conflicted man. He claimed that radio would be "a herald of peace and civilization between nations." At the same time, he aggressively peddled it to every military he could. He sold it to the British navy in 1901 and convinced the Belgian government to use it in the brutal colonization of the Congo. In the 1904–1905 Russo-Japanese War, both sides used Marconi radios.

Radio's promise, however, went well beyond connecting two points across land or sea. By eliminating the need for wire, radio freed communications in a manner akin to the printing press. One person could speak to thousands or even millions of people at once. Unlike the telegraph, which conveyed just dots and dashes, radio waves could carry the human voice and the entire musical spectrum, turning them into the conveyor of not only mass information but also mass entertainment.

The first radio "broadcast" took place in 1906, when an American engineer played "O Holy Night" on his violin. By 1924, there were an estimated 3 million radio sets and 20 million radio listeners in the United States alone. Quite quickly, radio waves collided with politics. Smart politicians began to realize that radio had shattered the old political norms. Speeches over the waves became a new kind of performance art crossed with politics. The average length of a political campaign speech in the United States fell from an hour to just ten minutes.

And there was no better performer than Franklin Delano Roosevelt, elected president in 1932. He used his weekly Fireside Chats to reach directly into the homes of millions of citizens. (After the December 7, 1941, Pearl Harbor attack, four-fifths of American households listened to his speech live.) In so doing, he successfully went over the heads of the political bosses and newspaper editors who fought to deny him a third and fourth term. So powerful were FDR's speeches that on the night of an important speech intended to rally listeners against Germany, the Nazis launched a heavy bombing raid against London to try to divert the news.

But radio also unleashed new political horrors. "It would not have been possible for us to take power or to use it in the ways we have without the radio," said Joseph Goebbels, who himself was something new to government, the minister of propaganda for Nazi Germany. Goebbels employed nearly a thousand propagandists to push Adolf Hitler's brutal, incendiary, enrapturing speeches. Aiding the effort was a giveaway with a catch: German citizens were gifted special radios marked with swastikas, which could only receive Nazi-broadcasted frequencies.

Just like the telegraph, radio would be used to foment war and become a new tool for fighting it. On the eve of the 1939 German invasion of Poland, Hitler told his generals, "I will provide a propagandistic *casus belli*. Its credibility doesn't matter. The victor will not be asked whether he told the truth." The following six years of World War II saw not just tanks, planes, and warships linking up by radio, but both sides battling back and forth over the airwaves to reshape what the opposing population knew and thought. As Robert D. Leigh, director of the Foreign Broadcast Intelligence Service, testified before Congress in 1944:

> Around the world at this hour and every hour of the 24, there is a constant battle on the ether waves for the possession of man's thoughts, emotions, and attitudes — influencing his will to fight, to stop fighting, to work hard, to stop working, to resist and sabotage, to doubt, to grumble, to stand fast in faith and loyalty . . . We estimate that by short wave alone, you as a citizen of this radio world are being assailed by 2,000 words per minute in 40–45 different languages and dialects.

Yet the reach and power of radio was soon surpassed by a technology that brought compelling imagery into broadcasts. The first working television in 1925 showed the face of a ventriloquist's dummy named Stooky Bill. From these humble beginnings, television soon rewired what people knew, what they thought, and even how they voted. By 1960, television sets were in nine of ten American homes, showing everything from *The Howdy Doody Show* to that infamous presidential debate between Richard M. Nixon and John F. Kennedy, won by the more "telegenic" candidate. In the United States, television forged a new sense of cultural identity. With a limited number of broadcasts to choose from, millions of families watched the same events and news anchors; they saw the same shows and gossiped eagerly about them the next day.

Television also changed what military victory and defeat looked like. In 1968, the Vietcong launched the Tet Offensive against South Vietnam and its American allies. The surprise operation quickly turned into a massive battlefield failure for the attackers. Half of the 80,000-strong Vietcong were killed or wounded; they captured little territory and held none of what they did. But that wasn't what American families watching in their dens back home saw. Instead, 50 million viewers saw clips of U.S. Marines in disarray; scenes of bloody retribution and bodies stacked deep. The most dramatic moment came when the U.S. embassy in Saigon was put under siege. Although the main building was never penetrated and the attackers were quickly defeated, the footage was mesmerizing — and, for many, deeply troubling.

Unfolding across a hundred South Vietnamese cities and towns, Tet was the biggest battle of the Vietnam War. But the war's real turning point came a month later and 8,000 miles away.

Legendary journalist Walter Cronkite was the anchor of the *CBS Evening News* and deemed "the most trusted man in America." In a three-minute monologue, Cronkite declared that the Vietnam War was never going to be the victory the politicians and generals had promised. In the White House, President Lyndon B. Johnson watched. Forlorn, he purportedly told his staff, "If I've lost Cronkite, I've lost Middle America." Such was the power of moving images and sound, interspersed with dramatic narration and beamed into tens of millions of households. Not only did video provide a new level of emotional resonance, but it was also hard to dispute. When the government claimed one thing and the networks showed another, the networks usually won.

As television ranged into even wider territory with the advent of real-time satellite coverage, the story seemed complete. From words pressed into Mesopotamian clay to broadcasts from the moon, the steady march of technological innovation had overcome the obstacles of time and distance. With each step, communications technology had altered the politics of the day, subverting some powers while crowning new ones in their place. Despite its originators' unfailing optimism about social promise and universal peace, each technology had also been effectively turned to the ends of war.

But one important limitation still bound all these technologies. Users could form a direct link, conversing one-to-one, as with the telegraph or telephone. Or one user could reach many at once, as with the printing press or radio and television broadcasts.

No technology could do it all — until ARPANET.

A SOCIAL SCIENCE FICTION

The first computer network grew quickly. Within weeks of the computers at UCLA and Stanford connecting in October 1969, a computer in Santa Barbara and then another in Utah joined the party. By 1971, fifteen university computer labs had been stitched together. In 1973, the network made its first international connection, incorporating computers at the Norwegian Seismic Array, which tracked earthquakes and nuclear tests.

With the bold idea of computer communication now proven workable, more and more universities and labs were linked together. But instead of joining the ARPANET, many forged their own mini-networks. One connected computers in Hawaii (delightfully called the ALOHAnet and MENEHUNE); another did so in Europe. These mini-networks presented an unexpected problem. Rather than forming one "galactic" network, computer communication was becoming isolated into a bunch of little clusters. It was even worse. Each network had its own infrastructure and governing authority. This meant that the networks couldn't easily link up. They were each setting their own rules about everything from how to maintain the network to how to communicate within it. Unless a common protocol could be established to govern a "network of networks" (or "internet"), the spread of information would be held back. This is when Vint Cerf entered the scene.

While figures like J.C.R. Licklider and Robert W. Taylor had conceived ARPANET, Cerf is rightfully known as the "father of the internet." As a teenager, he learned to code computer software by writing programs to test rocket engines. As a young researcher, he was part of the UCLA-Stanford team that connected the Pentagon's new network.

Recognizing the problem of compatibility would keep computerized communication from ever scaling, Cerf set out to find a solution. Working with his friend Robert Kahn, he designed the TCP/IP (transmission-control protocol/internet protocol), an adaptable framework that could track and regulate the transmission of data across an exponentially expanding network. Essentially, it is what allowed the original ARPANET to bind together all the mini-networks at universities around the world. It remains the backbone of the internet to this day.

Over the following years, Cerf moved to work at ARPA and helped set many of the rules and procedures for how the network would evolve. He was aware of the futuristic visions that his predecessors had laid out. But it was hard to connect that vision to what was still just a way for scientists to share computing time. Considerations of the internet's social or political impact seemed the stuff of fantasy.

This changed one day in 1979 when Cerf logged on to his workstation to find an unopened message from the recently developed "electronic mail" system. Because more than one person was using each computer, the scientists had conceived of "e-mail" (now commonly styled "email") as a way to share information, not just between computers but also from one person to another. But, just as with regular mail, they needed a system of "addresses" to send and receive the messages. The "@" symbol was chosen as a convenient "hack" to save typing time and scarce computer memory.

The message on Cerf's screen wasn't a technical request, however. The email subject was "SF-lovers." And it hadn't been sent just to him. Instead, Cerf and his colleagues scattered across the United States were all asked to respond with a list of their favorite science fiction authors. Because the message had gone out to the entire network, everybody's answers could then be seen and responded to by everybody else. Or users could send their replies to just one person or subgroup, generating scores of smaller discussions that eventually fed back into the whole.

Over forty years later, Cerf still recalls the moment he realized the internet would be something more than every other communications technology before it. "It was clear we had a social medium on our hands," he said.

The thread was a hit. After SF-lovers came Yumyum, a mailing list to debate the quality of restaurants in Silicon Valley. Soon the network was also used to share not just opinions but news about both science and science fiction, such as plans for a movie revival of the 1960s TV show *Star Trek*.

The U.S. military budgeters wanted to ban all this idle chatter from their expensive new network. However, they relented when engineers convinced them that the message traffic was actually a good stress test for ARPANET's machinery. Chain letters and freewheeling discussions soon proliferated across the network. ARPANET's original function had been remote computer use and file transfer, but soon email was devouring two-thirds of the available bandwidth. No longer was the internet simply improving the transfer of files from one database to another. Now it was creating those "interactive communities" that Licklider and Taylor had once envisioned, transforming what entire groups of people thought and knew. Soon enough, it would even change how they spoke to each other.

Perhaps no one — engineers included — understood by how much. At precisely 11:44 a.m. EST on September 19, 1982, computer scientist Scott Fahlman changed history forever. In the midst of an argument over a joke made on email, he wrote:

> I propose that [*sic*] the following character sequence for joke markers:
> :-)
> Read it sideways. Actually, it is probably more economical to mark things that are NOT jokes, given current trends. For this, use
> :-(

And so the humble emoticon was born. But it illustrated something more.

For all its promise, ARPANET was not the internet as we know it. It was a kingdom ruled by the U.S. government. And, as shown by the formal creation of the emoticon in the midst of an argument among nerds, its population was mostly PhDs in a handful of technical fields. Even the early social platforms these computer scientists produced were just digital re-creations of old and familiar things: the postal service, bulletin boards, and newspapers. The internet remained in its infancy.

But it was growing fast. By 1980, there were 70 institutions and nearly 5,000 users hooked up to ARPANET. The U.S. military came to believe that the computer network its budget was paying for had expanded too far beyond its needs or interests. After an unsuccessful attempt to sell

ARPANET to a commercial buyer (for the second time, AT&T said "no thanks"), the government split the internet in two. ARPANET would continue as a chaotic and fast-growing research experiment, while the military would use the new, secure MILNET. For a time, the worlds of war and the internet went their separate ways.

This arrangement also paved the way for the internet to become a civilian — and eventually a commercial — enterprise. The National Science Foundation took over from the Pentagon and moved to create a more efficient version of ARPANET. Called NSFNET, it proved faster by an order of magnitude and brought in new consortiums of users. The 28,000 internet users in 1987 grew to nearly 160,000 by 1989. The next year, the now outdated ARPANET was quietly retired. Vint Cerf was there to deliver the eulogy. "It was the first, and being first, was best, / But now we lay it down to ever rest. / . . . / Of faithful service, duty done, I weep. / Lay down thy packet, now, o friend, and sleep."

While the internet and the military were ostensibly dividing, other worlds were on the brink of colliding. Back in 1980, the British physicist Tim Berners-Lee had developed a prototype of something called "hypertext." This was a long-theorized system of "hyperlinks" that could bind digital information together in unprecedented ways. Called ENQUIRE, the system was a massive database where items were indexed based on their relationships to each other. It resembled a very early version of Wikipedia. There was a crucial difference, however. ENQUIRE wasn't actually part of the internet. The computers running this revolutionary indexing program couldn't yet talk to each other.

Berners-Lee kept at it. In 1990, he began designing a new index that could run across a network of computers. In the process, he and his team invented much of the digital shorthand still in use today. They wrote a new code to bind the databases together. Hypertext markup language (HTML) defined the structure of each item, could display images and video, and, most important, allowed anything to link to anything else. Hypertext transfer protocol (HTTP) determined how hypertext was sent between internet nodes. To give it an easy-to-find location, every item was then assigned a unique URI (uniform resource identifier), more commonly known as a URL (uniform resource locator). Berners-Lee dubbed his creation the World Wide Web.

Just as ARPANET had shaped the systems that made online communication possible and Cerf and Kahn's protocol had allowed the creation of a network of networks that spanned the globe, the World Wide Web — the layer on top that we now call the "internet" — shaped what this communication looked like. Forward-thinking entrepreneurs quickly set to building the first internet "browsers," software that translated the World Wide Web into a series of visual "pages." This helped make the internet usable for the masses; it could now be navigated by anyone with a mouse and a keyboard. During this same period, the U.S. government continued investment in academic research and infrastructure development, with the goal of creating an "information superhighway." The most prominent sponsor of these initiatives was Senator Al Gore, leading to the infamous claim that he "invented" the internet. More accurately, he valuably sped up its development.

The advent of the World Wide Web matched up perfectly with another key development that mirrored technology past: the introduction of profit-seeking. In 1993, early internet architects gathered to take their biggest step yet: privatizing the entire system and tying independent internet operators — of which there were thousands — into a single, giant network. At the same time, they also took steps to establish a common system of internet governance, premised on the idea that no one nation should control it. In 1995, NSFNET formally closed, and a longstanding ban on online commercial activity was lifted.

The internet took off like a rocket. In 1990, there were 3 million computers connected to the internet. Five years later, there were 16 million. That number reached 360 million by the turn of the millennium.

As with the technologies of previous eras, the internet's commercialization and rapid growth paved the way for a gold rush. Huge amounts of money were to be made, not just in owning the infrastructure of the network but also in all the new business that sprang from it. Among the earliest to profit were the creators of Netscape Navigator, the easy-to-use browser of choice for three-quarters of all internet users. When Netscape went public in 1995, the company was worth $3 billion by the end of its first day, despite having never turned a profit. At that moment, the internet ceased to be the plaything of academics.

Amid the flurry of new connections and ventures, the parallel world of the internet began to grow so fast that it became too vast for any one person to explore, much less understand. It was fortunate, then, that nobody *needed* to understand it. The explorers who would catalog the internet's most distant reaches would not be people but "bots," special programs built to "crawl" and index the web's endless expanse. The first bots were constructed by researchers as fun lab experiments. But as millions of users piled online, web search became the next big business. The most successful venture was born in 1996, created by two Stanford graduate students, Larry Page and Sergey Brin. Their company's name was taken from a mathematical term for the number 1 followed by 100 zeros. "Google" symbolized their idea to "organize a seemingly infinite amount of information on the web."

As the web continued its blistering growth, it began to attract a radically different user base, one far removed from the university labs and tech enclaves of Silicon Valley. For these new digital arrivals, the internet wasn't simply a curiosity or even a business opportunity. It was the difference between life and death.

In early 1994, a ragtag force of 4,000 disenfranchised workers and farmers rose up in Mexico's poor southern state of Chiapas. They called themselves the Zapatista National Liberation Army (EZLN). The revolutionaries occupied a few towns and vowed to march on Mexico City. The government wasn't impressed. Twelve thousand soldiers were deployed, backed by tanks and air strikes, in a swift and merciless offensive. The EZLN quickly retreated to the jungle. The rebellion teetered on the brink of destruction. But then, twelve days after it began — as the Mexican military stood ready to crush the remnant — the government declared a sudden halt to combat. For students of war, this was a head-scratcher.

But upon closer inspection, there was nothing conventional about this conflict. More than just fighting, members of the EZLN had been *talking* online. They shared their manifesto with like-minded leftists in other countries, declared solidarity with international labor movements protesting free trade (their revolution had begun the day the North American Free Trade Agreement, or NAFTA, went into effect), established contact with international organizations like the Red Cross, and urged every journalist they could find to come and observe the cruelty of the Mexican military firsthand. Cut off from many traditional means of communication, they turned en masse to the new and largely untested power of the internet.

The gambit worked. Their revolution was joined in solidarity by tens of thousands of liberal activists in more than 130 countries, organizing in 15 different languages. Global pressure to end the small war in Chiapas built swiftly on the Mexican government. Moreover, it seemed to come from every direction, all at once. Mexico relented.

Yet this new offensive didn't stop after the shooting had ceased. Instead, the war became a bloodless political struggle, sustained by the support of a global network of enthusiasts and admirers, most of whom had never even heard of Chiapas before the call to action went out. In the years that followed, this network would push and cajole the Mexican government into reforms the local fighters hadn't been able to obtain on their own. "The shots lasted ten days," lamented the Mexican foreign minister, José Ángel Gurría, in 1995, but "ever since the war has been a war of ink, of written word, a war on the internet."

Everywhere, there were signs that the internet's relentless pace of innovation was changing the social and political fabric of the real world. There was the invention of the webcam and the launch of eBay and Amazon; the birth of online dating; even the first internet-abetted scandals and crimes, one of which resulted in a presidential impeachment, stemming from a rumor first reported online. In 1996, Manuel Castells, among the world's foremost sociologists, made a bold prediction: "The internet's integration of print, radio, and audiovisual modalities into a single system promises an impact on society comparable to that of the alphabet."

Yet the most forward-thinking of these internet visionaries wasn't an academic at all. In 1999, musician David Bowie sat for an interview with the BBC. Rather than promote his albums, Bowie waxed philosophical about technology's future. The internet wouldn't just bring people together, he explained; it would also tear them apart. "Up until at least the mid-1970s, we really felt that we were still living under the guise of a single, absolute, created society — where there were known truths and known lies and there was no kind of duplicity or pluralism about the things that we believed in," the artist once known as Ziggy Stardust said. "[Then] the singularity disappeared. And that I believe has produced such a medium as the internet, which absolutely establishes and shows us that we are living in total fragmentation."

The interviewer was mystified by Bowie's surety about the internet's powers. "You've got to think that some of the claims being made for it are hugely exaggerated," he countered.

Bowie shook his head. "No, you see, I don't agree. I don't think we've even seen the tip of the iceberg. I think the potential of what the internet is going to do to society, both good and bad, is unimaginable. I think we're actually on the cusp of something exhilarating and terrifying . . . It's going to crush our ideas of what mediums are all about."

THE GOSPEL OF MARK

"The goal wasn't to create an online community, but a mirror of what existed in real life."

It's a grainy 2005 video of a college-age kid sitting on a sofa in a den, red plastic cup in hand. He's trying to describe what his new invention is and — more important — what it isn't. It isn't going to be just a place to hang out online, a young Mark Zuckerberg explains. It's going to be a lot more than that.

Zuckerberg was part of the first generation to be born into a world where the internet was available to the masses. By the age of 12, he had built ZuckNet, a chat service that networked his dad's dental practice with the family computer. Before he finished high school, he took a graduate-level course in computer science. And then one night in 2003, as a 19-year-old sophomore at Harvard, Zuckerberg began a new, ambitious project. But he did so with no ambition to change the world.

At the time, each Harvard house had "facebooks" containing student pictures. The students used them as a guide to their new classmates, as well as fodder for dorm room debates about who was hot or not. Originally printed as booklets, the facebooks had recently been posted on the internet. Zuckerberg had discovered that the online version could be easily hacked and the student portraits downloaded. So over a frenzied week of coding, he wrote a program that allowed users to rate which of two randomly selected student portraits was more attractive. He called his masterpiece "Facemash." Visitors to the website were greeted with a bold, crude proclamation: "Were we let in on our looks? No. Will we be judged on them? Yes."

Facemash appeared online on a Sunday evening. It spread like wildfire, with some 22,000 votes being cast in the first few hours. Student outrage spread just as quickly. As angry emails clogged his inbox, Zuckerberg was compelled to issue a flurry of apologies. Hauled before a university disciplinary committee, he was slapped with a stern warning for his poor taste and violation of privacy. Zuckerberg was embarrassed — but also newly famous.

Soon after, Zuckerberg was recruited to build a college dating site. He secretly channeled much of his energy in a different direction, designing a platform that would combine elements of the planned dating site with the lessons he'd learned from Facemash. On January 11, 2004, he formally registered TheFacebook.com as a new domain. Within a month, 20,000 students at elite universities around the country signed up, with tens of thousands more clamoring for Facebook to be made available at their schools.

For those fortunate enough to have it, the elegant mix of personal profiles, public postings, instant messaging, and common-friend groups made the experience feel both intimate and wholly unique. Early users also experienced a new kind of feeling: addiction to Facebook. "I've been paralyzed in front of the computer ever since signing up," one college freshman confessed to his student newspaper. That summer, Zuckerberg filed for a leave of absence from Harvard and boarded a plane to Silicon Valley. He would be a millionaire before he set foot on campus again and a billionaire soon thereafter.

Talented and entrepreneurial though Zuckerberg was, these qualities alone aren't enough to explain his success. What he had in greatest abundance — what *all* of history's great inventors have had — was perfect timing.

Facebook, after all, was hardly the first social network. From the humble beginnings of SF-lovers, the 1980s and early 1990s had seen all manner of online bulletin boards and message groups. As the internet went commercial and growth exploded, people started to explore how to profit from our willingness, and perhaps our need, to share. This was the start of "social media" — platforms designed around the idea that an expanding network of users might create and share the content in an endless (and endlessly lucrative) cycle.

The first company dedicated to creating an online platform for personal relationships was launched in 1997. Six Degrees was based on the idea first proposed in sociology studies that there were no more than six degrees of separation between any two people in the world. On this new site, you could maintain lists of friends, post items to a shared bulletin board, and even expand your network to second- and third-degree connections. At its peak, Six Degrees boasted 3.5 million registered members. But internet use was still too scattered for the network to grow at scale, and web browsers were still too primitive to realize many of the architects' loftiest ambitions.

Through the late 1990s, a wave of such new online services were created. Early dating sites like Match.com essentially applied the eBay model to the world of dating, where you would shop in a marketplace for potential mates. Massively multiplayer online role-playing games (MMORPGs) rose to prominence with the 1997 Ultima series, allowing users to form warring clans. And in 1999, a young programmer launched LiveJournal, which offered access to dynamic, online diaries. These journals would first be called "weblogs," which was soon shortened to "blogs." All of these networks would bloom (within ten years, there were more than 100 million active blogs), but across them all, the sociality was only a side effect of the main feature. The golden moment had yet to arrive.

And then came Armageddon. In 2000, the "dot-com" bubble burst, and $2.5 trillion of Silicon Valley investment was obliterated in a few weeks. Hundreds of companies teetered and collapsed. Yet the crash also had the same regenerative effect as a forest fire. It paved the way for a new generation of digital services, building atop the charred remains of the old.

Even as Wall Street retreated from Silicon Valley, the internet continued its extraordinary growth. The 360 million internet users at the turn of the millennium had grown to roughly 820 million in 2004, when Facebook was launched. Meanwhile, connection speed was improving by about 50 percent each year. The telephone-based modem, whose iconic dial-up sound had characterized and frustrated users' experiences, was dying a welcome death, replaced by fast broadband. Pictures and video, which had once taken minutes or even hours to download, now took seconds.

Most important of all was the steady evolution of HTML and other web development languages that governed the internet's fundamental capabilities. The first internet browsers were pieces of software that provided a gateway into an essentially static World Wide Web. Visitors could jump from page to page of a website, but rarely could they change what was written on those pages. Increasingly, however, websites could process user commands; access and update vast databases; and even customize users' experience based on hundreds or thousands of variables. To input a Google search, for instance, was essentially to borrow one of the world's mightiest supercomputers and have it spin up server farms across the world, just to help you find out who was the voice of Salem the cat in *Sabrina, the Teenage Witch* (it was Nick Bakay). The internet was becoming not just faster but more visual. It was both user-friendly and, increasingly, user-controlled. Media entrepreneur Tim O'Reilly dubbed this new, improved internet "Web 2.0."

An apt illustration of the Web 2.0 revolution in action was Wikipedia, launched in 2001. Since the very first encyclopedia was assembled in the first century CE by Pliny the Elder, these compendiums of knowledge had been curated by a single source and held in libraries or peddled door-to-door. By contrast, Wikipedia was an encyclopedia for the digital age, constructed of "wikis" — website templates that allowed anyone to edit pages or add new ones. The result was a user-administered, endlessly multiplying network of knowledge — essentially, a smaller version of the internet. By 2007, the English-language Wikipedia had amassed more than 2 million articles, making it the largest encyclopedia in history.

Wikipedia was just for knowledge, though. The first Web 2.0 site to focus on social networks of friends was launched in 2002. The name "Friendster" was a riff on Napster, the free (and notorious) peer-to-peer file-sharing system that gloriously let users swap music with each other. Following the same design, it linked peer groups of friends instead of music pirates. Within a few months, Friendster had 3 million users.

A series of social media companies seeking to profit from the new promise of online social networking quickly jumped in. Myspace offered a multimedia extravaganza: customizable profiles and embeddable music, peppered with all manner of other colorful options. Fronted by musicians and marketed to teenagers, Myspace unamicably knocked Friendster from its perch. Then there was LinkedIn, the staid professional social network for adults, which survives to this day. And finally came Photobucket, a "lite" social media service that offered the once-unthinkable: free, near-limitless image storage.

At this crucial moment, Facebook entered the scrum. Originally, a user had to be invited into the social network. Within a year, the service spread to 800 college campuses and had more than a million active accounts, buoyed by the kind of demand that could only come with the allure of exclusivity. When Facebook stripped away its original barriers to entry, more people stampeded into the online club. By the end of 2007, the service numbered 58 million users.

As exponential growth continued, Zuckerberg sought to make his creation ever more useful and indispensable. He introduced the News Feed, a dynamic scroll of status updates and "shares" that transformed Facebook from a static web service into a living, breathing world. It also made Facebook a driver of what the world knew about everything from a user's personal life to global news. For all the sharing Facebook brought to the world, however, little was known about the algorithm that governed the visibility, and thus the importance, of items in the News Feed. That was, like so much of Facebook's inner workings, shared only between Zuckerberg and his employees.

As Facebook's power and scale expanded beyond his wildest dreams, Zuckerberg began to grow more introspective about its (and his) potential place in history. Not only did Facebook offer a way to share news, he realized, but it also promised to shape and create mass narratives. As Zuckerberg explained in 2007, "You can start weaving together real events into stories. As these start to approach being stories, we turn into a massive publisher. Twenty to 30 snippets of information or stories a day, that's like 300 million stories a day. It gets to the point where we are publishing more in a day than most other publications have in the history of their whole existence."

Within a decade, Facebook would boast 2 billion users, a community larger than any nation on earth. The volume of conversation recorded each day on Facebook's servers would dwarf the accumulated writings of human history. Zuckerberg himself would be like William Randolph Hearst transposed to the global stage, entertaining visiting ministers and dignitaries from his beige cubicle in Menlo Park, California. He would show off a solar-powered Facebook drone to the pope and arbitrate the pleas of armed groups battling it out in Ukraine. In his hands lay more power and influence than that young teen or any of the internet's pioneers could have imagined.

But that future hadn't arrived quite yet. It would take one final revolution before Facebook and its ilk — the *new* face of the internet — could swallow the world.

THE WORLD WIDE WEB GOES MOBILE

On January 9, 2007, Apple cofounder and CEO Steve Jobs donned his signature black turtleneck and stepped onstage to introduce the world to a new technology. "Today, Apple is reinventing the phone!" Jobs gleefully announced. Although nobody knew it at the time, the introduction of the iPhone also marked a moment of destruction. Family dinners, vacations, awkward elevator conversations, and even basic notions of privacy — all would soon be endangered by the glossy black rectangle Jobs held triumphantly in his hand.

The iPhone wasn't the first mobile phone, of course. That honor belonged to a foot-long, three-pound Motorola monstrosity invented in 1973 and first sold a decade later for the modest price of (in today's dollars) $10,000. Nor was the iPhone the first internet-capable smartphone. Ericsson engineers had built one of those as far back as 1997, complete with touchscreen and full working keyboard. Only 200 were produced. The technology of the time was simply too clunky and slow.

By comparison, the iPhone was *sexy*. And the reason wasn't just the sleek design. Internet access was no longer a gimmick or afterthought; instead, it was central to the iPhone's identity. The packed auditorium at the 2007 Macworld Expo whooped with excitement as Jobs ran through the list of features: a touchscreen; handheld integration of movies, television, and music; a high-quality camera; and major advances in call reception and voicemail. The iPhone's most radical innovation, though, was a speedy, next-generation browser that could shrink and reshuffle websites, making the entire internet mobile-friendly.

A year later, Apple officially unveiled its App Store. This marked another epochal shift. For more than a decade, a smartphone could be used only as a phone, calculator, clock, calendar, and address book. Suddenly, the floodgates were thrown open to any possibility, as long as they were channeled through a central marketplace. Developers eagerly launched their own internet-enabled games and utilities, built atop the iPhone's sturdy hardware. (There are roughly 2.5 million such apps today.) With the launch of Google's Android operating system and competing Google Play Store that same year, smartphones ceased to be the niche of tech enthusiasts, and the underlying business of the internet soon changed.

By 2013, there were some 2 billion mobile broadband subscriptions worldwide; by 2018, 6 billion. By 2020, that number is expected to reach 8 billion. In the United States, where three-quarters of Americans own a smartphone, these devices have long since replaced televisions as the most commonly used piece of technology.

The smartphone combined with social media to clear the last major hurdle in the race started thousands of years ago. Previously, even if internet services worked perfectly, users faced a choice. They could be in *real life* but away from the internet. Or they could tend to their digital lives in quiet isolation, with only a computer screen to keep them company. Now, with an internet-capable device in their pocket, it became possible for people to maintain both identities simultaneously. Any thought spoken aloud could be just as easily shared in a quick post. A snapshot of a breathtaking sunset or plate of food (*especially* food) could fly thousands of miles away before darkness had fallen or the meal was over. With the advent of mobile livestreaming, online and offline observers could watch the same event unfold in parallel.

One of the earliest beneficiaries of the smartphone was Twitter. The company was founded in 2006 by Silicon Valley veterans and hardcore free speech advocates. They envisioned a platform with millions of public voices spinning the story of their lives in 140-character bursts (the number came from the 160-character limitation imposed by SMS mobile texting minus 20 characters for a URL). This reflected the new sense that it was the network, rather than the content on it, that mattered. As Twitter cofounder Jack Dorsey explained, "We looked in the dictionary . . . and we came across the word 'twitter,' and it was just perfect. The definition was a 'short burst of inconsequential information,' and 'chirps of birds.' And that's exactly what the product was."

As smartphone use grew, so did Twitter. In 2007, its users were sending 5,000 tweets per day. By 2010, that number was up to 50 million; by

2015, 500 million. Better web technology then offered users the chance to embed hyperlinks, images, and video in their updates.

Soon enough, Twitter was transforming the news — not just how it was experienced (as with Michael Jackson's death in 2009), but how it was reported. Journalists took to using social media to record notes and trade information, sharing the construction of their stories in real time. As they received instant feedback, Twitter became, in the words of tech reporter Farhad Manjoo, "a place." It was akin to a clubhouse for everyone, "where many journalists unconsciously build and gut-check a worldview." The social network had become where people decided what merited news coverage and what didn't.

Twitter also offered a means for those being reported on to bypass journalists. Politicians and celebrities alike turned to it to get their own messages out. Donald Trump likened his Twitter account to "owning your own newspaper," but drastically improved, by only featuring one perfect voice: his own. In just a few years, Twitter would become the engine driving political reporting across much of the world, even with a relatively "small" population of 330 million users.

Blistering advancements in smartphone camera quality and mobile bandwidth also began to change what a social network could look like. Instagram launched in 2010 — a next-generation photo-sharing service that combined user profiles, hashtags, and a range of attractive image filters. By 2017, Instagram was adding more than 60 million photographs to its archives each day. It was gobbled up by Zuckerberg's Facebook, just as its video counterpart, YouTube, had been scooped up by Google a decade earlier.

Before almost anyone realized it, mobile tech, carefully policed app stores, and corporate consolidation had effected another massive change in the internet — who controlled it. After decades of freewheeling growth, companies that had been startups just a few years earlier had rapidly risen to rule vast digital empires, playing host to hundreds of millions of virtual residents. Even more important, this handful of companies provided the pillars on which almost all the millions of other online services depended.

And it seems likely that things will stay this way for some time. Would-be rivals have been bought up with ungodly amounts of cash that only the titans can afford to throw around. WhatsApp, for instance, was bought by Facebook in 2014 for $19 billion, the largest acquisition of a venture-backed company in history. Even if smaller companies retain their independence, these titans now control the primary gateways through which hundreds of millions of people access the web. In countries like Thailand and the Philippines, Facebook literally *is* the internet. For all the internet's creative chaos, it has come to be ruled by a handful of digital kings.

The outcome is an internet that is simultaneously familiar but unrecognizable to its founders, with deep ramifications for not just the web's future, but for the future of politics and war as well. As Tim Berners-Lee has written, "The web that many connected to years ago is not what new users will find today. What was once a rich selection of blogs and websites has been compressed under the powerful weight of a few dominant platforms. This concentration of power creates a new set of gatekeepers, allowing a handful of platforms to control which ideas and opinions are seen and shared . . . What's more, the fact that power is concentrated among so few companies has made it possible to [weaponize] the web at scale."

It's an echo of how earlier tech revolutions created new classes of tycoons, as well as new powers deployed into conflict. But it differs, markedly, in the sheer breadth of the current companies' control. Guglielmo Marconi, for instance, invented radio and tried to monopolize it. But he was unable to contain the technology's spread or exert control over the emerging network of radio-based media companies. He could scarcely have imagined determining what messages politicians or militaries could send via the airwaves, nor seizing the entire global market for ads on them. Similarly, the inventions of Samuel Morse and then Alexander Graham Bell spawned AT&T, the most successful communications monopoly of the twentieth century. But neither they nor their corporate inheritors ever came close to exercising the political and economic influence wielded by today's top tech founders.

There is one more difference between this and earlier tech revolutions: not all of these new kings live in the West. WeChat, a truly remarkable social media model, arose in 2011, unnoticed by many Westerners. Engineered to meet the unique requirements of the enormous but largely isolated Chinese internet, WeChat may be a model for the wider internet's future. Known as a "super app," it is a combination of social media and marketplace, the equivalent of companies like Facebook, Twitter, Amazon, Yelp, Uber, and eBay all fused into one, sustaining and steering a network of nearly a billion users. On WeChat, one can find and review businesses; order food and clothing; receive payments; hail a car; post a video; and, of course, talk to friends, family, and everyone else. It is an app so essential to modern living that Chinese citizens quite literally can't do without it: they're not allowed to delete their accounts.

CHILDHOOD'S END

Put simply, the internet has left adolescence.

In the span of a generation, it has blossomed from a handful of scientists huddled around consoles in two university computer labs into a network that encompasses half the world's population. Behind this growth lies a remarkable expansion of the demographics of this new community. The typical internet user is no longer a white, male, American computer scientist from California. More than half of all users are now in Asia, with another 15 percent in Africa. As Jakob Nielsen, a pioneer in web interface and usability, once observed of the changes afoot in the "who" of the internet, "Statistically, we're likely talking about a 24-year-old-woman in Shanghai."

Half of the world's population is online, and the other half is quickly following. Hundreds of millions of new internet users are projected to join this vast digital ecosystem each year. This is happening for the most part in the developing world, where two-thirds of the online population now resides. There, internet growth has overtaken the expansion of basic infrastructure. In sub-Saharan Africa, rapid smartphone adoption will see the number of mobile broadband subscriptions double in the next five years. According to U.S. National Intelligence Council estimates, more people in sub-Saharan Africa and South Asia have access to the internet than to reliable electricity.

As a result, the internet is also now inescapable. Anyone seeking to go beyond its reach is essentially out of luck. Remote outposts in Afghanistan and Congo offer Wi-Fi. At the Mount Everest base camp, 17,500 feet above sea level, bored climbers can duck into a fully functional cybercafe. Meanwhile, hundreds of feet beneath the earth's surface, the U.S. Air Force has begun to rework the communications systems of its nuclear missile bunkers. Among the upgrades is making sure the men and women standing watch for Armageddon can access Facebook.

All these people, in all these places, navigate an online world that has grown unfathomably vast. While the number of websites passed 1 billion sometime in 2014, unknowable millions more lurk in the "deep web," hidden from the prying eyes of Google and other search indexes. If one counts all the pieces of content with a unique web address that have been created across all the various social networks, the number of internet nodes rises into the high trillions.

In some ways, the internet has gone the way of all communications mediums past. After decades of unbridled expansion, the web has fallen under the control of a few giant corporations that are essentially too big to fail, or at least too big to fail without taking down vast portions of global business with them.

But in other obvious ways, the internet is nothing like its precursors. A single online message can traverse the globe at the speed of light, leaving the likes of poor Pheidippides in the dust. It requires no cumbersome wires or operators. Indeed, it can leap language barriers at the press of a button.

Yet that very same message is also a tool of mass transmission, almost infinitely faster than the printing press and unleashed in a way that radio and television never were. And each and every one of those transmissions joins millions of others every minute, colliding and building upon each other in a manner that bears little resemblance to the information flow of centuries past.

This is what the internet has become. It is the most consequential communications development since the advent of the written word. Yet, like its precursors, it is inextricably tied to the age-old human experiences of politics and war. Indeed, it is bound more closely than any platform before it. For it has also become a colossal information battlefield, one that has obliterated centuries' worth of conventional wisdom about what is secret and what is known. It is to this revolution that we turn next.

3

The Truth Is Out There

Social Media and the End of Secrets

For nothing is secret, that shall not be made manifest;
neither any thing hid, that shall not be known . . .

— LUKE 8:17

OPERATION NEPTUNE SPEAR, the mission to get Osama bin Laden, was among the most secretive military missions in history. As the Navy SEAL team took off in the early morning of May 2, 2011, only a few dozen people around the world had been briefed on the operation. One group was clustered in a top-secret military tactical operations center, and the other was gathered around a table in the White House Situation Room. There, President Obama and his advisors tracked the SEALs' progress from 7,600 miles away via a direct video link that was the sole source of information about the mission.

Or at least it was supposed to be. No one had counted on @Really-Virtual.

@ReallyVirtual wasn't a spy. He wasn't a journalist either. His real name was Sohaib Athar, a Pakistani tech geek and café owner, whose social media handle described him as "an IT consultant taking a break from the rat-race by hiding in the mountains with his laptops."

A few years earlier, Athar had moved from the busy city of Lahore to the more pleasant town of Abbottabad, a mountain tourist hub and home of the Pakistan Military Academy — as well as, now, the most wanted man in the world. Crashing on a late-night software project, Athar was distracted by the sound of helicopters overhead. So he did what millions of people do each day: he took to social media to complain.

"Helicopter hovering above Abbottabad at 1AM (is a rare event)," he tweeted first. As the SEALs' mission played out over the next several minutes, Athar posted a litany of complaints that doubled as news reports. When the first helicopter took off, carrying away bin Laden's body and hard drives filled with data on Al Qaeda's networks, he tweeted, "Go away helicopter — before I take out my giant swatter :-/." As the remaining SEALs detonated a crashed helicopter and piled into a backup chopper, Athar shared the news of the explosion. "A huge window shaking bang here in Abbottabad . . . ," he tweeted. "I hope its not the start of something nasty :-S."

Eight hours later, the traditional news media finally caught up to one of the most important stories in a decade. On NBC, *The Celebrity Apprentice* was airing. Just as Donald Trump was explaining his rationale for "rehiring" singer La Toya Jackson, the network cut away. In a surprise prime-time address, President Obama announced that a top-secret raid had taken place in Pakistan and that Osama bin Laden was dead. "Justice has been done," he concluded. In cities around the United States, people danced in the streets. Thousands of miles away, Athar was coming to his own realization. "Uh oh," he tweeted, "now I'm the guy who liveblogged the Osama raid without knowing it."

It was lunchtime in Abbottabad when the messages began pouring in — a trickle that swiftly transformed into a torrent. Athar's Twitter follower count jumped from 750 to 86,000. He was deluged with calls for interviews and fan requests. Local journalists sped to his café to talk face-to-face. More worrying, a growing online mob accused him of spying for either the U.S. or Pakistani government. Surely, they argued, that was the only way that Athar could have known about such a top-secret military operation.

But the truth was simpler and more profound: Sohaib Athar was just a guy who happened to be near something newsworthy, with a computer and a social media account at hand.

When the internet first began to boom in the 1990s, internet theorists proclaimed that the networked world would lead to a wave of what they called "disintermediation." They described how, by removing the need for "in-between" services, the internet would disrupt all sorts of longstanding industries. Disintermediation soon remade realms ranging from retail stores (courtesy of Amazon) and taxi companies (courtesy of Uber) to dating (courtesy of Tinder). Athar's tale showed how the business of information gathering had undergone the same kind of disintermediation. No longer did a reporter need to be a credentialed journalist working for a major news organization. A reporter could be anyone who was in the right place at the right time. But this shift wasn't just about reporting the news. It was also changing all the people who make use of this information, be they citizens, politicians, soldiers, or spies.

There was one more lesson wrapped up in the surreal saga of Sohaib Athar, unnoticed by many at the time but painfully obvious to observers in the U.S. intelligence community. Operation Neptune Spear — among the most closely guarded operations in history — had nonetheless been documented in real time for anyone in the world to see. And this had happened accidentally, in a country where just 6 percent of the population had internet access at the time. What would the future hold as more and more people came online? Even more, how would intelligence agencies cope when it wasn't just a lone night owl inadvertently sharing secrets, but organized groups of analysts dedicated to parsing social media to find the operations hidden in plain sight?

"Secrets now come with a half-life," one CIA official told us, with more than a twinge of regret.

EVERYTHING IS ILLUMINATED

"Let's give a welcome to 'Macaca' here," puffed the fleshy man in the blue shirt, pointing toward the camera. "Welcome to America and the real world of Virginia."

It was August 2006, and Senator George Allen was barnstorming for votes at a rural Virginia park. A darling of the conservative wing of the Republican Party, he was already looking past this reelection bid. He'd recently made exploratory trips to Iowa and New Hampshire, testing the waters before a potential presidential run. But although Allen didn't know it yet, his political career had just ended in that single moment, all because of how the internet had changed.

The man behind the camera was S. R. Sidarth, a 20-year-old volunteer for Allen's opponent, who had taken to filming Allen's events. Sidarth was also Indian American — and the only brown face amid the rally's 100 attendees. And the name Allen had just called him, "Macaca," is Portuguese for "monkey," used as a racial slur for centuries.

The history of politicians saying and doing horrible or stupid things on the campaign trail is as old as democracy itself. But the trajectory from a bad moment to a fatal gaffe previously required a professional journalist to be on the scene to document it. Then the moment would have to be reported via a newspaper or radio or TV station. In order for the gaffe to build truly national momentum, other professional journalists and their outlets would then have to pick it up. Unfortunately for Allen, social media had altered this process, propelling his words beyond the control of any politician or journalist.

Sidarth's minute-long recording was quickly posted on YouTube, the new video-sharing platform, scarcely a year old in 2006. This was an unusual decision at the time, because the video clip was unedited and unattached to any broader story. It proved an ingenious move, however, as the very nature of the clip was part of its appeal. Easy to view and share, Sidarth's video went viral, with hundreds of thousands of people seeing it firsthand online, and the news media being able to report on and link to it.

Allen's advisors, skilled and experienced in the old model of political campaigning, were flummoxed. At first they denied that the incident had happened. Then they claimed that Allen had done nothing wrong, explaining that "Macaca" wasn't meant as a slur. And then they pivoted to claiming that Allen had actually said "Mohawk," referring to Sidarth's hair.

The problem with each explanation was that, unlike in the past, anyone who wanted to could now see the evidence for themselves. They could click "play" and hear the ugly word again and again. They could see that Allen was using it to describe the one brown-skinned person in a crowd of white people and suggesting that Sidarth wasn't a "real" American.

Allen's lead in the polls plummeted, and he went on to lose a race in which his victory had been all but guaranteed. Instead of making a run for president, he never served in elected office again. As for Sidarth, he was named Salon's person of the year: a "symbol of politics in the 21st century, a brave new world in which any video clip can be broadcast instantly everywhere and any 20-year-old with a camera can change the world."

What became known as the "Macaca moment" was a hint of the web-driven radical transparency that was just starting to change how information was gathered and shared — even the nature of secrecy itself.

The relatively new digital camera that Sidarth used to document Allen's fateful words has since been followed by some 9 billion digital devices that, importantly, are now linked online. By 2020, that number will soar to 50 billion, as devices ranging from smartphones to smart cars to smart toothbrushes all join in to feed the internet.

Most significantly, all of the new items coming online carry something that the computers used by ARPANET, and even the one used by Mark Zuckerberg to create Facebook, lacked: "sensors," devices for gathering information about the world beyond the computer. Some sensors are self-evident, like the camera of a smartphone. Others lurk in the background, like the magnometer and GPS that provide information about direction and location. These billions of internet-enabled devices, each carrying multiple sensors, are on pace to create a world of almost a trillion sensors. And any information put online comes with "metadata," akin to digital stamps that provide underlying details of the point of origin and movement of any online data. Each tweet posted on Twitter, for instance, carries with it more than sixty-five different elements of metadata.

This plethora of sensors and associated metadata is making real an idea that has long possessed (and frightened) humanity: the possibility of an ever-present watcher. The ancient Greeks imagined it as Argus Panoptes, a mythological giant with 100 eyes. During the Enlightenment, the English philosopher Jeremy Bentham turned the monster into the Panopticon — a hypothetical building in which all the occupants could be observed, but they never saw those watching them. Ominously, Bentham pitched his design as being useful for either a factory or a prison. George Orwell then gave the panoptic idea an even darker spin in his novel *1984*. His futuristic totalitarian world was filled with "telescreens," wall-mounted televisions that watched viewers while they watched the screens.

Today, the combination of mass sensors and social media has rendered these bizarre fantasies an equally bizarre reality. Yet rather than gods or rulers, *we* are collectively the ones doing the watching. A decade after Allen flamed out, any politician worth their salt, gazing into a crowd of a hundred people, might reasonably assume they are the subject of no less than a half dozen videos and many more photographs, texts, and audio snippets, any of which might prompt scores of social media reactions. Indeed, they would likely be upset if no one posted about the event online. To ensure that didn't happen, they would likely be doing it themselves. In the run-up to the 2018 U.S. midterm elections, some candidates pushed out more than a dozen Facebook videos a day.

None of this means that gaffes like Allen's no longer happen, nor that racist comments are no longer made; rather the opposite. When everything can be recorded, everything is on the record.

The amount of data being gathered about the world around us and then put online is astounding. In a minute, Facebook sees the creation of 500,000 new comments, 293,000 new statuses, and 450,000 new photos; YouTube the uploading of more than 500 hours of video; and Twitter the posting of more than 350,000 tweets. And behind this lies billions more dots of added data and metadata, such as a friend tagging who appeared in that Facebook photo or the system marking what cellphone tower the message was transmitted through. In the United States, the size of this "digital universe" doubles roughly every three years.

Each point of information might be from an observer consciously capturing a speech or a gun battle, or it might be unwittingly shared with the world, as was Athar's coverage of the bin Laden raid. The most valuable information might even lurk in the background. Snapping tourist photos of a harbor, Chinese civilians once accidentally revealed secrets of their navy's new aircraft carrier, under construction in the distance. Or an interesting tidbit might lie in the technical background. Exercise apps have inadvertently revealed everything from the movements of a murderer committing his crime to the location of a secret CIA "black site" facility in the Middle East. (A heat map made from tracing agents' daily jogs around the perimeter of their base provided a near-perfect outline of one installation.)

In 2017, General Mark Milley, chief of staff of the U.S. Army, summed up what this means for the military: "For the first time in human history, it is near impossible to be unobserved." Consider that in preparation for D-Day in June 1944, the Allies amassed 2 million soldiers and tens of thousands of tanks, cannons, jeeps, trucks, and airplanes in the British Isles. Although German intelligence knew that the Allied forces were there, they never figured out where or when they would strike. That information came only when the first Americans stormed Utah Beach. Today, a single soldier's or local civilian's Facebook account would be enough to give away the whole gambit. Indeed, even their digital *silence* might be enough to give it away, since a gap in the otherwise all-encompassing social media fabric would be conspicuous.

What's unveiled is not just the movement of armies. Such data can be used to geographically pinpoint people, even in circumstances where they'd rather not be found. For instance, Ashley Madison is a social network that links people thinking of cheating on their spouses. Its algorithms mine social media to detect when business travelers arrive at a hotel (and thus are more likely to stray from their marriages). In a similar fashion, in the fighting that began in Ukraine in 2014, Russian military intelligence pinpointed the smartphones of Ukrainian soldiers arriving on the front lines. Just as Ashley Madison uses geographically targeted data to fire off web ads to potentially philandering travelers, the Russians used it to send messages like "They'll find your bodies when the snow melts." Then their artillery began firing at the Ukrainians.

But what stands out about all this information is not just its massive scale and form. It is that most of it is *about* us, pushed out *by* us. This all arguably started back in 2006, when Facebook rolled out a design update that included a small text box that asked a simple question: "What's on your mind?" Since then, the "status update" has allowed people to use social media to share anything and everything about their lives they want to, from musings and geotagged photos to live video feeds and augmented-reality stickers.

The result is that we are now our own worst mythological monsters—not just watchers but chronic over-sharers. We post on everything, from events small (your grocery list) to momentous (the birth of a child, which one of us actually live-tweeted). The exemplar of this is the "selfie," a picture taken of yourself and shared as widely as possible online. At the current pace, the average American millennial will take around 26,000 selfies in their lifetime. Fighter pilots take selfies during combat missions. Refugees take selfies to celebrate making it to safety. In 2016, one victim of an airplane hijacking scored the ultimate millennial coup: taking a selfie with his hijacker.

These postings are revelatory of our personal experiences, of course. But they also now convey the weightiest issues of public policy. The first sitting world leader to use social media was Canadian prime minister Stephen Harper in 2008, followed quickly by U.S. president Barack Obama. A decade later, the leaders of 178 countries have joined in. Even former Iranian president Mahmoud Ahmadinejad, who famously banned Twitter during a brutal crackdown, has since changed his mind on the morality — and utility — of social media. He debuted online with a friendly English-language video as he stood next to the Iranian flag. "Let's all love each other," he tweeted.

It is not just world leaders, though. Agencies at every level and in every type of government now share their own news, from some 4,000 national embassies to the fifth-grade student council of the Upper Greenwood Lake Elementary School. National militaries have gotten in the game as well. In the United States, the Army, Navy, Air Force, and Marine Corps all have an official social media presence. So do their bases, combat units, generals, and admirals. Even the status of individual military operations is now updated. When the U.S. military's Central Command expanded Operation Inherent Resolve against ISIS in 2016, Twitter users could follow along directly via the hashtag #TalkOIR. A U.S. military officer even popped up on the discussion forum Reddit for its signature Ask Me Anything series. He answered dozens of questions on the state of anti-ISIS operations, but delicately declined to offer the U.S. military's thoughts on the latest season of the *Archer* television show.

The result of all this sharing is an immense, endlessly multiplying churn of information and viewpoints. And it matters not just for the here and now. Nothing truly disappears online. Instead, the data builds and builds, just waiting to reemerge at any moment. According to law professor Jeffrey Rosen, the social media revolution has essentially marked "the end of forgetting."

This massive accumulation of updates, snapshots, and posts over time offers revelations of its own. The clearest exemplar of this phenomenon is the first president to have used social media before running for office. As both a television celebrity and social media addict, Donald Trump entered politics with a vast digital trail behind him. The Internet Archive has a fully perusable, downloadable collection of more than a thousand hours of Trump-related video, while his Twitter account has generated some 40,000 messages. Never has a president shared so much of himself — not just words, but even neuroses and particular psychological tics — for all the world to see. Trump is a man — the most powerful in the world — whose very essence has been imprinted on the internet.

Tom Nichols is a professor at the U.S. Naval War College who worked with the intelligence community during the Cold War. He explains the unprecedented value of this vault of information: "It's something you never want the enemy to know. And yet it's all out there . . . It's also a window into how the President processes information — or how he doesn't process info he doesn't like. Solid gold info." Russian intelligence services reportedly came to the same conclusion, using Trump's Twitter account as the basis on which to build a psychological profile of the forty-fifth president.

And exhaustive though it is, Trump's online dossier encompasses only a single decade and started only when he was in his 60s. Nearly every future politician or general or soldier or voter will have a much bigger dataset, from much earlier in life. Indeed, this inescapable record may well change the prospects of those who wish to become leaders in the future. As Barack Obama said after he left office, "If you had pictures of everything I'd done in high school, I probably wouldn't have been President of the United States."

The consequences of such extensive online sharing go beyond broadcasting our day-to-day activities and thoughts. Social media can also provide a surprisingly clear window into our psychological and neurological states. Luke Stark, a researcher in the sociology department at Dartmouth College, explains that accumulated online postings provide "something much more akin to medical data or psychiatric data." Even the most trivial details can be unexpectedly revealing. Consistent use of black-and-white Instagram filters and single-face photos, for instance, has proven a fairly good identifier of clinical depression.

As unprecedented as all this information may be, it matters little unless there's someone on the other end to appreciate — or exploit — it.

Just as the internet has changed the volume, source, and availability of content, it also has wrought dramatic changes in *how* this information is used.

THE ELECTRIC BRAIN AWAKENS

The ten killers snuck into Mumbai's port aboard inflatable boats on November 26, 2008. Once ashore, they split up, fading into the megacity of some 18 million people. The attacks started soon after: a staccato of massacres at a railway station, a tourist café, a luxury hotel, and a synagogue. Over the next three days, 164 civilians and police officers would be killed. Another 300 would be injured. The tragedy would mark the deadliest Indian terror attack in a generation. It also signaled a radical change in how the news was both parsed and spread.

Although there were just 6 million Twitter users worldwide at the time, Mumbai's booming IT sector had helped build a small, vocal network of early adopters. The tweets began minutes after the first attack: 140-character reports, observations, cries for help, and words of warning that captured each explosion and gunshot. "I have just heard 2 more loud blasts around my house in colaba," wrote @kapilb, blocks from one attack. "Grenades thrown at colaba," added @romik a few minutes later.

Users living thousands of miles from the attack also served as unexpected conduits for those trapped inside it. When the terrorists took hostages in two Mumbai hotels, a Silicon Valley–based venture capitalist helped spread the news. "I just spoke with my friends at the Taj and Oberoi," tweeted @skverma, "people have been evacuated or are barricaded in their Rooms."

With Indian authorities reeling and journalists restricted as Mumbai became an instant and unexpected war zone, much of the useful reporting originated from a network that emerged spontaneously. Mumbai's online community kicked into gear, sharing riveting stories that quickly spread across the digital ecosystem. One brave resident took to the streets, snapping dozens of pictures. He posted them to the image-sharing service Flickr, originally created for video gamers. In a reversal of journalistic practice, these amateur photographs filled the front pages of newspapers the next day.

Of course, as this network of amateur reporters recounted the Mumbai attacks, it also spread all the spurious rumors that we know now accompany such events. There were fake reports of nonexistent attacks, which then begat more falsehoods as people reacted to them. So new communities arose to sift through the mountain of data, networks of online analysts separating fact from fiction.

In a move that would soon become the norm, the Mumbai attacks got their own Wikipedia page — roughly four hours after the first shots had been fired. Dozens of volunteer editors pitched in, debating everything from serious allegations of external support (rumors of Pakistani government involvement were already swirling) to tricky issues of phrasing (were the attackers "Muslim militants" or "Muslim terrorists"?). All told, before the last terrorist had been cornered and shot, the Wikipedia entry was edited more than 1,800 times.

Another important new tool was put to work. Google Maps, launched in 2005, had made it possible for people scattered across the world to locate and share precise coordinates — a capability previously reserved for only the most advanced militaries. The location of each bomb blast and firefight was plotted as soon as it was reported. This revealed something else unexpected. It became possible not just to track the latest news, but to build the operation's history. You could trace where the terrorists had disembarked from their boats; where the first car bomb had detonated; where each gun battle had taken place. Eventually, Google Maps would even plot the location of victims' funerals.

Anyone in the world could now monitor a battle unfolding in real time. This included even the people who had sent the attackers. Nesting in a control room in Karachi, Pakistan, the operation's commanders reportedly kept in cellphone contact with the gunmen on the ground, guiding them from target to target. Rather than relying on some secret intelligence network for updates, they could track the same social media platforms as everyone else. Just as a tweet about a possible explosion warned others to stay away, it helped the militants predict the focus of attention and the paths of emergency responders.

Eventually, the online crowd began realizing this and urged others to stop talking about the movement of Indian security forces. Some took it upon themselves to police the news. One widely shared message declared, "Indian government asks for live Twitter updates from Mumbai to cease immediately. ALL LIVE UPDATES — PLEASE STOP TWEETING." The Indian government had said no such thing; it was fake news. Others began to fight back the only way they knew how, sending a flurry of tweets at the terrorists. "Die, die, die, if you're reading this," one wrote.

The online activity also begat a new kind of emergency coordination. Messages poured in begging for blood donations, directing donors to the hospital where most of the victims had been taken. Other users spread word of tip lines, publicizing the information further and faster than the Indian government could on its own. This was a radical departure from past emergencies, which had depended on sluggish public broadcast systems and word of mouth.

As the smoke cleared, the Mumbai attack left several legacies. It was a searing tragedy visited upon hundreds of families. It brought two nuclear powers to the brink of war. And it foreshadowed a major technological shift. Hundreds of witnesses — some on-site, some from afar — had generated a volume of information that might previously have taken months of diligent reporting to gather. By stitching these individual accounts together, the online community had woven seemingly disparate bits of data into a cohesive whole. It was like watching the growing synaptic connections of a giant electric brain.

There is a word for this: "crowdsourcing." An idea that had danced excitedly on the lips of Silicon Valley evangelists for years, crowdsourcing had been originally conceived as a new way to outsource software programming jobs, the internet bringing people together to work collectively, more quickly and cheaply than ever before. As social media use had skyrocketed, the promise of crowdsourcing had extended into a space beyond business. Mumbai proved an early, powerful demonstration of the concept in action. The incidents would swiftly multiply from there.

At its core, crowdsourcing is about redistributing power — vesting the many with a degree of influence once reserved for the few. Sometimes, crowdsourcing might be about raising awareness, other times about money (also known as "crowdfunding"). It can kick-start new businesses or throw support to people who might once have languished in the shadows. It was through crowdsourcing, for instance, that septuagenarian socialist Bernie Sanders became a fundraising juggernaut in the 2016 U.S. presidential election, raking in $218 million online.

Of course, like any useful tool, crowdsourcing has also been bent to the demands of war. A generation ago, Al Qaeda was started by the son of a Saudi billionaire. By the time of the Syrian civil war and the rise of ISIS, the internet was the "preferred arena for fundraising" for terrorism, for the same reasons it has proven so effective for startup companies, nonprofits, and political campaigns. It doesn't just allow wide geographic reach. It expands the circle of fundraisers, seemingly linking even the smallest donor with their gift target on a personal level. As *The Economist* explained, this was, in fact, one of the key factors that fueled the years-long Syrian civil war. Fighters sourced needed funds by learning "to crowdfund their war using Instagram, Facebook and YouTube. In exchange for a sense of what the war was really like, the fighters asked for donations via PayPal. In effect, they sold their war online."

Just as any digital marketing guru would advise, Syrian fighting groups have molded their message to reflect the donor pool's interest. Many of Syria's early rebel fighters sought to establish a free, secular democracy. But this prospect didn't excite the fundamentalist donors from the wealthy Arab states. So, to better sell their effort online, even secular fighters grew impressively long beards and made sure to pepper their battle videos with repetitions of *"Allahu Akbar"* (God is great).

Fundraisers also got creative through what became known as "financial jihad." Some clerics argued that online pledges allowed donors to fulfill their religious duties in the same way they would if they had actually served in battle. Much as you might sponsor your cousin's "fun run" to fight cancer, you could sponsor a rocket-propelled grenade (RPG) launcher for a Syrian rebel (it went for $800). Or you could back the rebels' opponents. Hezbollah, the Iranian-sponsored terror group that allied with the Syrian regime, ran an "equip a mujahid" campaign on Facebook and Twitter. It similarly allowed online supporters to fulfill their religious obligations by buying weapons and ammunition for the war.

As radical transparency merges with crowdsourcing, the result can wander into the grotesque. In 2016, a hard-line Iraqi militia took to Instagram to brag about capturing a suspected ISIS fighter. The militia then invited its 75,000 online fans to vote on whether to kill or release him. Eager, violent comments rolled in from around the world, including many from the United States. Two hours later, a member of the militia posted a follow-up selfie; the body of the prisoner lay in a pool of blood behind him. The caption read, "Thanks for vote." In the words of Adam Linehan, a blogger and U.S. Army veteran, this represented a bizarre evolution in warfare: "A guy on the toilet in Omaha, Nebraska, could emerge from the bathroom with the blood of some 18-year-old Syrian on his hands."

The rapidity with which these bloodthirsty votes came in illustrates how the speed at which these spontaneous online collectives form is accelerating apace with the information environment. In 2008, the online mosaic of the Mumbai terror attacks was pieced together in a matter of hours. Five years later, at the Boston Marathon Bombing of April 15, 2013, this timeline had shifted by an order of magnitude.

It took about thirty seconds for Boston's emergency coordination center to learn of the attack that killed 3 people and wounded nearly 300 more — a fact that it would proudly document in an after-action report commissioned a year later. But the news was already online. The very moment when Boston police officers and firefighters were shouting into their radios, @KristenSurman tapped out a frantic message to her Twitter followers: "Holy shit! Explosion!" Seconds later, @Boston_to_a_T uploaded the first photo of the attack, taken mid-fireball. It took three minutes for the terror attack to be reported by a professional media outlet. The coverage came via a quick tweet from Fox Sports Radio — but not in Boston; in Washington State. By contrast, it took nearly an hour for Boston police to formally confirm the bombing.

Since then, the size and reach of this online audience has continued to grow exponentially, while the number of global smartphone users has more than doubled. Virtually any event leaves a digital trail that can be captured, shared, and examined by hungry internet users. As the audience eagerly rushes from one development to the next, it drives new developments of its own. Less and less is there a discrete "news cycle." Now there is only the news, surrounding everyone like the Force in *Star Wars,* omnipresent and connected to all.

The best way to describe the feeling that results is a term from the field of philosophy: "presentism." In presentism, the past and future are pinched away, replaced by an incomprehensibly vast *now.* If you've ever found yourself paralyzed as you gaze at a continually updating Twitter feed or Facebook timeline, you know exactly what presentism feels like. Serious reflection on the past is hijacked by the urgency of the current moment; serious planning for the future is derailed by never-ending distraction. Media theorist Douglas Rushkoff has described this as "present shock." Buffeted by a constant stream of information, many internet users can feel caught in a struggle just to avoid being swept away by the current.

With the marriage of radical transparency and ever faster and more feverish crowdsourcing, the line between *observer* and *participant* has been irrevocably blurred. "News" originates not just with journalists, but with anyone at the scene with a smartphone; any soldier with an Instagram account; any president tweeting away while watching Fox News in his bedroom. In a sense, *everyone* has become part of the news. And while people who serve to make sense of the madness still exist, the character and identity of these gatekeepers have transformed as well.

THE NEW MIDDLE

The word "media" comes from the Latin word for "middle." For the past century, "media" has been used to refer to the professional journalists and news organizations paid to serve as the conduit between the public and the news. Today, social media has put new voices in the middle.

At the ripe old age of 11, Rene Silva joined this once-august profession. He started his first newspaper, printing it on his school's photocopier. Soon Silva was running a full-fledged online news agency, dedicated to catching the stories that fell through the cracks. He built a website, he launched a Facebook page and YouTube channel, and he maintained an active Twitter feed. All this before he left high school.

Familiar though his story might sound, Silva had none of the advantages of a typical tech-savvy American teenager from suburbia. Instead, his *Voz das Comunidades* (Voice of the Community) reported on life in Complexo do Alemão, a cluster of Brazilian favelas infested with drug traffickers, with 70,000 people squeezed into a single square mile. Although most of Rene's stories resembled those of any local newspaper — articles on illegal parking and profiles of local leaders — sometimes circumstances caught up with him. In 2010, when the Brazilian government's security forces launched a series of rolling firefights against the gangs, the 17-year-old Silva was there to capture it all. He live-tweeted the location of each battle, set up a video feed, and enlisted friends to help hunt for more stories. At times he took to correcting the accounts of his rivals — professional, adult journalists, who didn't know the neighborhood and so often misreported the street names. Silva's efforts won him international acclaim, and he was named an honorary torchbearer in both the 2012 and 2016 Olympic Games. Today, he seeks to apply his hyperlocal reporting approach to neighborhoods across Brazil.

Just as social media has altered the people who witness and report the news, it has also changed those who deliberately go out and gather it. Across the world, there is a new breed of journalist, empowered by the web, often referred to as the "citizen reporter."

Sometimes, these reporters fill in the gaps, using social media to report on areas the traditional media finds unprofitable to report from. Selinsgrove, Pennsylvania, population 5,500, is far more bucolic than the Alemão favelas, but it, too, lacks a truly local paper. Into the void stepped 9-year-old Hilde Kate Lysiak, whose *Orange Street News* has covered everything from vandalism to a fire department corruption scandal. Hilde's journalistic chops are perhaps best illustrated by how she responded when police tried to keep a murder investigation quiet, asking her not to report it. She refused. "I may be nine, but I have learned that my job as a reporter is to get the truth to the people. I work for them, not the police."

Other times, they cover stories in places where the job of reporting has become too dangerous. For instance, Mexico's decade-long drug war between the government, paramilitaries, and cartels has taken a terrible toll on society — and journalists. There were nearly 800 documented attacks on reporters between 2006 and 2016, killing dozens.

As they seek to avoid this fate, media organizations face a terrible set of choices. Some have become criminal mouthpieces, employing cartel-affiliated reporters known as "links." Others have become more selective in the kinds of news they cover. The *Norte* newspaper in Juárez took the second route, avoiding stories that might anger the state's secret cartel overseers. As one editor explained, "You do it or you die, and nobody wants to die. *Auto Censura* — self-censorship — that's our shield." But even that wasn't enough to hold the cartels at bay. In 2017, the *Norte* shuttered with a final, bitter "Adiós."

Into this void stepped Felina, a woman whose Twitter handle and image were inspired by Catwoman, of comic book legend. Her home state of Tamaulipas had been split between two gangs, the Gulf Cartel and the Zetas, whose war had killed over 15,000 people. Faced with this ongoing horror, she and a small group of online vigilantes banded together to form an organization to inform and protect their fellow citizens. They called it Valor por Tamaulipas (Courage for Tamaulipas). With Felina serving as administrator, the group built a crowdsourced news service that gathered and distributed the information that residents of Tamaulipas needed to know to stay safe. The reports ranged from notifications of ongoing shootouts to photos of trigger-happy cartel members, so citizens would know which streets and people to avoid.

Felina was soon curating one of the most popular social media channels in the state, with over half a million followers on Facebook and 100,000 on Twitter. This success, though, marked Felina as a target. A cartel offered a reward of 600,000 pesos (about $48,000 at the time) for the identities of the site's anonymous administrators. It even posted its own message to the citizen journalists' feed: "We're coming very close to many of you . . . watch out felina."

Felina was undeterred. The threats attracted even more attention — the internet's most powerful currency. Web traffic quadrupled, and fans from near and far celebrated their work. Inspired, Felina threw herself into a grander vision of transforming Tamaulipas. She labored to raise money for poor people, organizing blood donor drives and even helping to find a needy person a pair of orthopedic shoes.

This generosity of spirit may have been Felina's undoing, seeding clues that the cartels used to track her down. On October 16, 2014, Felina's Twitter account posted the following message: "FRIENDS AND FAMILY, MY REAL NAME IS MARÍA DEL ROSARIO FUENTES RUBIO. I AM A PHYSICIAN. TODAY MY LIFE HAS COME TO AN END."

The feed that had long sought to help the citizens of Tamaulipas then posted two photos in rapid succession. The first showed a middle-aged woman looking directly into the camera. The second showed the same woman lying on the floor, a bullet hole in her head. As journalist Jason McGahan put it, "She tweeted against the Mexican cartels. They tweeted her murder."

With such risks, the work of these brave citizen reporters becomes something more than mere reporting. Perhaps the most notable example unfolded in the midst of the terrible group ISIS, whose rise also personified social media's new power. In 2013, the city of Raqqa, Syria, fell to ISIS, becoming its capital. Soon, Raqqa became an epicenter of horror, from the brainwashing of children to public crucifixions. Yet, because ISIS had a practice of murdering professional journalists, there was no one from the international media to document and report this reign of terror.

So a group of seventeen citizens banded together to tell the story of their city's destruction. They did so via an online news network they called Raqqa Is Being Slaughtered Silently. It was as much an act of resistance as reporting. Their belief, as one member put it, was that "truthtelling" would prove to be more powerful than ISIS's weapons.

For years, these citizen reporters were the primary source of information on life in a city ruled by ISIS. The group's work was incredibly dangerous, with ISIS on the constant hunt for them and anyone they loved. Soon after the network's launch, ISIS began broadcasting counterprogramming titled "They Are the Enemy So Beware of Them." It showed a group, whom ISIS claimed to be the reporters and their families, being paraded in front of cameras and then hanged from a tree. Raqqa Is Being Slaughtered Silently then reported these executions, sadly noting that ISIS had killed the wrong people. Eventually, ten members of the network were found and killed. One woman, Ruqia Hassan, dashed off a quick Facebook post shortly before ISIS police captured and executed her. "It's okay because they will cut my head," she wrote. "And I have dignity, it's better than I live in humiliation." For both its bravery and novelty, the group was awarded a 2015 International Press Freedom Award.

A common thread runs through all of these stories. From favela life to cartel bloodlettings to civil wars, social media has erased the distinction between citizen, journalist, activist, and resistance fighter. Anyone with an internet connection can move seamlessly between these roles. Often, they can play them all at once.

This communications revolution has been complemented by another shift, easier to miss but perhaps even more consequential. Just as the people who document and reveal secrets to the world have changed, so, too, have the people who traditionally work behind the scenes collecting and analyzing this information. They're called "intelligence analysts," or, more colloquially, "spies." And today, they look (and work) quite a bit different indeed.

WORK-FROM-HOME SHERLOCK HOLMES

As he boarded his long-haul flight from the Netherlands to Malaysia on July 17, 2014, the Dutch musician Cor Pan snapped a picture of the waiting Boeing 777 and uploaded it to Facebook. "In case we go missing, here's what it looks like," he wrote. It was supposed to be a joke. Instead, it would become one of the last digital echoes of a terrible tragedy.

A few hours later, the plane was flying over eastern Ukraine, a region divided between the local government and Russian-backed separatists. Many of the 298 passengers and crew aboard were drowsing with the shades drawn. They were a mix of vacationers, business travelers, and a group of scientists bound for an HIV/AIDS conference. In the cockpit, the mood was similarly placid. The pilots' biggest worry was light turbulence. We know this because all the sounds and activity on the cockpit voice recorder were perfectly normal, up until the thirty milliseconds before it stopped working.

The pilots never saw the missile as it pierced the cloud cover to their left. In a fraction of a second, over 7,600 pieces of superheated shrapnel tore through the cockpit, ripping the pilots to shreds. The blast cleaved the front of the plane from the rest of the fuselage. As the aircraft jerked and began to fall, it separated into three pieces. Many passengers in the cabin remained alive through the plane's plummet, struggling to understand what was happening. For ninety seconds, they were overwhelmed with deafening sounds; stomach-turning acceleration; a cyclone of serving trays and carry-on luggage; fierce winds and extreme cold.

There was no surviving the final impact. All 298 would die.

As flaming wreckage fell just outside the town of Hrabove, it took under five minutes for the first reports to appear online. A local witness described the moment she would never forget: the bodies "just fell very, very hard to the ground."

As the story rocketed across the internet, each side in the war zone that the plane had fallen into — the Ukrainian government and Russian-backed separatists — quickly blamed the other for the tragedy. Although social media exploded with theories, actual facts were hidden in a swirling fog. The rebels barred international investigators from visiting the crash site, preventing any independent examination for the next two weeks. It seemed that whoever had just killed 298 civilians would have ample time to disappear.

What they hadn't counted on was a soft-spoken former World of Warcraft addict. Sitting 2,000 miles away at his computer, he already had access to all the evidence he would need.

Three years earlier, Eliot Higgins had been a stay-at-home dad, doting on his infant daughter in their cozy Leicester, England, home. Deciding he was spending too much time online playing video games and commenting on news stories, he turned to channel his interests into something more useful — starting a blog about the Syrian civil war that had just begun. He took the handle "Brown Moses," from one of the iconoclastic rocker Frank Zappa's more obscure songs, which asked, "What wickedness is dis?"

Yet Higgins had never even been to Syria. He didn't speak Arabic. By his own admission, his knowledge of any type of conflict was limited to what he'd seen in Rambo movies. Indeed, he rarely left his house. He didn't have to. Thanks to millions of social media accounts, the Syrian civil war came to him.

His gifts were patience and diligence. His weapons were YouTube and Google Maps. Higgins taught himself how to find and track weapon serial numbers; how to use landmarks and satellite imaging to trace someone's steps; how to combine and catalog a stream of tens of thousands of videos. Soon enough, Higgins was charting each new development in a twisting, chaotic conflict. He uncovered hidden rebel weapon supply lines. He built a mountain of evidence to show that the Syrian dictator Bashar al-Assad had used nerve gas on his own people. From modest Blogspot beginnings, Brown Moses was soon rivaling not just the professional news media in his reporting, but some government intelligence agencies as well.

But this was only his first act. As interest in his unusual methods grew, Higgins launched a crowdfunding campaign for a new kind of online project dedicated to "citizen investigative journalists." The organization was dubbed Bellingcat. The name was taken from the old fable in which a group of mice conspire to place a bell around a cat's neck, so that they might always be warned of its approach.

Bellingcat had barely launched when flight MH17 fell from the sky. The first cat had come stalking.

One day after the tragedy, with international media awash in speculation and finger-pointing, Bellingcat published its first report. It was a straightforward summary of social media evidence to date, focused on sightings of a Russian Buk surface-to-air missile launcher that had been prowling near the crash site at the time of the tragedy. On July 22, Higgins posted a follow-up, superimposing images of wreckage against pictures of the intact plane, tracing the pattern of shrapnel damage. He drew no firm conclusions, pushing back against sketchy accounts posted by both Russians and Ukrainians. Bellingcat would only report the facts, as the digital bread crumbs revealed them.

It took a certain kind of person to excel in this work. Social media forensics requires endless focus and an attention to detail that sometimes borders on unhealthy. "I played a lot of role-player games," Higgins explained. "Believe me, there are a lot of obsessive people out there who could probably put their passions to a more productive use." A diverse crew of obsessive volunteers joined the hunt, an international online collective forming. They ranged from a Finnish military officer who knew Russian weapons, to Aric Toler, an American volunteer from North Carolina, who sheepishly told us how his main qualification was just being "good at the internet." He'd spend hours each day tumbling through obscure Russian social media channels, surfacing only for occasional coffee breaks and visits to the Chipotle next door. His in-laws thought he was wasting his time online. They didn't realize Toler was investigating war crimes.

The Bellingcat team soon tracked down multiple images and videos that showed a Buk missile launcher in the vicinity of MH17's flight path on the day of the tragedy, clearly within separatist territory. But the team noticed something even more telling. In photographs posted from before the time of the plane crash, the vehicle carried four missiles; in photographs taken soon after the crash, only three. They'd found their smoking gun.

But then the trail seemed to go cold. Although they could find plenty of other online photos of Russian-manned Buks deployed to aid the rebels, they couldn't find any additional matches for this particular vehicle. The images also showed that a shell game was being played. The painted vehicle number had been changed both before and after the July 17 event, and it could easily be changed again.

The breakthrough came when the analysts shifted their gaze lower. They realized that every Buk vehicle had a rubber skirt to help stop its treads from throwing up mud and dirt. Because each vehicle has its own particular driving history, each rubber skirt has a unique pattern of wear and tear. Now the Bellingcat team had the equivalent of a fingerprint to hunt for across every photograph and snippet of video that came out of eastern Ukraine.

They soon located photos of the Buk that had shot the missile in a convoy filmed crossing from Russia into Ukraine on June 23 — and leaving again on July 20. They were then able to trace backward, finding the unit to which the Buk in the convoy belonged: the 2nd Battalion of the 53rd Anti-Aircraft Missile Brigade of the Russian army. Bellingcat posted their findings online, mapping out the odyssey of the weapon that had killed 298 people, as well as showing its Russian origin.

The weapon and even its unit had been found, but who had pulled the trigger? Here the answer was provided by the shooters themselves. Searching through Russian soldiers' profiles on VKontakte, or VK (essentially a Russian version of Facebook), the Bellingcat team found images of military equipment, dour group photographs, and hundreds of angsty selfies. One conscript had even snapped a picture of an attendance sheet for a 2nd Battalion drill shortly before its deployment to Ukraine.

It wasn't just the soldiers who opened the lens to their war, but also their friends and families. The Bellingcat team found particular value in an online forum frequented by Russian soldiers' wives and mothers. Worried about their loved ones, they traded gossip about the deployments of specific units, which also proved an intelligence gold mine.

After nearly two years of research, Bellingcat presented its findings to the Dutch tribunal that had been tasked with bringing the killers to justice. It included the names, photographs, and contact information of the twenty soldiers who the data showed had been manning the missile system that had shot down flight MH17. It was an extraordinary feat, accomplished using only what was available on the internet. It was also damning evidence of Russian participation in a war crime.

Through their stubborn and breathtakingly focused investigation of the MH17 case, Eliot Higgins and Bellingcat displayed the remarkable new power of what is known as "open-source intelligence" (OSINT).

With today's OSINT, anyone can gather and process intelligence in a way that would have been difficult or impossible for even the CIA or KGB a generation ago. One OSINT analyst explained to us just how simple it can be, through a story of his pursuit of Iranian arms smuggling. He began by searching for some common weapons-related words in Farsi, courtesy of Google Translate. Soon he discovered an article that profiled a young Iranian CEO, who had launched a company specializing in drones. The journalist also found a video online that showed the CEO's face. This allowed him to locate the CEO in a registry of Iranian aerospace professionals, which, in turn, yielded an email address as well as phone and fax numbers. Translating the CEO's name into Farsi and searching Facebook (cross-referencing it with the email address), he quickly found the CEO's account. It was confirmed with a perfect facial match. He was then able to trace the CEO's movements, including a trip to Malaysia, where there is a market for drone parts. He also found that, despite working for a theocratic state, the Iranian CEO had a notable friendship with a charming young woman there, who liked to wear bunny ears and a six-inch skirt. Her own posts, in turn, revealed her to be an alumna of the "International Sex School."

He ended the search there. In just one hour of digging, he'd been able to compile a list of leads that might once have taken an intelligence agency months to find, as well as to tease out a fertile opportunity for blackmail. Although he'd begun his work as a reporter, in the end it was tough to shake the feeling that he'd become something else entirely: a spy.

In some cases, these new OSINT analysts don't even have to be human. GVA Dictator Alert is an algorithm whose sole purpose is tracking the flights of dictators into and out of Geneva, Switzerland — a favorite destination for money laundering and other shady business dealings. The program does its job unerringly, scanning real-time aviation data for aircraft registered to repressive governments. Whenever an autocrat arrives to check on their money, GVA Dictator Alert instantly squawks the information across Twitter. As the bot's creator explained, "It's a cool idea to know that every time the front wheel of their plane touches the tarmac in Geneva, there's a tweet saying 'hi, you are here, and now it's public.'"

In revealing these secrets, OSINT showed how its ability to unveil what was once hidden can be a powerful potential force for good. It can not only catch people cutting corners (quite literally, OSINT analysts found that one of every five racers in the 2017 Mexico City Marathon cheated, including several politicians hoping to tout their endurance), but also shine a light on the world's worst crimes. The manner in which Hitler and Pol Pot were able to kill en masse, unbeknownst to the wider world, is simply not possible today. The Bellingcat team, for instance, would use the very same approach it had brought to detecting war crimes in Ukraine to documenting the use of chemical weapons in Syria. Indeed, both technology and international law crossed a new frontier in 2017, when the International Criminal Court indicted Mahmoud Al-Werfalli for a series of mass killings in Libya. He was the first person ever to be charged with war crimes based solely on evidence found on social media.

Yet the very same techniques can also aid evil causes. Terrorists can reconnoiter potential targets without ever visiting them in person; they can tap into a global network of bomb makers and weapons experts without leaving their homes. The OSINT revolution also enables wholly new categories of crime. In Colombia, Mexico, and Venezuela, targets for kidnapping are selected through information gathered from their social media accounts. In some "virtual kidnappings," the actual kidnapping is skipped entirely as the criminals extort ransom from friends and family, knowing when the "victim" is out of contact and unable to respond.

OSINT can even offer a glimpse into previously impenetrable worlds. After decades of dark secrecy, the modern "Lords of War" — arms dealers — have embraced social media along with everybody else. Scouring Libyan Facebook groups, you can track hundreds of arms trades each month, as dealers advertise and negotiate shipments of everything from anti-aircraft missiles to heavy machine guns. By monitoring these sales, you can spot not just looming battles (evidenced by who is stocking up) but also evidence of policy failure. Many of the thousands of weapons traded online by Middle Eastern Facebook users, for instance, can be traced to stocks of weapons given to the Iraqi army and Syrian rebels by the CIA and U.S. military.

At its most promising, the OSINT revolution doesn't just help people parse secrets from publicly accessible information; it may also help them predict the future. Predata is a small company founded by James Shinn, a former CIA agent. Shinn modeled his unique service on saber-metrics, the statistics-driven baseball analysis method popularized by Michael Lewis's book *Moneyball*. "By carefully gathering lots and lots of statistics on their past performance from all corners of the internet, we are predicting how a large number of players on a team will bat or pitch in the future," Shinn explains. In this case, however, the statistics his firm mines are tens of millions of social media feeds around the world. But instead of predicting hits and strikeouts, it predicts events like riots and wars.

Predata uses such mass monitoring to discern online patterns that might be used to project real-world occurrences. Each Sunday, it sends out a "Week Ahead" mailer, breaking down the statistical likelihood of particular contingencies based on web monitoring. Billion-dollar Wall Street hedge funds are interested in any hints of unrest that might move markets. U.S. intelligence agencies are interested in signs of looming terror attacks or geopolitical shifts. For instance, North Korean experts were able to use the firm's vast quantity of social media data to identify "signal spikes" that predicted major diplomatic developments, military exercises, and even ballistic missile tests. The world of social media is becoming so revelatory that it can even help someone to anticipate what will happen next.

TRUE BELIEVER

"The exponential explosion of publicly available information is changing the global intelligence system . . . It's changing how we tool, how we organize, how we institutionalize — everything we do."

This is how a former director of the U.S. Defense Intelligence Agency (DIA) explained to us how the people who had once owned and collected secrets — professional spies — were adjusting to this world *without* secrets.

OSINT has a long history, being first separated from the classic spy craft of coaxing and interrogation (known as human intelligence, or HUMINT) and the intercept of confidential communications (signals intelligence, or SIGINT) during World War II. The breakthrough came when Allied analysts with the Office of Strategic Services (the predecessor of today's CIA) discovered that they could figure out the number of Nazi casualties by reading the obituary sections of German newspapers that were available in neutral Switzerland. By war's end, these analysts were cataloging roughly 45,000 pages of European periodicals each week. America also launched the Foreign Broadcast Monitoring Service (renamed the Foreign Broadcast Information Service, or FBIS), which transcribed 500,000 words of radio broadcasting each day.

Through much of the Cold War, U.S. intelligence agencies collected OSINT on a massive scale. The U.S. embassy in Moscow maintained subscriptions to over a thousand Soviet journals and magazines, while the FBIS stretched across 19 regional bureaus, monitoring more than 3,500 publications in 55 languages, as well as nearly a thousand hours of television each week. But intelligence chiefs traditionally put little faith in what this mass of free data yielded, rarely giving it the same weight as the other sources of information. Part of their skepticism arose because the information was so readily available (if it could be acquired so easily, how could it be valuable?), and part of it was because they suspected trickery (anything willingly shared by the Soviet Union must be a lie).

Ultimately, the FBIS was undone by the sheer volume of OSINT the internet produced. In 1993, the FBIS was creating 17,000 reports a month; by 2004, that number had risen to 50,000. In 2005, the FBIS was shuttered. Information was simply spreading too quickly online, in too many forms, for it to keep up. There was also no reason for the U.S. government to work so hard. For years, analysts had labored to maintain a sprawling, updated encyclopedia on the regions of the Soviet Union. Now there was Wikipedia.

However, a few forward-thinking intelligence officers dared to take the next big cognitive leap. What if OSINT wasn't losing its value, they asked, but was instead becoming the new coin of the realm? The question was painful because it required setting aside decades of training and established thinking. It meant envisioning a future in which the most valued secrets wouldn't come from cracking intricate codes or the whispers of human spies behind enemy lines — the sort of information that only the government could gather. Instead, they would be mined from a vast web of open-source data, to which everyone else in the world had access. If this was true, it meant changing nearly every aspect of every intelligence agency, from shifting budget priorities and programs to altering the very way one looked at the world. But the intelligence expert we interviewed felt it was a crucial change that had to be made.

"Publicly available information is now probably the greatest means of intelligence that we could bring to bear," he told us. "Whether you're a CEO, a commander in chief, or a military commander, if you don't have a social media component . . . you're going to fail."

The expert we consulted was Michael Thomas Flynn.

Flynn joined the U.S. military in 1981, at the height of the Cold War. He built his career in Army intelligence, rising through the ranks. After 9/11, he was made director of intelligence for the task force that deployed to Afghanistan. He then assumed the same role for the Joint Special Operations Command (JSOC), the secretive organization of elite units like the bin Laden–killing Navy SEAL team that had been outed by social media in Abbottabad. In this position, trying to track down the terror cells of Al Qaeda in Iraq (AQI), Flynn realized that his operatives had to look elsewhere, out in the open, for clues to where the enemy was hiding. And they had to do it faster than ever before.

As he explained, U.S. special operations forces were commandos without equal, "the best spear fishermen in the world." But in order to beat an adversary that recruited so rapidly and blended so easily with the civilian population, the commandos would have to become "net fishermen." They would eschew individual nodes and focus instead on taking down the entire network, hitting it before it could react and reconstitute itself. As Flynn's methods evolved, JSOC got better, capturing or killing dozens of terrorists in a single operation, gathering up the intelligence, and then blasting off to hit another target before the night was done. Eventually, the shattered remnants of AQI would flee Iraq for Syria, where they would ironically later reorganize themselves into the core of ISIS.

Flynn's career took off. He was promoted to three-star general and, in 2012, appointed to lead the DIA, the agency charged with centralizing intelligence across the entire U.S. military. Although he had no experience commanding such a large organization (the DIA numbered some 17,000 employees), Flynn was eager to translate his ideas into action. He envisioned not just reform of the DIA, but a wider reorganization of how the entire intelligence system worked in the twenty-first century. It was time to shift what was being collected and how. Before the rise of social media, he explained, 90 percent of useful intelligence had come from secret sources. Now it was the exact opposite, with 90 percent coming from open sources that anyone could tap.

Flynn sought to steer the agency in a new direction, boosting OSINT capabilities and prioritizing the hiring of computational analysts, who could put the data gushing from the digital fire hose to good use. He expected an uphill battle. OSINT, he explained, had only recently stopped being the "unwanted pregnancy" of military intelligence. Now, at best, it was a "redheaded stepchild."

He didn't realize the shake-up would prove a bridge too far. Flynn's aggressive moves alarmed the DIA's bureaucracy — not least because it threatened their own jobs. The agency was soon mired in chaos. Flynn's leadership was also questioned, his grand vision undermined by poor management. Just a year and a half after his term began, Flynn was informed that he was being replaced. He was forced into retirement, leaving the Army after thirty-three years of service.

If that were the whole story, Flynn's legacy might be one of a forward-thinking prophet of the social media revolution who paid the price for seeking change. But his tale didn't end there.

Flynn didn't take his dismissal well. After he left the military, he channeled his energy into media appearances and speaking engagements, as well as building a consulting business. He was originally marketed as the general who had seen the future. But he quickly became better known as a critic of the Obama administration that had fired him, angrily denouncing it for betraying him and the nation. This brought celebrity and money well beyond what Flynn had made in military service, but also new entanglements. His firm signed a shady $530,000 deal with a company linked to the Turkish government, which became doubly questionable when Flynn failed to register as a foreign lobbying agent. He accepted $45,000 to speak at a glitzy Russian government–

sponsored gala in Moscow. Photos of the ex-general sitting next to Vladimir Putin at the dinner shocked many in the U.S. security establishment.

Most important, Flynn's rising celebrity came to the attention of Donald Trump, who had just announced his run for office. Their first meeting was supposed to last thirty minutes. When it ended ninety minutes later, the former intelligence officer walked away with new insight into the future. "I knew he was going to be President of the United States."

Now the angry general became Trump's fiercest campaign surrogate, bestowing the inexperienced candidate with much-needed national security credibility. He used his old Army rank as a weapon, relentlessly attacking Trump's rivals. In doing so, however, Flynn began to dive deeply into the online world that he'd previously just observed. The result wasn't pretty.

Flynn had started his personal Twitter account, @GenFlynn, in 2011 with a tweet linking to a news article on Middle East politics. Not a single person had replied or retweeted it. But as he entered politics, Flynn's persona changed dramatically. His feed pushed out messages of hate ("Fear of Muslims is RATIONAL," he fumed in one widely shared tweet), anti-Semitism ("Not anymore, Jews. Not anymore," referring to the news media), and one wild conspiracy theory after another. His postings alleged that Obama wasn't just a secret Muslim, but a "jihadi" who "laundered" money for terrorists, that Hillary Clinton was involved with "Sex Crimes w Children," and that if she won the election, she would help erect a one-world government to outlaw Christianity. To wild acclaim from his new Twitter fans, Flynn even posted on #spiritcooking, an online conspiracy theory that claimed Washington, DC, elites regularly gathered at secret dinners to drink human blood and semen. That message got @GenFlynn over 2,800 "likes."

It was a remarkable turn for the once-respected intelligence officer. Just a few months earlier, he had cautioned us about the internet, "Now it's a matter of making sure that the accuracy matters . . . You combine your judgment, your experience, your analysis along with the valuable data you get."

Despite the online madness that violated his advice (or perhaps because of it), things seemed to work out well for the general. When

Trump won the election, Flynn was named to the position of national security advisor, one of the most powerful jobs in the world. His first tweet in the new role proclaimed, "We are going to win and win and win at everything we do."

The winning didn't last long.

Within a few weeks, Flynn would be fired, done in by a web of mistruths regarding his contact with Russian government officials. He was the shortest-serving national security advisor in American history. Within the year, in a plea bargain with the U.S. Department of Justice, Flynn would admit to making, in his words, "false, fictitious, and fraudulent statements."

As it all played out, we were reminded of one more piece of wisdom Flynn had imparted to us before his downfall. He'd spoken of the importance of piercing through the "fog" of the modern information environment; of getting to the "golden nuggets" of actionable intelligence that lurked in the mists of social media. The right bit of data was already out there, he explained. You just had to know where to look.

The general was right. The internet has indeed exposed the golden nuggets — the truth — for anyone to find. But, as his story also shows, scattered among these bits of truth is pyrite, "fool's gold," cleverly engineered to distract or even destroy us. It is harder than ever to keep a secret. But it is also harder than ever to separate the truth from lies. In the right hands, those lies can become powerful weapons.

4

The Empires Strike Back

Censorship, Disinformation, and the Burial of Truth

"Truth" is a lost cause and . . . reality is essentially malleable.

— PETER POMERANTSEV AND MICHAEL WEISS,
"The Menace of Unreality"

"INFORMATION WANTS TO BE FREE," declared web pioneer and counterculture icon Stewart Brand at the world's first Hackers Conference in 1984. This freedom wouldn't just sound the death knell of censorship; it would also mark the end of authoritarian regimes that relied on it. After all, what government could triumph against a self-multiplying network of information creators and consumers, where any idea might mobilize millions in a heartbeat? John Gilmore, an early cyberactivist and cofounder of the Electronic Frontier Foundation, put it simply in a 1993 interview: "The Net interprets censorship as damage and routes around it."

For many years, this seemed to be the case. In a dispatch for the newly launched *Wired* magazine, reporter Bruce Sterling described the key role of an early freedom fighter. In 1989, a mysterious digital Johnny Appleseed appeared in Czechoslovakia. Activists would credit him with helping to spark the uprisings that spread across Soviet-ruled Eastern Europe. But at the time, he was known simply as "the Japanese guy."

Without any warning or fanfare, some quiet Japanese guy arrived at the university with a valise full of brand-new and unmarked 2400-baud Taiwanese modems. The astounded Czech physics and engineering students never did quite get this gentleman's name. He just deposited the modems with them free of charge, smiled cryptically, and walked off diagonally into the winter smog of Prague, presumably in the direction of the covert-operations wing of the Japanese embassy. They never saw him again.

The Czech students distributed the new networking technology, using it to circulate manifestos and disseminate daily news updates. They were able to expand their revolutionary circles in a way never before possible, while evading the old methods of monitoring and censorship.

As the internet continued its blistering growth, the power of democratic dissidents followed. The first so-called internet revolution shook Serbia in 1996. Cut off from state media, young people used mass emails to plan protests against the regime of President Slobodan Milošević. Although the initial protests failed, they returned stronger than ever in 2000, being organized even more online. Serbia's youth won out and kicked off a series of "color revolutions," which soon spread throughout the former Soviet bloc, toppling rulers in Georgia, Ukraine, and Kyrgyzstan.

Then, in 2009, anger against a rigged election swept across theocratic Iran. While the front pages of Iranian newspapers were full of blank spaces (where government censors had blotted out reports), young people took to social media to organize and share the news. An astounding 98 percent of the links posted on Twitter that week were about Iran. Photos showed tens of thousands of Iranian youth pouring into the streets, a smartphone in nearly every hand. "The Revolution Will Be Twittered," declared one excited headline. *Wired* magazine's Italian edition nominated the internet for a Nobel Peace Prize.

In 2010, Mohamed Bouazizi, a 26-year-old Tunisian, touched off the next outbreak of web-powered freedom. Each morning for ten years, he had pushed a cart to the city marketplace, selling fruit to support his widowed mother and five siblings. Every so often, he had to navigate a shakedown from the police — the kind of petty corruption that had festered under the two-decade-long rule of dictator Zine el-Abidine Ben Ali. But on December 17, 2010, something inside Bouazizi snapped. After police confiscated his wares and he was denied a hearing to plead his case, Bouazizi sat down outside the local government building, doused his clothes with paint thinner, and lit a match.

Word of the young man's self-immolation spread quickly through the social media accounts of Tunisians. His frustration with corruption was something almost every Tunisian had experienced. Dissidents began to organize online, planning protests and massive strikes. Ben Ali responded with slaughter, deploying snipers who shot citizens from rooftops. Rather than retreat, however, some protesters whipped out their smartphones. They captured grisly videos of death and martyrdom. These were shared tens of thousands of times on Facebook and YouTube. The protests transformed into a mass uprising. On January 14, 2011, Ben Ali fled the country.

The conflagration soon leapt across national borders. While the Egyptian dictator Hosni Mubarak ordered censorship of the events in Tunisia, Wael Ghonim, a 30-year-old Google executive, used Facebook to organize similar protests in Cairo. When the first 85,000 people pledged online to march with him, *Time* magazine asked, "Is Egypt about to have a Facebook Revolution?" It was and it did. The trickle of pro-democracy protests turned to a raging torrent. Hundreds of thousands of demonstrators braved tear gas and bullets to demand Mubarak's resignation. His thirty-year reign ended in a matter of days. In the geopolitical equivalent of the blink of an eye, Egypt became a free nation.

A euphoric Ghonim gave credit where credit seemed due. "The revolution started on Facebook," he said. "We would post a video on Facebook that would be shared by 60,000 people . . . within a few hours. I've always said that if you want to liberate a society, just give them the Internet." Elsewhere, he said, "I want to meet Mark Zuckerberg one day and thank him." Another Egyptian revolutionary gave thanks in a more unorthodox way, naming his firstborn baby girl "Facebook."

Political unrest soon rocked Syria, Jordan, Bahrain, and a dozen more nations. In Libya and Yemen, dictators who had ruled for decades through the careful control of their population and its sources of information saw their regimes crumble in a matter of days. Tech evangelists hailed what was soon called the Arab Spring as the start of a global movement that would end the power of authoritarian regimes around the world, perhaps forever.

The Arab Spring seemed the perfect story of the internet's promise fulfilled. Social media had illuminated the shadowy crimes through which dictators had long clung to power, and offered up a powerful new means of grassroots mobilization. In the words of technology writer Clay Shirky, online social networks gave activists a way to "organize without organizations." Through Facebook events and Twitter hashtags, protests grew faster than the police could stamp them out. Each time the autocrats reacted violently, they created new online martyrs, whose deaths sparked further outrage. Everywhere, it seemed, freedom was on the march, driven by what Roger Cohen of the *New York Times* extolled as "the liberating power of social media."

Yet not everyone felt so sure. The loudest dissenter was Evgeny Morozov. Born in 1984 in the former Soviet bloc nation of Belarus, Morozov had been raised in an environment where a strongman had clung to power for nearly three decades. Like others his age, Morozov had enthusiastically embraced the internet as a new means to strike back against authoritarianism. "Blogs, social networks, wikis," he remembered. "We had an arsenal of weapons that seemed far more potent than police batons, surveillance cameras, and handcuffs."

But it never seemed to be enough. Not only did the activists fail to sustain their movement, but they noticed, to their horror, that the government began to catch up. Tech-illiterate bureaucrats were replaced by a new generation of enforcers who understood the internet almost as well as the protesters. They no longer ignored online sanctuaries. Instead, they invaded them, not just tracking down online dissidents, but using the very same channels of liberation to spread propaganda. More alarming, their tactics worked. Years after the first internet revolutions had sent shivers down dictators' spines, the Belarusian regime actually seemed to be strengthening its hand.

Morozov moved to the United States and set his sights squarely on the Silicon Valley dreamers, whom he believed were leading people astray. In a scathing book titled *The Net Delusion,* he coined a new term, "cyber-utopianism." He decried "an enthusiastic belief in the liberating power of technology," made worse by a "stubborn refusal to acknowledge its downside." When his book was released at the height of the Arab Spring, those he attacked as "cyber-utopians" were happy to laugh him off. If newly freed populations were literally naming their kids after social media, who could doubt its power for good?

As it turned out, the Arab Spring didn't signal the first steps of a global, internet-enabled democratic movement. Rather, it represented a high-water mark. The much-celebrated revolutions began to fizzle and collapse. In Libya and Syria, digital activists would soon turn their talents to waging internecine civil wars. In Egypt, the baby named Facebook would grow up in a country that quickly turned back to authoritarian government, the new regime even more repressive than Mubarak's.

Around the world, information had been freed. But so had a countering wave of authoritarianism using social media itself, woven into a pushback of repression, censorship, and even violence. The web's unique strengths had been warped and twisted toward evil ends. In truth, democratic activists had no special claim to the internet. They'd simply gotten there first.

CONTROL THE SIGNAL

Liu was a new arrival to the city of Weifang, China. He was new, too, to the city's traditions. One balmy August evening, he stumbled upon a neighborhood square dance. It looked like fun, and Liu — tired from another day of hunting for work — decided to join in the festivities.

Too late, Liu noticed the laughter and pointed fingers from the audience, the smartphones snapping his photo. He realized that nearly all the other dancers were middle-aged women. Liu fled the scene, flushed with embarrassment. He became petrified that his picture would be shared online for others to mock. So he did the only thing that made sense to him: he decided to destroy the internet.

Liu prowled the city looking for optical cable receivers, the big boxes of coiled wire that relay internet data to individual households. Each time he found one, he forced it open and tore the receiver apart by hand. By the time Liu was caught, he'd caused $15,000 worth of damage. Liu was sent to prison, but the internet — although temporarily disrupted across parts of Weifang — kept right on chugging. We know this because we read about him in an online report that made its way around the world.

While Liu failed in his mission, he actually had the right idea. After all, the internet isn't really a formless, digital "cloud." It is made up of physical things. His problem was that these "things" include billions of computers and smartphones linked to vast server farms that play host to all the world's online services. These are then bound together through an ever-growing network of everything from fiber-optic cable that runs twenty-five times the circumference of the earth, to some 2,000 satellites that circle the planet.

No one human could hope to control so monumental a creation. But governments are a different story.

For all the immensity of today's electronic communications network, the system remains under the control of only a few thousand internet service providers (ISPs), the firms that run the backbone, or "pipes," of the internet. Just a few ISPs supply almost all of the world's mobile data. Indeed, because two-thirds of all ISPs reside in the United States, the average number per country across the rest of the globe is relatively small. Many of these ISPs hardly qualify as "businesses" at all. They are state-sanctioned monopolies or crony sanctuaries directed by the whim of local officials. Liu would never have been able to "destroy" the internet. Neither can any one government. But regimes can control when the internet goes on (or off) and what goes on it.

Designed as an open system and built on trust, the web remains vulnerable to governments that play by different rules. In less-than-free nations around the world, internet blackouts are standard practice. All told, sixty-one countries so far have created mechanisms that allow for national-level internet cutoffs. When the Syrian uprising began, for instance, the government of Bashar al-Assad compelled Syria's main ISP to cut off the internet on Fridays, as that was the day people went to mosques and organized for protests. It doesn't just happen in wartime. In 2016, the exam questions for a national high school test in Algeria were leaked online, spreading across kids' social media. In response, government officials cut off the entire nation's access to the internet for three days while students took the test. Many Algerians suspected their government was actually using the scandal over the exam as a way to test its new tools of mass censorship.

These blackouts come at a cost. A 2016 study of the consequences of eighty-one instances of internet cutoffs in nineteen countries assessed the economic damage. Algeria's economy lost at least $20 million during that three-day shutdown, while a larger economy like Saudi Arabia lost $465 million from an internet shutdown in May 2016.

With this in mind, governments are investing in more efficient ways to control internet access, targeting particular areas of a country. For example, India is the world's largest democracy, but when violent protests started in the district of Rohtak in 2016, everyone in the district had their mobile connections cut for a week. (Even this limited focus cost the Indian economy $190 million.) Yet even more finely tuned censorship is possible. That same year, Bahrain instituted an "internet curfew" that affected only a handful of villages where antigovernment protests were brewing. When Bahrainis began to speak out against the shutdown, authorities narrowed their focus further, cutting access all the way down to specific internet users and IP addresses.

A variant of this cutoff strategy is "throttling." Whereas internet blocks cut off access completely, throttling slows down connections. It allows vital online functions to continue while making mass coordination more difficult. It's also harder to detect and prove. (Your Facebook posts on the evils of the government might not be loading because of a web slowdown or simply because your neighbor is downloading a video game.) Web monitoring services, for instance, have noticed that every time a protest is planned in Iran, the country's internet coincidentally and conveniently slows to a crawl.

A corollary to this strategy is the effort by governments to bring more of the internet's infrastructure under their direct control. Apologists call this "data localization," but it is better known as "balkanization," breaking up the internet's global network into a series of tightly policed national ones. The Islamic Republic of Iran, for instance, has poured billions of dollars into its National Internet Project. It is intended as a web replacement, leaving only a few closely monitored connections between Iran and the outside world. Iranian officials describe it as creating a "clean" internet for its citizens, insulated from the "unclean" web that the rest of us use. Of course, with each new stride in censorship, human ingenuity finds ways to get around it. Identity-masking technologies can circumvent even the strongest government controls, while communications satellites can beam data into neighboring nations as easily as into their own. Despite the regime's best efforts, for instance, Syrian rebel fighters were able to maintain active social media profiles by using solar-powered phone chargers and tapping into the mobile data network of neighboring Turkey.

But outside of the absolute-authoritarian state of North Korea (whose entire "internet" is a closed network of about thirty websites), the goal isn't so much to stop the signal as it is to weaken it. If one has to undertake extensive research and buy special equipment to circumvent government controls, the empowering parts of the internet are no longer for the masses. The potential network shrinks. The flow of information slows. And authoritarians' greatest fear — the prospect of spontaneous, widespread political mobilization — becomes harder to realize.

Yet governments' reach extends beyond the internet's infrastructure. They also have the police and the courts, all the mechanisms of state-sanctioned violence. As the internet has magnified the power of speech, these authoritarians haven't hesitated to use their own unique powers to control it.

CONTROL THE BODY

Does a retweet actually mean endorsement? For Dion Nissenbaum, the answer to this question landed him in a Turkish prison.

A soft-spoken man with a neat, gray-flecked goatee, Nissenbaum is a journalist who has spent years reporting from the most dangerous places in the world. He's been abducted by masked gunmen in the Gaza Strip, shot at by Israeli soldiers, rocketed by Hezbollah militants, and forced to ditch a broken-down car in the midst of Taliban-controlled Afghanistan. When the *Wall Street Journal* sent him on assignment to Turkey, Nissenbaum assumed the situation would be comparatively tame. He was wrong.

In July 2016, Turkey was roiled by an attempted military coup. The plotters followed the classic playbook, rounding up politicians in the middle of the night, setting up armed checkpoints at key locations in the major cities, and seizing control of newspaper printing presses and

TV stations. The idea was that the Turkish public would wake up the next morning to a fait accompli.

Instead, the coup became a story of the internet at its very best: a tale of mass mobilization that wouldn't have been possible without social media. The first rallying cry came from the mayor of Ankara, taking to Twitter as he evaded antigovernment forces. "RT HERKES SOKAGA," he wrote. "RETWEET: EVERYONE ON THE STREETS."

Hundreds of thousands of Turkish citizens streamed from their homes. They engulfed city squares and surrounded military positions, chanting slogans. In almost every hand was a smartphone, inviting friends and family to join them, and the world to cheer them on. When armed soldiers took control of the printing press of the nation's largest newspaper, with a daily circulation of just over 300,000 print copies, it didn't matter. Its 34-year-old digital content coordinator reported the news on the newspaper's Facebook page, allowing him to reach the ten times as many subscribers instantly. When soldiers tried to track him through the office building, he kept up a running commentary via Facebook, livestreaming his dangerous game of hide-and-seek.

As the online furor grew, even more protesters hit the streets. Meanwhile, the soldiers were beset by doubt. Many had been told by their officers that this was a routine training exercise. Staring at the faces of their furious countrymen and reading the online reports, they began to realize the truth. By dawn, the coup's architects had been captured or killed. The confused soldiers surrendered.

Instead of celebrating the triumph of online people power, Turkish president Recep Tayyip Erdoğan, the target of the coup, saw a different opportunity. "This insurgency is a blessing from Allah, because it will allow us to purge the military," he declared. Within three days, over 45,000 people suspected of links to his political opponents were pushed out of public service or marched before kangaroo courts. Among those arrested were 103 admirals and generals, 15,200 teachers, even 245 staffers at the Ministry of Youth and Sports. With the rebellious soldiers already in jail, few of these subsequent arrests had any connection to the coup. They were just people Erdogan wanted to get rid of.

Within months, over 135,000 civil servants were purged, and 1,058 schools and universities, 16 television stations, 23 radio stations, 45 newspapers, 15 magazines, and 29 book publishers were shut down. As part of this crackdown, Facebook, Twitter, and YouTube — services whose unfettered access had been crucial in stopping the coup — were increasingly restricted. Journalists saw their accounts suspended at the behest of the government. Freedom of speech was curtailed, the consequences demonstrated in a series of arrests of prominent figures. A satirical Instagram caption, penned by a former Miss Turkey, was enough to net her a fourteen-month jail sentence.

As conditions worsened, Dion Nissenbaum kept doing his job reporting the news. A few months after the coup, Nissenbaum was reading his Twitter feed, where he came across a report from an OSINT social media tracker — one of the same sources used by Eliot Higgins and the Bellingcat team. The report revealed that two Turkish soldiers being held captive by ISIS had been burned alive in a gruesome propaganda video. Nissenbaum thought it was newsworthy, as the Turkish government had been claiming that its operations in Syria were going well. He clicked the "retweet" button, sharing someone else's OSINT news with his few thousand followers. He thought little of it, as he regularly retweeted tidbits of news that came across his feed, mixed in with stories he found amusing, like of a new robotic waitress at a pizza shop.

As Nissenbaum explained, he quickly learned that "Twitter is a bare-knuckle battleground." A network of Turkish nationalists circulated screenshots of Nissenbaum's online profile, overlaid with threats. Another person turned his picture into a mug shot and urged people in Istanbul to be on the lookout for "this son of a whore." A popular Turkish newspaper editor, meanwhile, called for him to be deported. A friend quickly sent him a message warning him to check out the furor building online. After seeing the reaction, Nissenbaum took down his retweet, which had been up for just a few minutes. It was too late. As anger continued to swell, the Turkish government called his office, warning of unspecified "consequences."

Those consequences soon arrived in the form of three Turkish police officers, who showed up at Nissenbaum's apartment that night. They explained that he needed to pack a bag and come with them. There was no room for discussion. As he was driven away in a police van, Nissenbaum assumed he was going to be deported from the country. He grew alarmed when the van passed the airport and kept on going.

Nissenbaum was taken to a detention center, where he found himself strip-searched and thrown into a windowless isolation cell. For three days, he was denied all contact with the outside world. He played tic-tac-toe with himself and read the one book he'd been allowed to bring, a guide for new parents (he had just become a father).

And then, just as abruptly, he was yanked out of the jail, put in another van, and driven to a gas station parking lot. There waited his *Wall Street Journal* colleagues, who had been working around the clock to get him released. He didn't waste any time. Within hours, he and his family were on a one-way flight out of Istanbul.

Afterward, Nissenbaum reflected on the experience. If he could turn back the clock, he admitted, he would do things differently. "The cost of the retweet was so high," he said, "and the news value of putting it out was modest at best." There was a broader lesson, he added. Social media was a "volatile political battleground." What was said and shared — even a hasty retweet — carried "real-world consequences."

In retrospect, Nissenbaum was lucky. As an American citizen, he had powerful advocates on his side. He also had the power to leave. For thousands of Turks jailed for online speech, as well as tens of thousands more "under investigation," there was no such protection.

Nissenbaum's story shows how the internet's ultrafast and vast reach can spread information as never before. Yet it also shows how written (and unwritten) laws still vest immense power in government authorities, who determine the consequences for what is shared online.

Often, these restrictions are wrapped in the guise of religion or culture. But almost always they are really about protecting the government. Iran's regime, for example, polices its "clean" internet for any threats to "public morality and chastity," using such threats as a reason to arrest human rights activists. In Saudi Arabia, the harshest punishments are reserved for those who challenge the monarchy and the competence of the government. A man who mocked the king was sentenced to 8 years in prison, while a wheelchair-bound man was given 100 lashes and 18 months in jail for complaining about his medical care. In 2017, Pakistan became the first nation to sentence someone to death for online speech after a member of the Shia minority got in a Facebook argument with a government official posing as someone else.

Such codes are not limited to the Muslim world. Thailand has strictly enforced its law of "lèse majesté," promising years of prison for anyone who insults a member of the royal family. The scope can be incredibly expansive. In 2017, unflattering photos of the king wearing a crop-top shirt and (especially) low-rise jeans appeared on Facebook. The government threatened to punish not just anyone who posted the images, but anyone who looked at them.

These regimes are also proactive in searching for online dissent. "We'll send you a friend request," a Thai government official explained. "If you accept the friend request, we'll see if anyone disseminates [illegal] information. Be careful: we'll soon be your friend." The regime's eyes are many, extending into the ranks of the very young. Since 2010, Thai police have administered a "Cyber Scouts" program for children, encouraging them to report on the online activity of friends and family — and promising $15 for each report of wrongdoing.

More than religion or culture, this new generation of censors relies on appeals to national strength and unity. Censorship is not for *their* sake, these leaders explain, but rather for the good of the country. A Kazakh visiting Russia and criticizing Russian president Vladimir Putin on his Facebook page was sentenced to three years in a penal colony for inciting "hatred." A Russian woman who posted negative stories about the invasion of Ukraine was given 320 hours of hard labor for "discrediting the political order."

The state can wield this power not only against users but also against the companies that run the networks. They may seem like faceless organizations, but there are real people behind them, who can be reached by the long arm of the law — or other means. VKontakte is the most popular social network in Russia. After anti-Putin protesters used VK in the wake of the Arab Spring, the regime began to take a greater interest in it and the company's young, progressive-minded founder, Pavel Durov. When the man once known as "the Mark Zuckerberg of Russia" balked at sharing user data about his customers, armed men showed up at his apartment. He was then falsely accused of driving his Mercedes over a traffic cop's foot, a ruse to imprison him. Getting the message, Durov sold his shares in the company to a Putin crony and fled the country.

Over time, such harsh policing of online speech actually becomes less necessary as self-censorship kicks in. Communications scholars call it the "spiral of silence." Humans continually test their beliefs against those of the perceived majority and often quietly moderate their most extreme positions in order to get along better with society as a whole. By creating an atmosphere in which certain views are stigmatized, governments are able to shape what the majority opinion appears to be, which helps steer the direction of *actual* majority opinion.

Although plenty of dissenters still exist in authoritarian states, like those seeking to circumvent web bans and throttling, they now have to work harder. Their discussions have migrated from open (and easily monitored) social media platforms to secure websites and encrypted message applications, where only true believers can find them.

Yet there is more. Through the right balance of infrastructure control and enforcement, digital-age regimes can exert remarkable control over not just computer networks and human bodies, but the minds of their citizens as well. No nation has pursued this goal more vigorously — or successfully — than China.

CONTROL THE SPIRIT

"Across the Great Wall we can reach every corner in the world."

So read the first email ever sent from the People's Republic of China, zipping 4,500 miles from Beijing to Berlin. The year was 1987. Chinese scientists celebrated as their ancient nation officially joined the new global internet.

Other milestones soon followed. In 1994, China adopted the same TC/IP system that powered the World Wide Web. Almost overnight, the dour research tool of Chinese scientists became a digital *place,* popping with colorful websites and images. Two years later, the internet was opened to Chinese citizens, not just research institutions. A trickle of new users turned into a flood. In 1996, there were just 40,000 Chinese online; by 1999, there were 4 million. In 2008, China passed the United States in number of active internet users: 253 million. Today, that figure has tripled again to nearly 800 million (over a quarter of all the world's netizens), and, as we saw in chapter 2, they use some of the most vibrant and active forms of social media.

Yet it was also clear from the beginning that for the citizens of the People's Republic of China, the internet would not be — *could* not be

— the freewheeling, crypto-libertarian paradise pitched by its American inventors. China has remained a single, cohesive political entity for 4,000 years. The country's modern history is defined by two critical periods: a century's worth of embarrassment, invasion, and exploitation by outside nations, and a subsequent series of revolutions that unleashed a blend of communism and Chinese nationalism. For these reasons, Chinese authorities treasure harmony above all else. Harmony lies at the heart of China's meteoric rise and remains the underlying political doctrine of the Chinese Communist Party (CCP), described by former president Hu Jintao as the creation of a "harmonious society." Dissent, on the other hand, is viewed as only harmful to the nation, leaving it again vulnerable to the machinations of foreign powers.

Controlling ideas online has thus always been viewed as a vital, even natural, duty of the Chinese state. Unity must be maintained; harmful ideas must be stamped out. Yuan Zhifa, a former senior government propagandist, described this philosophy in 2007. "The things of the world must have cadence," he explained. His choice of words was important. Subtly different from "censorship," "cadence" means managing the "correct guidance of public opinion."

From the beginning, the CCP made sure that the reins of the internet would stay in government hands. In 1993, when the network began to be seen as something potentially important, officials banned all international connections that did not pass through a handful of state-run telecommunications companies. The Ministry of Public Security was soon tasked with blocking the transmission of all "subversive" or "obscene" information, working hand in hand with network administrators. In contrast to the chaotic web of international connections emerging in the rest of the globe, the Chinese internet became a closed system. Although Chinese internet users could build their own websites and freely communicate with other users *inside* China, only a few closely scrutinized strands of cable connected them to the wider world. Far from surmounting the Great Wall, the "Chinese internet" had become defined by a new barrier: the Great Firewall.

Chinese authorities also sought to control information *within* the nation. In 1998, China formally launched its Golden Shield Project, a feat of digital engineering on a par with mighty physical creations like the Three Gorges Dam. The intent was to transform the Chinese internet into the largest surveillance network in history — a database with records of every citizen, an army of censors and internet police, and automated systems to track and control every piece of information transmitted over the web. The project cost billions of dollars and employed tens of thousands of workers. Its development continues to this day. Notably, the design and construction of some of the key components of this internal internet were outsourced to American companies — particularly Sun Microsystems and Cisco — which provided the experience gained from building vast, closed networks for major businesses.

The most prominent part of the Golden Shield Project is its system of keyword filtering. Should a word or phrase be added to the list of banned terms, it effectively ceases to be. As Chinese internet users leapt from early, static websites to early-2000s blogging platforms to the rise of massive "microblogging" social media services starting in 2009, this system kept pace. Today, it is as if a government censor looms over the shoulder of every citizen with a computer or smartphone. Web searches won't find prohibited results; messages with banned words will simply fail to reach the intended recipient. As the list of banned terms updates in real time, events that happen on the rest of the worldwide web simply never occur inside China.

In 2016, for instance, the so-called Panama Papers were dumped online and quickly propelled to virality. The documents contained 2.6 terabytes of once-secret information on offshore bank accounts used by global elites to hide their money — a powerful instance of the internet's radical transparency in action. Among the disclosures were records showing that the families of eight senior CCP leaders, including the brother-in-law of President Xi Jinping, were funneling tens of millions of dollars out of China through offshore shell companies.

The information in all its details was available for anyone online — unless you lived in China. As soon as the news broke, an urgent "Delete Report" was dispatched by the central State Council Internet Information Office. "Find and delete reprinted reports on the Panama Papers," the order read. "Do not follow up on related content, no exceptions. If material from foreign media attacking China is found on any website, it will be dealt with severely." With that, the Panama Papers and the information in them was rendered inaccessible to all Chinese netizens. For a time, the entire nation of Panama briefly disappeared from search results in China, until censors modulated the ban to delete only if the post contained "Panama" and leaders' names or related terms like "offshore."

So ubiquitous is the filter that it has spawned a wave of surreal word-play to try to get around it. For years, Chinese internet users referred to "censorship" as "harmony" — a coy reference to Hu Jintao's "harmonious society." To censor a term, they'd say, was to "harmonize" it. Eventually, the censors caught on and banned the use of the word "harmony." As it happens, however, the Chinese word for "harmony" sounds similar to the word for "river crab." When a word had been censored, savvy Chinese internet users then took to calling it "river crab'd." And, as social media has become more visual, the back-and-forth expanded to image blocks. In 2017, the lovable bear Winnie-the-Pooh was disappeared from the Chinese internet. Censors figured out "Pooh" was a reference to President Xi, as he walks with a similar waddle.

History itself (or rather people's knowledge and awareness of it) can also be changed through this filtering, known as the "cleanse the web" policy. Billions of old internet postings have been wiped from existence, targeting anything from the past that fails to conform to the regime's "harmonious" history. Momentous events like the 1989 Tiananmen Square protests have been erased through elimination of nearly 300 "dangerous" words and phrases. Baidu Baike, China's equivalent of Wikipedia, turns up only two responses to a search on "1989": "the number between 1988 and 1990" and "the name of a computer virus." The result is a collective amnesia: an entire generation ignorant of key moments in the past and unable to search out more information if they ever do become aware.

Chinese censorship extends beyond clearly political topics to complaints that can be seen as challenging the state in any way. In 2017, a man in Handan was arrested for "disturbing public order" after he posted a negative comment online about hospital food.

As we've seen, many nations muzzle online discussion. But there is a key difference in China: the content of the story is sometimes irrelevant to the perceived crime. Unlike in other states where the focus is on banning discourse on human rights or calls for democratization, Chinese censorship seeks to suppress any messages that receive too much grassroots support, even if they're apolitical — or even complimentary to the authorities. For example, what seemed like positive news of an environmental activist who built a mass movement to ban plastic bags was harshly censored, even though the activist started out with support from local government officials. In a truly "harmonious society," only the central government in Beijing should have the power to inspire and mobilize on such a scale. Spontaneous online movements challenge the state's authority — and, by extension, the unity of the Chinese people. Or, as China's state media explained, "It's not true that 'everyone is entitled to their own opinion.'"

From the first days of the Chinese internet, authorities have ruled that websites and social media services bear the legal responsibility to squelch any "subversive" content hosted on their networks. The definition of this term can shift suddenly. Following a spate of corruption scandals in 2016, for instance, the government simply banned all online news reporting that did not originate with state media. It became the duty of individual websites to eliminate such stories or suffer the consequences.

Ultimately, however, the greatest burden falls on individual Chinese citizens. Although China saw the emergence of an independent blogging community in the early 2000s, the situation abruptly reversed in 2013 with the ascendancy of President Xi Jinping. That year, China's top court ruled that individuals could be charged with defamation (and risk a three-year prison sentence) if they spread "online rumors" seen by 5,000 internet users or shared more than 500 times. Around the same time, China's most popular online personalities were "invited" to a mandatory conference in Beijing. They received notebooks stamped with the logo of China's internet security agency and were treated to a slide show presentation. It showed how much happier a blogger had become after he'd switched from writing about politics to exploring more "appropriate" subjects, like hotel reviews and fashion. The message was clear: Join us or else.

The government soon took an even harder line. Charles Xue, a popular Chinese American blogger and venture capitalist, was arrested under suspicious circumstances. He appeared on state television in handcuffs shortly afterward, denouncing his blogging past and arguing for state control of the internet. "I got used to my influence online and the power of my personal opinions," he explained. "I forgot who I am."

The pace of internet-related detainments soon spiked dramatically.

Since Xi came to power, tens of thousands of Chinese citizens have been charged with "cybercrimes," whose definition has expanded from hacking to pretty much anything digital that authorities don't like. In 2017, for instance, Chinese regulators determined that the creator of a WeChat discussion group wasn't responsible just for their own speech, but also for the speech of each group member.

In China, it's not enough simply to suppress public opinion; the state must also take an active hand in shaping it. Since 2004, China's provincial ministries have mobilized armies of bureaucrats and college students in publishing positive stories about the government. As a leaked government memo explained, the purpose of these commenters is to "promote unity and stability through positive publicity." In short, their job is to act as cheerleaders, presenting an unrelentingly positive view of China, and looking like real people as they do so.

Where this phalanx of internet commenters differs from a traditional crowdsourcing network is the level of organization that comes from a state bureaucracy, boasting its own pay scales, quotas, and guidelines, as well as examinations and official job certifications. Critics quickly labeled these commenters the "50-Cent Army," for the 50 Chinese cents they were rumored to be paid for each post. (Eventually, China would simply ban the term "50 cents" from social media entirely.) One early advertisement for the 50-Cent Army promised that "performance, based on the number of posts and replies, will be considered for awards in municipal publicity work." By 2008, the 50-Cent Army had swelled to roughly 280,000 members. Today, there are as many as 2 million members, churning out at least 500 million social media postings each year. This model of mass, organized online positivity has grown so successful and popular that many members no longer have to be paid. It has also been mimicked by all sorts of other organizations in China, from public relations companies to middle schools.

All of these firewalls, surveillance, keyword censorship, arrests, and crowdsourced propagandists are intended to merge the consciousness of 1.4 billion people with the consciousness of the state. While some may see it as Orwellian, it actually has more in common with what China's Communist Party founder, Mao Zedong, described as the "mass line." When Mao broke with the Soviet Union in the 1950s, he criticized Joseph Stalin and the Soviet version of communism for being too concerned with "individualism." Instead, Mao envisioned a political cycle in which the will of the masses would be refracted through the lens of Marxism and then shaped into policy, only to be returned to the people for further refinement. Through this process, diverse opinions would be hammered into a single vision, shared by all Chinese people. The reality proved more difficult to achieve, and, indeed, such thinking was blamed for the Cultural Revolution that purged millions through the 1960s and 1970s, until it was repudiated after Mao's death in 1976.

Through the possibilities offered by the Chinese internet, this mass-line philosophy has made a comeback. President Xi Jinping has lauded these new technologies for offering the realization of Mao's vision of "condensing" public opinion into one powerful consensus.

To achieve this goal, even stronger programs of control lurk on the horizon. In the restive Muslim-minority region of Xinjiang, residents have been forced to install the Jingwang (web-cleansing) app on their smartphones. The app not only allows their messages to be tracked or blocked, but it also comes with a remote-control feature, allowing authorities direct access to residents' phones and home networks. To ensure that people were installing these "electronic handcuffs," the police have set up roving checkpoints in the streets to inspect people's phones for the app.

The most ambitious realization of the mass line, though, is China's "social credit" system. Unveiled in 2015, the vision document for the system explains how it will create an "upward, charitable, sincere and mutually helpful social atmosphere" — one characterized by unwavering loyalty to the state. To accomplish this goal, all Chinese citizens will receive a numerical score reflecting their "trustworthiness . . . in all facets of life, from business deals to social behavior."

Much like a traditional financial credit score, each citizen's "social credit" is calculated by compiling vast quantities of personal information and computing a single "trustworthiness" score, which measures, essentially, someone's usefulness to society. This is possible thanks to Chinese citizens' near-universal reliance on mobile services like We-Chat, in which social networking, chatting, consumer reviews, money transfers, and everyday tasks such as ordering a taxi or food delivery are all handled by one application. In the process, users reveal a staggering amount about themselves — their conversations, friends, reading lists, travel, spending habits, and so forth. These bits of data can form the basis of sweeping moral judgments. Buying too many video games, a program director explained, might suggest idleness and lower a person's score. On the other hand, regularly buying diapers might suggest recent parenthood, a strong indication of social value. And, of course, one's political proclivities also play a role. The more "positive" one's online contributions to China's cohesion, the better one's score will be. By contrast, a person who voices dissent online "breaks social trust," thus lowering their score.

In an Orwellian twist, the system's planning document also explains that the "new system will reward those who report acts of breach of trust." That is, if you report others for bad behavior, your score goes up. Your score also depends on the scores of your friends and family. If they aren't positive enough, you get penalized for their negativity, thus motivating everyone to shape the behavior of the members of their social network.

What gives the trustworthiness score its power is the rewards and risks, both real and perceived, that underpin it. Slated for deployment throughout China in 2020, the scoring system is already used in job application evaluations as well as doling out micro-rewards, like free phone charging at coffee shops for people with good scores. If your score is too low, however, you can lose access to anything from reserved beds on overnight trains to welfare benefits. The score has even been woven into China's largest online matchmaking service. Value in the eyes of the Chinese government thus will also shape citizens' romantic and reproductive prospects.

Luckily, no other nation has enjoyed China's level of success in subordinating the internet to the will of the state, because of both its head start and its massive scale of investment. But other nations are certainly jealous. The governments of Thailand, Vietnam, Zimbabwe, and Cuba have all reportedly explored establishing a Chinese-style internet of their own. Russian president Vladimir Putin has even gone so far as to sign a pact calling for experienced Chinese censors to instruct Russian engineers on building advanced web control mechanisms. All told, Chinese officials have held "information management" training for the governments of at least thirty-six countries. Just as U.S. tech companies once helped China erect its Great Firewall, so China has begun to export its hard-won censorship lessons to the rest of the world.

Programs like these make it clear that the internet has not loosened the grip of authoritarian regimes. Instead, it has become a new tool for maintaining their power. Sometimes, this occurs through visible controls on physical hardware or the people using it. Other times, it happens through sophisticated social engineering behind the scenes. Both build toward the same result: controlling the information and controlling the people.

Yet the web has also given authoritarians a tool that has never before existed. In a networked world, they can extend their reach across borders to influence the citizens of *other* nations just as easily as their own.

This is a form of censorship that hardly seems like censorship at all.

DAZE AND CONFUSE

"It was difficult to get used to at first," the young man confessed. "Why was I sitting in a stuffy office for eight hours a day, doing what I did? But I was tempted by easy work and good money."

On the surface, his story is familiar. A philosophy major in college, he was short on job options and found himself sucked into the corporate grind. But this young man didn't become a bored paralegal or a restless accountant. Instead, his job was causing chaos on the internet, to the benefit of the Russian government. He did this by writing more than 200 blog posts and comments each day, assuming fake identities, hijacking conversations, and spreading lies. He joined a war of global censorship by means of disinformation.

It is not surprising that Russia would pioneer this strategy. From its birth, the Soviet Union relied on the clever manipulation and weaponization of falsehood (called *dezinformatsiya*), both to wage ideological battles abroad and to control its population at home. One story tells how, when a forerunner of the KGB set up an office in 1923 to harness the power of *dezinformatsiya*, it invented a new word — "disinformation" — to make it sound of French origin instead. In this way, even the origin of the term was buried in half-truths.

During the Cold War, the Soviet Union turned disinformation into an assembly-line process. By one count, the KGB and its allied agencies conducted more than 10,000 disinformation operations. These ranged from creating front groups and media outlets that tried to amplify political divisions in the West, to spreading fake stories and conspiracy theories to undermine and discredit the Soviet Union's foes.

These operations often used "black propaganda," in which made-up sources cleverly laundered made-up facts. Perhaps the most notorious was Operation INFEKTION, the claim that the U.S. military invented AIDS, a lie that echoes through the internet to this day. The campaign began in 1983, launched via an article the KGB planted in the Indian newspaper *Patriot,* which itself was created as a KGB front in 1967. Its purported author was presented as a "well-known American scientist and anthropologist." It was given further academic validation by another article in which two East Germans posed as French scientists and confirmed the findings reported in the fake article by the fake author. This subsequent article was the subject of no less than forty reports in Soviet newspapers, magazines, and radio and television broadcasts. At this point, the reports began to be distributed into the West through pro-Soviet, left-leaning media outlets and extreme right-wing ones prone to conspiracy theories (such as the fringe Lyndon LaRouche movement). The operation was a remarkable success, but it took four years to reach fruition.

The fall of the Soviet Union brought a seeming end to such initiatives. In article 29 of its newly democratic constitution, the Russian Federation sought to close the door on the era of state-controlled media and shadowy propaganda campaigns. "Everyone shall have the right to freely look for, receive, transmit, produce and distribute information by any legal way," the document declared.

In reality, the Cold War's end didn't mean the end of disinformation. With new means of dissemination via social media, the prospect of spreading lies became all the more attractive, especially after the ascension of Vladimir Putin, a former KGB officer once steeped in them.

By way of crony capitalism and forced buyouts, Russia's large media networks soon lay in the hands of oligarchs, whose finances are deeply intertwined with those of the state. Today, the Kremlin makes its positions known through press releases and private conversations, the contents of which are then dutifully reported to the Russian people, no matter how much spin it takes to make them credible.

Of course, this modern spin differs considerably from the propaganda of generations past. In the words of *The Economist,* old Soviet propagandists "spoke in grave, deliberate tones, drawing on the party's lifelong wisdom and experience." By contrast, the new propaganda is colorful and exciting, reflecting the tastes of the digital age. It is a cocktail of moralizing, angry diatribes, and a celebration of traditional values, constantly mixed with images of scantily clad women. A pop star garbed like a teacher in a porn video sings that "freedom, money and girls — even power" are the rewards for living a less radical lifestyle, while a rapper decries human rights protesters as "rich brats." Running through it all is a constant drumbeat of anxiety about terrorism, the CIA, and the great specter of the West. Vladimir Milov, a former Russian energy minister turned government critic, explained it best. "Imagine you have two dozen TV channels," he said, "and it is all Fox News."

Milov's freedom to say this, though, shows another twist on the traditional model. Unlike the Soviet Union of the past, or how China and many other regimes operate today, Russia doesn't prevent political opposition. Indeed, opposition makes things more interesting — just so long as it abides by the unspoken rules of the game. A good opponent for the government is a man like Vladimir Zhirinovsky, an army colonel who premised his political movement on free vodka for men and better underwear for women. He once proposed beating the bird flu epidemic by shooting all the birds from the sky. Zhirinovsky was entertaining, but he also made Putin seem more sensible in comparison. By contrast, Boris Nemtsov was not a "good" opponent. He argued for government reform, investigated charges of corruption, and organized mass protests. In 2015, he was murdered, shot four times in the back as he crossed a bridge. The government prefers caricatures to real threats. Nemtsov was one of at least thirty-eight prominent opponents of Putin who died under dubious circumstances between 2014 and 2017 alone, from radioactive poisonings to tumbling down an elevator shaft.

Dissent is similarly allowed among the few journalists at news outlets independent of the state, but again, only within certain boundaries. Those who become too vocal or popular will experience a backlash. It might be through low-level harassment to make their life gratingly tenuous (such as by raising their taxes or instructing their landlord to suddenly break their lease). Or it might be through disinformation efforts to undermine their reputation. A favorite tactic is the state-linked media accusing them of being terrorists or arranging "scandals" using *kompromat,* a tactic whereby compromising material, like a sex tape, is dumped online. There are also more forceful methods of ensuring silence. Since Putin consolidated power in 1999, dozens of independent journalists have been killed under circumstances as suspicious as those that have befallen his political opponents.

The outcome has been an illusion of free speech within a newfangled Potemkin village. "The Kremlin's idea is to own all forms of political discourse, to not let any independent movements develop outside its walls," writes Peter Pomerantsev, author of *Nothing Is True and Everything Is Possible.* "Moscow can feel like an oligarchy in the morning and a democracy in the afternoon, a monarchy for dinner and a totalitarian state by bedtime."

But importantly, the village's border no longer stops at Russia's frontier. After the color revolutions roiled Eastern Europe and the Arab Spring swept the Middle East, a similar wave of enthusiasm in late 2011 inspired tens of thousands of young Russians to take to the streets, mounting the most serious protests of Putin's reign. Perceiving the combined forces of liberalization and internet-enabled activism as an engineered attack by the West, the Russian government resolved to fight back.

The aim of Russia's new strategy, and its military essence, was best articulated by Valery Gerasimov, the country's top-ranking general at the time. He channeled Clausewitz, declaring in a speech reprinted in the Russian military's newspaper that "the role of nonmilitary means of achieving political and strategic goals has grown. In many cases, they have exceeded the power of force of weapons in their effectiveness." In contrast to the haphazard way that Western governments have conceived of the modern information battlefield, Gerasimov proposed restructuring elements of the Russian state to take advantage of the "wide asymmetrical possibilities" that the internet offered.

These observations, popularly known as the Gerasimov Doctrine, have been enshrined in Russian military theory, even formally written into the nation's military strategy in 2014. Importantly, Russian theorists saw this as a fundamentally *defensive* strategy — essentially a "war on information warfare against Russia."

Such a power for Russia would only arise through strategic investment and organization, a stark contrast to the way most people in the West think about what happens on the internet as inherently chaotic and "organic." A conglomerate of nearly seventy-five education and research institutions was devoted to the study and weaponization of information, coordinated by the Federal Security Service, the successor of the KGB. It was a radical new way to think about conflict (and one we'll return to in chapter 7), premised on defanging adversaries abroad before they are able to threaten Russia at home. Ben Nimmo, who has studied this issue for NATO and the Atlantic Council, has described the resultant strategy as the "4 Ds": dismiss the critic, distort the facts, distract from the main issue, and dismay the audience. Just as Western radio and television signals once ranged into the Soviet Union, Russian propagandists began to return the favor — with interest.

The most visible vehicle for this effort is Rossiya Segodnya (Russia Today, or RT), a state news agency founded in 2005 with the declared intention of sharing Russia with the world. Initially, it was a fairly boring, traditional broadcasting outlet. But when Russia reforged its information warfare strategy, the organization's identity and mission shifted. Today, RT is a glitzy and contrarian media empire, whose motto can be found emblazoned everywhere from Moscow's airport to bus stops adjacent to the White House: "Question More."

RT was originally launched with a Russian government budget of $30 million per year in 2005. By 2015, the budget had jumped to approximately $400 million, an investment more in line with the Russian view of the outlet as a "weapons system" of influence. That support, and the fact that its long-serving editor in chief, Margarita Simonyan, simultaneously worked on Putin's election team, belies any claims of RT's independence from the Russian government. Indeed, on her desk sits a yellow landline phone with no dial or buttons — a direct line to the Kremlin. When asked its purpose, she answered, "The phone exists to discuss secret things."

The reach of the RT network is impressive, broadcasting across the world in English, Arabic, French, and Spanish. Its online reach is even more extensive, pushing out digital content in these four languages plus Russian and German. RT is also popular; it has more YouTube subscribers than any other broadcaster, including the BBC and Fox News.

The network's goal is no longer sharing Russia with the world, but rather showing why all the other countries are wrong. It does so by publishing harsh, often mocking stories about Russia's political opponents, along with attention-grabbing pieces designed to support and mobilize divisive forces inside nations Russia views as its adversaries (such as nationalist parties in Europe or the Green Party and extreme right-wing in the United States). Any content that grabs eyeballs and sows doubt represents a job well done. Snarky videos designed to go viral (*Animated Genitals* and *Lawnmower Explodes* were major hits) are intermingled with eye-popping conspiracy theories (RT has promoted everything from Trump's "birther" claims about Barack Obama to regular reporting on UFO sightings). As Matt Armstrong, a former member of the U.S. Broadcasting Board of Governors, has explained, "'Question More' is not about finding answers, but fomenting confusion, chaos, and distrust. They spin up their audience to chase myths, believe in fantasies, and listen to faux ... 'experts' until the audience simply tunes out."

After RT's initial success, a supplementary constellation of Russian government–owned or co-opted outlets was organized, allowing stories and scoops to be shared from one mouthpiece to the next, building more and more online momentum. Sputnik International is a "news service" modeled on savvy web outlets like BuzzFeed, claiming to "[cover] over 130 cities and 34 countries." Meanwhile, the news service Baltica targets audiences in the Baltic (and NATO-member) nations of Estonia, Latvia, and Lithuania. These well-funded Russian propaganda outlets can often outgun and overwhelm their local media competitors.

This modern network of disinformation can quickly rocket a falsehood around the world. In 2017, for instance, the U.S. Army announced that it would be conducting a training exercise in Europe involving 87 tanks. That nugget of truth was transformed into an online article headlined "US Sends 3,600 Tanks Against Russia — Massive NATO Deployment Under Way." The first source of this false report was Donbass News International (DNI), the official media of the unofficial Russian separatist parts of Ukraine. That DNI Facebook page article was then distributed through nineteen different outlets, ranging from a Norwegian communist news aggregator to far-left activist websites to seemingly reputable outlets like the "Centre for Research on Globalization." However, the "Centre" was actually an online distribution point for conspiracy theories on everything from "chemtrails" (the idea that the air is secretly being poisoned by mysterious aircraft) to claims that Hillary Clinton was behind a pedophile ring at a Washington, DC, pizzeria. That second cascade of reports was read by tens of thousands. The reports were then used as inspiration for further reporting, under different titles, by official Russian media like RT, which extended the story's reach by orders of magnitude more.

This was exactly how Operation INFEKTION worked during the Cold War, except for two key differences. Through the web, a process that once took four years now takes mere hours, and it reaches millions more people.

The strategy also works to blunt the impact of any news that is harmful to Russia, spinning up false and salacious headlines to crowd out the genuine ones. Recall how Eliot Higgins and Bellingcat pierced the fog of war surrounding the crash of flight MH17, compiling open-source data to show — beyond a reasonable doubt — that Russia had supplied and manned the surface-to-air missile launcher that stole 298 lives. The first response from Russia was a blanket denial of any role in the tragedy, accompanied by an all-out assault on the Wikipedia page that had been created for the MH17 investigation, seeking to erase any mention of Russia. Then came a series of alternative explanations pushed out by the official media network, echoed by allies across the internet. First the Ukrainian government was to blame. Then the Malaysian airline was at fault. ("Questions over Why Malaysia Plane Flew over Ukrainian Warzone," one headline read, even though the plane flew on an internationally approved route.) And then it was time to play the victim, claiming Russia was being targeted by a Western smear campaign.

Mounting evidence of Russia's involvement in the shootdown proved little deterrent. Shortly after the release of the Bellingcat exposé showing who had shot the missiles, Russian media breathlessly announced that, actually, a newfound satellite image showed the final seconds of MH17. Furthermore, it could be trusted, as the image had both originated with the Russian Union of Engineers and been confirmed by an independent expert.

The photo was indeed remarkable, showing a Ukrainian fighter jet in the act of firing at the doomed airliner. It was a literal smoking gun.

It was also a clear forgery. The photo's background revealed it had been stitched together from multiple satellite images. It also pictured the wrong type of attack aircraft, while the airliner said to be MH17 was just a bad photoshop job. Then it turned out the engineering expert validating it did not actually have an engineering degree. The head of the Russian Union of Engineers, meanwhile, explained where he'd found it: "It came from the internet."

All told, Russian media and proxies spun at least a half dozen theories regarding the MH17 tragedy. It hardly mattered that these narratives often invalidated each other. (In addition to the fake fighter jet photos, another set of doctored satellite images and videos claimed to show it hadn't been a Russian, but rather a Ukrainian, surface-to-air missile launcher in the vicinity of the shootdown, meaning now the airliner had somehow been shot down from both above and below.) The point of this barrage was to instill doubt — to make people wonder how, with so many conflicting stories, one could be more "right" than any other.

It is a style of censorship akin to the twist in Edgar Allan Poe's "The Purloined Letter." In the famous short story, Parisian police hunt high and low for a letter of blackmail that they know to be in their suspect's possession. They comb his apartment for months, searching under the floorboards and examining the joints of every piece of furniture; they probe each cushion and even search the moss between the bricks of the patio. Yet they come up empty-handed. In desperation, they turn to an amateur detective, C. Auguste Dupin, who visits the suspect's apartment and engages him in pleasant conversation. When the suspect is distracted, Dupin investigates a writing desk strewn with papers — and promptly finds the missing letter among the suspect's other mail. The very best way to hide something, Dupin explains to the shocked police, is to do so in plain sight. So, too, is it with modern censorship. Instead of trying to hide information from prying eyes, it remains in the open, buried under a horde of half-truths and imitations.

Yet, for all the noise generated by Russia's global media network of digital disinformation sites, there's an even more effective, parallel effort that lurks in the shadows. Known as "web brigades," this effort entails an army of paid commenters (among them our charming philosophy major), who manage a vast network of online accounts. Some work in the "news division," others as "social media seeders," still others tasked with creating "demotivators": visual content designed to spread as far and quickly as possible. Unlike the 50-Cent Army of China, however, the Russian version isn't tasked with spreading positivity. In the words of our philosophy student's boss, his job was to sow "civil unrest" among Russia's foes. "This is information war, and it's official."

While these activities have gained much attention for their role in the 2016 U.S. presidential election and the UK's Brexit vote the same year, Russia's web brigades actually originated almost a decade earlier in a pro-Kremlin youth group known as Nashi. When government authorities (firmly in control of traditional media) struggled to halt the fierce democratic activism spreading through Russian social media circles after the color revolutions and Arab Spring, the group stepped in to pick up the slack, praising Putin and trashing his opponents. The Kremlin, impressed with these patriotic volunteers, used the engine of capitalism to accelerate the process. It solicited Russian advertisers to see if they could offer the same services, dangling fat contracts as a reward. Nearly a dozen major companies obliged. And so the "troll factories" were born. (In 2018, several Russian oligarchs associated with these companies would be indicted by Special Counsel Robert Mueller in his investigation of Russian interference in the U.S. election.)

Each day, our hapless Russian philosophy major and hundreds of other young hipsters would arrive for work at organizations like the innocuously named Internet Research Agency, located in an ugly neo-Stalinist building in St. Petersburg's Primorsky District. They'd settle into their cramped cubicles and get down to business, assuming a series of fake identities known as "sockpuppets." The job was writing hundreds of social media posts per day, with the goal of hijacking conversations and spreading lies, all to the benefit of the Russian government. For this work, our philosophy major was paid the equivalent of $1,500 per month. (Those who worked on the "Facebook desk" targeting foreign audiences received double the pay of those targeting domestic audiences.) "I really only stayed in the job for that," he explained. "I bought myself a Mazda Six during my time there."

Like any job, that of being a government troll comes with certain expectations. According to documents leaked in 2014, each employee is required, during an average twelve-hour day, to "post on news articles 50 times. Each blogger is to maintain six Facebook accounts publishing at least three posts a day and discussing the news in groups at least twice a day. By the end of the first month, they are expected to have won 500 subscribers and get at least five posts on each item a day. On Twitter, they might be expected to manage 10 accounts with up to 2,000 followers and tweet 50 times a day."

The hard work of a sockpuppet takes three forms, best illustrated by how they operated during the 2016 U.S. election. One is to pose as the organizer of a trusted group. @Ten_GOP called itself the "unofficial Twitter account of Tennessee Republicans" and was followed by over 136,000 people (ten times as many as the official Tennessee Republican Party account). Its 3,107 messages were retweeted 1,213,506 times. Each retweet then spread to millions more users, especially when it was disseminated by prominent Trump campaign figures like Donald Trump Jr., Kellyanne Conway, and Michael Flynn. On Election Day 2016, it was the seventh most retweeted account across *all* of Twitter. Indeed, Flynn followed at least five such documented accounts, sharing Russian propaganda with his 100,000 followers at least twenty-five times.

The second sockpuppet tactic is to pose as a trusted news source. With a cover photo image of the U.S. Constitution, @tpartynews presented itself as a hub for conservative fans of the Tea Party to track the latest headlines. For months, the Russian front pushed out anti-immigrant and pro-Trump messages and was followed and echoed out by some 22,000 people, including Trump's controversial advisor Sebastian Gorka.

Finally, sockpuppets pose as seemingly trustworthy individuals: a grandmother, a blue-collar worker from the Midwest, a decorated veteran, providing their own heartfelt take on current events (and who to vote for). Another former employee of the Internet Research Agency, Alan Baskayev, admitted that it could be exhausting to manage so many identities. "First you had to be a redneck from Kentucky, then you had to be some white guy from Minnesota who worked all his life, paid taxes and now lives in poverty; and in 15 minutes you have to write something in the slang of [African] Americans from New York." Baskayev waxed philosophic about his role in American politics. "It was real postmodernism. Postmodernism, Dadaism and Sur[realism]."

Yet, far from being postmodern, sockpuppets actually followed the example of classic Cold War "active measures" by targeting the extremes of both sides of American politics during the 2016 election. The fake accounts posed as everything from right-leaning Tea Party activists to "Blacktivist," who urged those on the left to "choose peace and vote for Jill Stein. Trust me, it's not a wasted vote." A purported African American organizer, Blacktivist, was actually one of those Russian hipsters sitting in St. Petersburg, whose Facebook posts would be shared an astounding 103.8 million times before the company shut the account down after the election.

By cleverly leveraging readers' trust, these engineers of disinformation induced thousands — sometimes millions — of people each day to take their messages seriously and spread them across their own networks via "shares" and retweets. This sharing made the messages seem even more trustworthy, since they now bore the imprimatur of whoever shared them, be it a distinguished general or a family friend. As the Russians moved into direct advertising, this tactic enabled them to achieve an efficiency that digital marketing firms would kill for. According to a dataset of 2016 Facebook advertisements purchased by Russian proxies, the messages received engagement rates as high as 24 percent — far beyond the single digits to which marketing firms usually aspire.

The impact of the operation was further magnified by how efforts on one social media platform could complement (and amplify) those on another. Russian sockpuppets ran rampant on services like Instagram, an image-sharing platform with over 800 million users (larger than Twitter and Snapchat combined) and more popular among youth than its Facebook corporate parent. Here, the pictorial nature of Instagram made the disinformation even more readily shareable and reproducible. In 2017, data scientist Jonathan Albright conducted a study of just twenty-eight accounts identified as having been operated by the Russian government. He found that this handful of accounts had drawn an astounding 145 *million* "likes," comments, and plays of their embedded videos. They'd also provided the visual ammunition subsequently used by other trolls who stalked Facebook and Twitter.

These messages gained even greater power as they reached beyond social media, taking advantage of how professional news outlets — feeling besieged by social media — had begun embedding the posts of online "influencers" in their own news stories. In this, perhaps no one matched the success of @Jenn_Abrams. A sassy American teen, who commented on everything from Kim Kardashian's clothes to the need to support Donald Trump, her account amassed nearly 70,000 Twitter followers. That was impressive, but not nearly as impressive as the ripple effect of her media efforts. "Jenn" was quoted in articles in the BBC News, BET, Breitbart, Business Insider, BuzzFeed, CNN, The Daily Caller, The Daily Dot, the *Daily Mail*, Dallas News, Fox News, France24, Gizmodo, HuffPost, IJR, the *Independent*, Infowars, Mashable, the *National Post*, the *New York Daily News, New York Times, The Observer*, Quartz, Refinery29, Sky News, the *Times of India, The Telegraph, USA Today, U.S. News and World Report*, the *Washington Post*, Yahoo Sports, and (unsurprisingly) Russia Today and Sputnik. Each of these articles was then read and reacted to, spreading her views even further and wider. In 2017, "Jenn" was outed by Twitter as yet another creation of Russia's Internet Research Agency.

The Russian effort even turned the social media firms' own corporate strategies against their customers. As a way to draw users deeper into its network, Facebook automatically steered people to join groups, where they could find new friends who "share their common interests and express their opinion." The Russian sockpuppets learned to create and then manipulate these online gatherings. One of the more successful was Secured Borders, an anti–Hillary Clinton Facebook group that totaled over 140,000 subscribers. It, too, was actually run out of the St. Petersburg office of the Internet Research Agency. By combining online circulation with heavy ad buys, just one of its posts reached 4 million people on Facebook and was "liked" more than 300,000 times.

Much like the harassment campaigns inside Russia, sockpuppets also targeted Putin critics abroad. The most extreme efforts were reserved for those who investigated the disinformation campaigns themselves. After journalist Jessikka Aro published an exposé of the fake accounts, sockpuppets attacked her with everything from posts claiming she was a Nazi and drug dealer to messages pretending to be from her father, who had died twenty years earlier. When another group of Western foreign affairs specialists began to research the mechanics of disinformation campaigns, they found themselves quickly savaged on the professional networking site LinkedIn. One was labeled a "pornographer," and another was accused of harassment. Such attacks can be doubly effective, not only silencing the direct targets but also discouraging others from doing the sort of work that earned such abuse.

While the sockpuppets were extremely active in the 2016 election, it was far from their only campaign. In 2017, data scientists searched for patterns in accounts that were pushing the theme of #UniteTheRight, the far-right protests that culminated in the killing of a young woman in Charlottesville, Virginia, by a neo-Nazi. The researchers discovered that one key account in spreading the messages of hate came to life each day at 8:00 a.m. Moscow time. Realizing they'd unearthed a Russian sockpuppet, they dug into its activities before the Charlottesville protests. For four years, it had posted around a hundred tweets a day, more than 130,000 messages in all. At first, the chief focus was support for UKIP, a far-right British party. Then it shifted to pushing Russia's stance on the Ukraine conflict. Then it pivoted to a pro-Brexit stance, followed by support for Trump's candidacy. After his election, it switched to white nationalist "free speech" protests. The efforts of these networks continue to this day, ever seeking to sow anger and division within Russia's foes.

Indeed, a full three years after the flight MH17 tragedy, we tested the strength of the Russian disinformation machine for ourselves by setting what's known as a "honeypot." The term traditionally referred to a lure — in fiction, usually a sexy female agent — which enemy operatives couldn't resist. Think Vesper Lynd's seduction of James Bond in *Casino Royale,* or her real-life counterpart, Anna Chapman, the redheaded KGB agent who worked undercover in New York and then, after she was caught by the FBI and deported back to Russia, began a second career as a Facebook lingerie model. We posted something even more enticing on Twitter: one of Bellingcat's reports. Within minutes, an account we'd had no prior link with reached out, inundating us with images disputing the report as "#Bellingcrap." The account's history showed it, day after day, arguing against Russia's role in MH17, while occasionally mixing things up with anti-Ukrainian conspiracy theories and tweets in support of far-right U.S. political figures. In trying to persuade us, our new online friend had instead provided a window into a fight over "truth" that will likely continue to rage for as long as the internet exists.

Success breeds imitators. Just as some nations have begun to study China's internet engineering, many others are copying Russian techniques. In Venezuela, the nominally elected "president," Nicolas Maduro, enjoys an online cult of personality in which loyal (and paid) supporters quickly suppress critical headlines. In Azerbaijan, "patriotic trolls" launch coordinated attacks to discredit pro-democratic campaigners. Even in democratic India, rumors fly of shadowy online organizations that exist to defend the party of Prime Minster Narendra Modi. They applaud each new government policy and circulate "hit lists" to dig up dirt on opponents and pressure them into silence. If no incriminating material exists, they simply invent it.

A 2018 study from Oxford University's Computational Propaganda Research Project found that, all told, at least forty-eight regimes have followed this new model of censorship to "steer public opinion, spread misinformation, and undermine critics." Even more worrisome, between 2017 and 2018 at least thirty national-level elections were targeted by such social media manipulation. As more governments become attuned to the internet's dark possibilities, this figure will only grow.

Perhaps the most pernicious effect of these strategies, however, is how they warp our view of the world around us. It is a latter-day incarnation of the phenomenon explored in *Gaslight,* a 1938 play that was subsequently turned into a movie. In the story, a husband seeks to convince his new wife that she's going mad (intending to get her committed to an asylum and steal her hidden jewels). He makes small changes to her surroundings — moving a painting or walking in the attic — then tells her that the things she is seeing and hearing didn't actually occur. The play's title comes from the house's gas lighting, which dims and brightens as he prowls the house late at night. Slowly but surely, he shatters his wife's sense of reality. As she says of her mounting self-doubt and resulting self-censorship, "In the morning when the sun rises, sometimes it's hard to believe there ever was a night."

Since the 1950s, the term "gaslighting" has been used to describe relationships in which one partner seeks control over another by manipulating or even denying the truth. We're now seeing a new form of gaslighting, perpetrated repeatedly and successfully through social media on the global stage. In the words of writer Lauren Ducca, "Facts . . . become interchangeable with opinions, blinding us into arguing amongst ourselves, as our very reality is called into question." All the while, a new breed of authoritarians tighten their grip on the world.

Yet sinister as they might be, even the strongest dictators cannot force someone to believe that the earth is flat. Nor can the accumulated weight of 100,000 online comments so much as bend a blade of grass unless someone chooses to act on them. There's another piece of the puzzle still unaccounted for, perhaps the information battlefield's most dangerous weapon of all.

Our own brains.

5

The Unreality Machine

The Business of Veracity vs. Virality

When all think alike, no one thinks very much.

— WALTER LIPPMANN, *The Stakes of Diplomacy*

NEVER BEFORE COULD THESE TEENAGE BOYS have afforded the $100 bottles of Moët champagne that they sprayed across the nightclub floor. But that was before the gold rush, before their lives were flooded with slick wardrobes and fancy cars and newly available women. In the rusted old industrial town of Veles, Macedonia, they were the freshly crowned kings.

They worked in "media." More specifically, they worked in American social media. The average U.S. internet user was basically a walking bag of cash, worth four times the advertising dollars of anyone else in the world — and they were very gullible. In a town with 25 percent unemployment and an annual income of under $5,000, these young men had discovered a way to monetize their boredom and decent English-language skills. They set up catchy websites, peddling fad diets and weird health tips, and relying on Facebook "shares" to drive traffic. With each click, they got a small slice of the pie from the ads running along the side. Soon the best of them were pulling in tens of thousands of dollars a month.

But there was a problem. As word got out, competition swelled. More and more Veles teens launched websites of their own.

Fortunately, these young tycoons had timed their business well. The American political scene soon brought them a virtually inexhaustible source of clicks and resulting fast cash: the 2016 U.S. presidential election.

The Macedonians were awed by Americans' insatiable thirst for political stories. Even a sloppy, clearly plagiarized jumble of text and ads could rack up hundreds of thousands of "shares." The number of U.S. politics–related websites operated out of Veles ballooned into the hundreds. As U.S. dollars poured into the local economy, one nightclub even announced that it would hold special events the same day Google released its advertising payouts.

"Dmitri" (a pseudonym) was one of the successful entrepreneurs. He estimated that in six months, his network of fifty websites attracted some 40 million page views driven there by social media. It made him about $60,000. The 18-year-old then expanded his media empire. He outsourced the writing to three 15-year-olds, paying each $10 a day. Dmitri was far from the most successful of the Veles entrepreneurs. Several became millionaires. One even rebranded himself as a "clickbait coach," running a school where he taught dozens of others how to copy his success.

Some 5,000 miles from actual American voters, this small Macedonian town had become a cracked mirror of what Mark Zuckerberg had pulled off just a decade earlier. Its entrepreneurs had pioneered a new industry that created an unholy amount of cash and turned a legion of young computer nerds into rock stars. As one 17-year-old girl explained at the nightclub, watching the teen tycoons celebrate from her perch at the bar, "Since fake news started, girls are more interested in geeks than macho guys."

The viral news stories pumped out by these young, hustling Macedonians weren't just exaggerations or products of political spin; they were flat-out lies. Sometimes, the topic was the long-sought "proof" that Obama had been born in Kenya or revelations that he was planning a military coup. Another report warned that Oprah Winfrey had told her audience that "some white people have to die." In retrospect, such articles seem unbelievable, but they were read on a scale that soared past reports of the truth. A study of the top election news–related stories found that false reports received more engagement on Facebook than the top stories from all the major traditional news outlets combined.

As with their peddling of fad diets, the boys turned to political lies for the sole reason that this was what their targets seemed to want. "You see they like water, you give water," said Dmitri. "[If] they like wine, you give wine." There was one cardinal rule in the business, though: target the Trumpkins. It wasn't that the teens especially cared about Trump's political message, but, as Dmitri explained, "nothing [could] beat" his supporters when it came to clicking on their made-up stories.

Of the top twenty best-performing fake stories spread during the election, seventeen were unrepentantly pro-Trump. Indeed, the single most popular news story of the entire election — "Pope Francis Shocks World, Endorses Donald Trump for President" — was a lie fabricated in Macedonia before blasting across American social networks. Three times as many Americans read and shared it on their social media accounts as they did the top-performing article from the *New York Times.* Pope Francis didn't mince words in his reaction to such articles: "No one has a right to do this. It is a sin and it is hurtful."

Dmitri and his colleagues, though, were unrepentant. "I didn't force anyone to give me money," he said. "People sell cigarettes, they sell alcohol. That's not illegal, why is my business illegal? If you sell cigarettes, cigarettes kill people. I didn't kill anyone." If anything, the fault lay with the traditional news media, which had left so much easy money on the table. "They're not allowed to lie," Dmitri noted scornfully.

At the same time that governments in Turkey, China, and Russia sought to obscure the truth as a matter of policy, the monetization of clicks and "shares" — known as the "attention economy" — was accomplishing much the same thing. Social media provided an environment in which lies created by anyone, from anywhere, could spread everywhere, making the liars plenty of cash along the way.

When the work of these Macedonian media moguls came to light, President Obama himself huddled with advisors on Air Force One. The most powerful man in the world dwelled on the absurdity of the situation and his own powerlessness to fight back. He could dispatch Navy SEALs to kill Osama bin Laden, but he couldn't alter this new information environment in which "everything is true and nothing is true." Even in the absence of digitally empowered censorship, the free world had still fallen victim to the forces of disinformation and unreality.

When the social media revolution began in earnest, Silicon Valley evangelists enthused about the possibilities that would result from giving everyone "access to their own printing press." It would break down barriers and let all opinions be heard. These starry-eyed engineers should have read up on their political philosophy. Nearly two centuries earlier, the French scholar of democracy Alexis de Tocqueville — one of the first foreigners to travel extensively in the new United States of America — pondered the same question. "It is an axiom of political science in the United States," he concluded, "that the only way to neutralize the influence of newspapers is to multiply their number." The greater the number of newspapers, he reasoned, the harder it would be to reach public consensus about a set of facts.

Tocqueville was worried about the number of newspapers expanding past the few hundred of his time. Today, the marvels of the internet have created the equivalent of several billion newspapers, tailored to the tastes of each social media user on the planet. Consequently, there is no longer one set of facts, nor two, nor even a dozen. Instead, there exists a set of "facts" for every conceivable point of view. All you see is what you *want* to see. And, as you'll learn how it works, the farther you're led into this reality of your own creation, the harder it is to find your way out again.

THE ECHO CHAMBERS OF ME

"Imagine a future in which your interface agent can read every newswire and newspaper and catch every TV and radio broadcast on the planet, and then construct a personalized summary."

This is what MIT media professor Nicholas Negroponte prophesied in 1995. He called it the "Daily Me." A curated stream of information would not only keep people up to date on their own personal interests, but it would cover the whole political spectrum, exposing people to other viewpoints. His vision aligned with most internet pioneers. The internet didn't just mean the end of censorship and authoritarians. Access to *more* information would also liberate democracies, leading to a smarter, wiser society.

As the web exploded in popularity and the first elements of the "Daily Me" began to take shape, some pondered whether the *opposite* might actually be true. Rather than expanding their horizons, people were just using the endless web to seek out information with which they already agreed. Harvard law professor Cass Sunstein rebranded it as the "Daily We."

> Imagine . . . a system of communications in which each person has unlimited power of individual design. If some people want to watch news all the time, they would be entirely free to do exactly that. If they dislike news, and want to watch football in the morning and situation comedies at night, that would be fine too . . . If people want to restrict themselves to certain points of view, by limiting themselves to conservatives, moderates, liberals, vegetarians, or Nazis, that would be entirely feasible with a simple point-and-click. If people want to isolate themselves, and speak only with like-minded others, that is feasible too . . . The implication is that groups of people, especially if they are like-minded, will end up thinking the same thing that they thought before — but in more extreme form.

With the creation of Facebook just a few years later, the "Daily We" — the algorithmically curated newsfeed — became a fully functioning reality. However, the self-segregation was even worse than Sunstein had predicted. So subtle was the code that governed user experience on these platforms, most people had no clue that the information *they* saw might differ drastically from what *others* were seeing. Online activist Eli Pariser described the effect, and its dangerous consequences, in his 2011 book, *The Filter Bubble*. "You're the only person in your bubble," he wrote. "In an age when shared information is the bedrock of shared experience, the filter bubble is the centrifugal force, pulling us apart."

Yet, even as social media users are torn from a shared reality into a reality-distorting bubble, they rarely want for company. With a few keystrokes, the internet can connect like-minded people over vast distances and even bridge language barriers. Whether the cause is dangerous (support for a terrorist group), mundane (support for a political party), or inane (belief that the earth is flat), social media guarantees that you can find others who share your views. Even more, you will be steered to them by the platforms' own algorithms. As groups grow, it becomes possible for even the most far-flung of causes to coordinate and organize, to gain visibility and find new recruits.

Flat-earthers, for instance, had little hope of gaining traction in a post–Christopher Columbus, pre-internet world. It wasn't just because of the silliness of their views, but they couldn't easily find others who shared them.

Today, the World Wide Web has given the flat-earth belief a dramatic comeback. Proponents now have an active online community and an aggressive marketing scheme. They spread stories that claim government conspiracy, and produce slick videos that discredit bedrock scientific principles. Pushing back at the belief only aids it, giving proponents more attention and more followers. "YouTube cannot contain this thing," declared one flat-earther. "The internet cannot contain it. The dam is broken; we are everywhere."

Flat-earthism may sound amusing, but substitute it for any political extreme and you can see the very same dynamics at play. As groups of like-minded people clump together, they grow to resemble fanatical tribes, trapped in echo chambers of their own design. The reason is basic human nature. In numerous studies, across numerous countries, involving millions of people, researchers have discovered a cardinal rule that explains how information disseminates across the internet, as well as how it shapes our politics, media, and wars. The best predictor is not accuracy or even content; it is the number of friends who share the content first. They are more likely to believe what it says — and then to share it with others who, in turn, will believe what *they* say. It is all about us, or rather our love of ourselves and people like us.

This phenomenon is called "homophily," meaning "love of the same." Homophily is what makes humans social creatures, able to congregate in such large and like-minded groups. It explains the growth of civilization and cultures. It is also the reason an internet falsehood, once it begins to spread, can rarely be stopped.

Homophily is an inescapable fact of online life. If you've ever shared a piece of content after seeing it on a friend's newsfeed, you've become part of the process. Most people don't ponder deeply when they click "share." They're just passing on things that they find notable or that might sway others. Yet it shapes them all the same. As users respond positively to certain types of content, the algorithms that drive social media's newsfeeds ensure that they see more of it. As they see more, they share more, affecting all others in their extended network. Like ripples in a pond, each of these small decisions expands outward, altering the flow of information across the entire system.

But there's a catch: these ripples also reverberate back toward you. When you decide to share a particular piece of content, you are not only influencing the future information environment, you are also being influenced by any information that has passed your way already. In an exhaustive series of experiments, Yale University researchers found that people were significantly more likely to believe a headline ("Pope Francis Shocks World, Endorses Donald Trump for President") if they had seen a similar headline before. It didn't matter if the story was untrue; it didn't even matter if the story was preceded by a warning that it might be fake. What counted most was *familiarity.* The more often you hear a claim, the less likely you are to assess it critically. And the longer you linger in a particular community, the more its claims will be repeated until they become truisms — even if they remain the opposite of the truth.

Homophily doesn't just sustain crazy online echo chambers; its effects can sow deadly consequences for society. A prime example is the anti-vaccine movement, which claims that one of the most important discoveries in human history is actually a vast conspiracy. The movement got its start in the 1960s but exploded in popularity along with social media. People with radical but seemingly disparate views — those on the far left suspicious of pharmaceutical companies, the far right suspicious of the government, and religious fundamentalists suspicious of relying on anything but prayer — found common cause online. Across Facebook groups and alternative-health websites, these "anti-vaxxers" shared made-up stories about the links between childhood vaccination and autism, reveling in conspiracy theories and claims they were the ones who faced a second "Holocaust."

In an endless feedback loop, each piece of content shared within the anti-vaxxer community leaves them only more convinced that *they* are the sane ones, defending their children against blasphemous, corporate-enriching, government-induced genetic engineering. In the process, the personalization afforded by social media also becomes a weapon. Whenever they are challenged, anti-vaxxers target not just the counterargument but also the person making it. Any critic becomes part of the conspiracy, transforming a debate over "facts" into one over motivations.

Their passion has also made them a potent online force. In turn, this has made them an attractive movement for others to leverage to their own ends. Beginning in the late 2000s, this cadre of true believers was joined by a series of lower-tier celebrities whose popularity had diminished, like Jenny McCarthy and Donald Trump (who tweeted, "Healthy young child . . . gets pumped with massive shot of many vaccines, doesn't feel good and changes — AUTISM. Many such cases!"). These failing stars used the attention-getting power of the anti-vaxxers to boost their personal brands — magnifying the reach of the conspiracy in the process.

In the United States, the net result of this internet-enabled movement is that — after more than two centuries of proven, effective use and hundreds of millions of lives saved — vaccines have never faced so much public doubt. That might be just as funny as the flat-earthers, claiming that the earth is flat while coordinating via satellites that circle the globe, except for the real costs borne by the most vulnerable members of society: children. In California, the percentage of parents applying a "personal belief exception" to avoid vaccinating their kindergartners quadrupled between 2000 and 2013, and disease transmittal rates among kids soared as a result. Cases of childhood illnesses like whooping cough reached a sixty-year high, while the Disneyland resort was rocked by an outbreak of measles that sickened 147 children. Fighting an infectious army of digital conspiracy theorists, the State of California eventually gave up arguing and passed a law *requiring* kindergarten vaccinations, which only provided more conspiracy theory fodder.

Tempting as it may be to blame the internet for this, the real source of these digital echo chambers is again deeply rooted in the human brain. Put simply, people like to be right; they *hate* to be proven wrong. In the 1960s, an English psychologist isolated this phenomenon and put a name to it: "confirmation bias." Other psychologists then discovered that trying to fight confirmation bias by demonstrating people's errors often made the problem worse. The more you explain with facts that someone is mistaken, the more they dig in their heels.

What the internet does do is throw this process into overdrive, fueling the brain's worst impulses and then spreading them to countless others. Social media transports users to a world in which their every view seems widely shared. It helps them find others just like them. After a group is formed, the power of homophily then knits it ever closer together. U.S. Army colonel turned historian Robert Bateman summarizes it pointedly: "Once, every village had an idiot. It took the internet to bring them all together."

Thanks to this combination of internet-accelerated homophily and confirmation bias, civil society can be torn into fragments. Each group comes to believe that only *its* members know the truth and that all others are ignorant or, even worse, evil. In fragile states, the situation can become untenable. A 2016 study from George Washington University's Institute for Public Diplomacy and Global Communication explored this phenomenon in the context of the Arab Spring (which, as we saw earlier, marked the height of optimism about the power of social media), helping to explain how these democratic uprisings were so quickly exploited by authoritarianism.

When the researchers pored over nearly 63 million Twitter and Facebook posts that followed the initial uprisings, a pattern became clear. The availability of information and ease of organizing online had catalyzed masses of disparate people into action. But then came the division. "As time went on, social media encouraged political society to self-segregate into communities of the like-minded, intensifying connections among members of the same group while increasing the distance among different groups."

Once the shared enemy was gone, wild allegations demonized former allies and drove people farther apart. As the researchers explained elsewhere, "The speed, emotional intensity and echo-chamber qualities of social media content make those exposed to it experience more extreme reactions. Social media is particularly suited to worsening political and social polarization because of its ability to spread violent images and frightening rumors extremely quickly and intensely."

Although the main case study was Egypt, they could well have been describing the plight of any nation on earth.

The outcome is a cruel twenty-first-century twist on one of the classic quotes of the twentieth century. "Everyone is entitled to his own opinion, but not his own facts," declared the legendary sociologist and New York senator Daniel Patrick Moynihan in a widely attributed axiom. He was born in the age of radio and rose to power in the age of television. He died in 2003 — the same year Mark Zuckerberg was mucking around in his Harvard dorm room. In Moynihan's time, such noble words rang true. Today, they're a relic.

Fact, after all, is a matter of consensus. Eliminate that consensus, and fact becomes a matter of opinion. Learn how to command and manipulate that opinion, and you are entitled to reshape the fabric of the world. As a Trump campaign spokesperson famously put it in 2016, "There's no such thing, unfortunately, anymore as facts." It was a preposterous claim, but in a certain way, it is true.

And yet there's another disturbing phenomenon at work. On social media, everyone may be entitled to their own facts, but rarely do they form their own opinions. There's someone else manufacturing the beliefs that go viral online.

THE SUPER SPREAD OF LIES

The families were just sitting down for lunch on December 4, 2016, when the man with the scraggly beard burst through the restaurant door. Seeing him carrying a Colt AR-15 assault rifle, with a Colt .38 revolver strapped to his belt, parents shielded their terrified children. But Edgar Welch hardly noticed. After all, he was a man on a mission. The 28-year-old part-time firefighter knew for a fact that the Comet Ping Pong pizza restaurant was just a cover for Hillary Clinton's secret pedophilia ring, and, as a father of two young girls, he was going to do something about it.

As the customers made a run for it (and, of course, started posting on social media about it), Welch headed to the back of the pizza place. He expected to find the entrance to the vast, cavernous basement that he knew to hold the enslaved children. Instead, he found an employee holding pizza dough. For the next forty-five minutes, Welch hunted for the secret sex chambers, overturning furniture and testing the walls. Eventually, his attention turned to a locked door. This was surely it, he thought. He fired his weapon, destroying the lock, and flung open the door. It was a tiny computer room little larger than a storage closet. There were no stairs to a secret underground sex chamber. Indeed, there was no basement at all. Dejected and confused, Welch dropped his weapons and surrendered to police.

In the subsequent trial, neither side would suggest Welch was insane. Indeed, the prosecution wrote that Welch "was lucid, deadly serious, and very aware." He'd sincerely believed he was freeing children from captivity, that he was embarking on a one-way mission for which he was prepared to give his life. On his 350-mile journey, he'd recorded a tearful farewell to his family on his smartphone; a martyrdom message to broadcast on social media if he died in a hail of bullets. Welch would be sentenced to four years in prison.

For James Alefantis, the founder and owner of the pizza café, the conviction was small comfort. "I do hope one day, in a more thoughtful world, every one of us will remember this day as an aberration," he said. "When the world went mad and fake news was real."

But it was no aberration. Welch's ill-fated odyssey could be traced to a flurry of viral conspiracy theories known collectively as #Pizzagate. Arising in the final days of the 2016 U.S. election, the hoax claimed Hillary Clinton and her aides were involved in satanic worship and underage sex trafficking at a DC-area pizza parlor. Their "evidence" was a picture of the owner, Alefantis, hosting a fundraiser for Clinton, and a heart-shaped logo on the restaurant's website. Working through a crowdsourced "investigation" that was a perverse reflection of the sort conducted by Bellingcat, these far-right sleuths had determined the heart was a secret sign for child predators. It was actually the symbol of a fundraiser for St. Jude Children's Research Hospital.

#Pizzagate blazed across social media, garnering 1.4 million mentions on Twitter alone. On the Infowars YouTube channel, conspiracy theorist Alex Jones told his 2 million subscribers, "Something's being covered up. All I know, God help us, we're in the hands of pure evil." Spying opportunity, the Russian sockpuppets working in St. Petersburg also latched onto the #Pizzagate phenomenon, their posts further boosting its popularity. #Pizzagate not only dominated far-right online conversation for weeks, but actually *increased* in power following Clinton's electoral defeat. When polled after the election, nearly half of Trump voters affirmed their belief that the Clinton campaign had participated in pedophilia, human trafficking, and satanic ritual abuse.

Yet, as Welch ruefully admitted after his arrest, "the intel on this wasn't 100 percent." Welch's use of the word "intel" was eerily appropriate. Among the key voices in the #Pizzagate network was Jack Posobiec, a young intelligence officer in the U.S. Navy Reserve. Although Posobiec's security clearance had been revoked and he'd been reassigned by his commanding officers to such duties as "urinalysis program coordinator," on Twitter he was a potent force. Posobiec was relentless in pushing #Pizzagate to his more than 100,000 followers. He'd even livestreamed his own "investigation" of the restaurant, barging in on a child's birthday party and filming until he was escorted from the premises.

For Posobiec, social media offered a path to popularity that eluded him in real life and a way to circumvent the old media gatekeepers. "They want to control what you think, control what you do," he bragged. "But now we're able to use our own platforms, our own channels, to speak the truth."

The accuracy of Posobiec's "truth" was inconsequential. Indeed, Welch's violent and fruitless search didn't debunk Posobiec's claims; it only encouraged him to make new ones. "False flag," Posobiec tweeted as he heard of Welch's arrest. "Planted Comet Pizza Gunman will be used to push for censorship of independent news sources that are not corporate owned." Then he switched stories, informing his followers that the DC police chief had concluded, "Nothing to suggest man w/ gun at Comet Ping Pong had anything to do with #pizzagate." It was, like the rest of the conspiracy, a fabrication. The only thing real was the mortal peril and psychological harm that opportunists like Posobiec had inflicted on the workers of the pizza place and the families dining there.

Yet Posobiec suffered little for his falsehoods. Indeed, they only increased his online fame and influence. They also brought other rewards. Just a few months after he'd trolled a pizza parlor into near tragedy, he was livestreaming from the White House press briefing room, as a specially invited guest. And then came the ultimate validation. Posobiec and his messages were retweeted multiple times by the most powerful social media platform in all the world, that of President Donald Trump.

#Pizzagate shows how online virality — far from a measure of sincere popularity — is a force that can be manipulated and sustained by just a few influential social media accounts. In internet studies, this is known as "power law." It tells us that, rather than a free-for-all among millions of people, the battle for attention is actually dominated by a handful of key nodes in the network. Whenever they click "share," these "super-spreaders" (a term drawn from studies of biologic contagion) are essentially firing a Death Star laser that can redirect the attention of huge swaths of the internet. This even happens in the relatively controlled parts of the web. A study of 330 million Chinese Weibo users, for instance, found a wild skew in influence: fewer than 200,000 users had more than 100,000 followers; only about 3,000 accounts had more than 1 million. When researchers looked more closely at how conversations *started,* they found that the opinions of these hundreds of millions of voices were guided by a mere 300 accounts.

The internet may be a vast, wild, and borderless frontier, but it has its monarchs all the same. Vested with such power, these super-spreaders often have little regard for the truth. Indeed, why should they? The information of truth is less likely to draw eyes.

In the past several years, episodes like #Pizzagate have become all too common, as have fabulists like Posobiec. These conspiracy-mongers' influence has been further reinforced by the age-old effects of homophily and confirmation basis. Essentially, belief in one conspiracy theory ("Global warming is a hoax") increases someone's susceptibility to further falsehoods ("Ted Cruz's dad murdered JFK"). They're like the HIV of online misinformation: a virus that makes its victims more vulnerable to subsequent infections.

The combination of conspiracy theories and social media is even more toxic than that, however. As psychologist Sander van der Linden has written, belief in online conspiracy theories makes one more supportive of "extremism, racist attitudes against minority groups (e.g., anti-Semitism) and even political violence."

Modest lies and grand conspiracy theories have been weapons in the political arsenal for millennia. But social media has made them more powerful and more pervasive than ever before. In the most comprehensive study of its kind, MIT data scientists charted the life cycles of 126,000 Twitter "rumor cascades" — the first hints of stories before they could be verified as true or false. The researchers found that the fake stories spread about six times faster than the real ones. "Falsehood diffused significantly farther, faster, deeper, and more broadly than the truth in all categories of information," they wrote.

Ground zero for the deluge, however, was in politics. The 2016 U.S. presidential election released a flood of falsehoods that dwarfed all previous hoaxes and lies in history. It was an online ecosystem so vast that the nightclubbing, moneymaking, lie-spinning Macedonians occupied only one tiny corner. There were thousands of fake websites, populated by millions of baldly false stories, each then shared across people's personal networks. In the final three months of the 2016 election, more of these *fake* political headlines were shared on Facebook than real ones. Meanwhile, in a study of 22 million tweets, the Oxford Internet Institute concluded that Twitter users, too, had shared more "misinformation, polarizing and conspiratorial content" than actual news stories.

The Oxford team called this problem "junk news." Like junk food, which lacks nutritional value, these stories lacked news value. And also like junk food, they were made of artificial ingredients and infused with sweeteners that made them hard to resist. This was the realization of a danger that internet sociologist danah boyd had warned about as far back as 2009:

> Our bodies are programmed to consume fats and sugars because they're rare in nature ... In the same way, we're biologically programmed to be attentive to things that stimulate: content that is gross, violent, or sexual and that [sic] gossip which is humiliating, embarrassing, or offensive. If we're not careful, we're going to develop the psychological equivalent of obesity. We'll find ourselves consuming content that is least beneficial for ourselves or society as a whole.

What the Oxford researchers called "junk news" soon became more commonly known as "fake news." That term was originally created to describe news that was verifiably untrue. However, President Trump quickly co-opted it (using it more than 400 times during his first year in office), turning "fake news" into an epithet to describe information that someone doesn't like. That is, even the term used to describe untruths went from an objective measure of accuracy to a subjective statement of opinion.

Whatever the term, in the United States, as in Macedonia, many people saw dollar signs in this phenomenon. Posobiec, for instance, would market his expertise as an online conspiracy theorist in a book that promised to explain "how social media was weaponized." As with Operation INFEKTION and the anti-vaxxers, however, the right-wing had no monopoly on driving lies viral, or making money along the way. One example could be seen in Jestin Coler, a self-described family man in his early 40s. With a degree in political science and an avid interest in propaganda, Coler claimed to have gotten into the fake news business as an experiment, testing the gullibility of right-wing conspiracy theorists. "The whole idea from the start," he explained, "was to build a site that could kind of infiltrate the echo chambers of the alt-right, publish blatantly or fictional stories and then be able to publicly denounce those stories and point out the fact that they were fiction." But then the money began to pour in — sometimes tens of thousands of dollars in a single month. Any high-minded purpose was forgotten.

Coler expanded his operation into a full-fledged empire: twenty-five websites, manned by a stable of two dozen freelance writers, each of whom took a cut of the profits. The wilder the headline, the more clicks it got and the more money everyone made. One of Coler's most popular pieces told the tragic and wholly false story of an FBI agent and his wife who, amid an investigation of Hillary Clinton, had died in a suspicious murder-suicide. In a ten-day period, 1.6 million readers were drawn to the real-sounding but fake newspaper (the *Denver Guardian*) that had posted the fake story. On Facebook, the damning headline would be glimpsed at least 15 million times.

Coler was unmasked when an intrepid reporter for National Public Radio pierced through his shell of web registrations and tracked him to his home. Asked why he'd stayed hidden, Coler was blunt about the people who were making him rich. "They're not the safest crowd," he said. "Some of them I would consider domestic terrorists. So they're just not people I want to be knocking on my door." Their beliefs were crazy; their cash was good.

Yet these individual for-profit purveyors of lies were just the small fry. More significant was the new media business environment surrounding them, which meshed profit and partisan politics. When researchers at the *Columbia Journalism Review* broke down the readership of some 1.25 million news stories published during the 2016 election cycle, they found that liberal and conservative news consumers were both relying more on social media than on traditional media outlets, but both groups essentially existed in their own parallel universes. This finding confirmed what we've seen already. Homophily and virality combined to increase users' exposure to information they agreed with, while insulating them from information they found distasteful.

But the research revealed something else. The drivers of conversation in the left-leaning social media universe were divided across multiple hubs that included old media mainstays like the *New York Times* and avowedly liberal outlets like the Huffington Post. In contrast, the right-leaning universe was separate but different. It had just one central cluster around the hyperpartisan platform Breitbart, which had been launched in 2005 (the year after Facebook) with the new media environment deliberately in mind. As founder Andrew Breitbart explained, "I'm committed to the destruction of the old media guard . . . and it's a very good business model."

After Breitbart's death in 2012, the organization was run by Steve Bannon, a former investment banker turned Hollywood producer, who intimately understood both markets and the power of a good viral headline. Bannon embraced social media as a tool to dominate the changing media marketplace, as well as to remake the right-wing. The modern internet wasn't just a communications medium, he lectured his staff, it was a "powerful weapon of war," or what he called "#War."

Through Breitbart, Bannon showered favorable coverage on the "alt-right," an emerging online coalition that could scarcely have existed or even been imagined in a pre–social media world. Short for "alternative right" (the term popularized by white supremacist leader Richard Spencer), the alt-right fused seemingly disparate groups ranging from a new generation of web-savvy neo-Nazis to video gamer collectives using online harassment campaigns to battle perceived "political correctness." All these groups found unity in two things. The first was a set of beliefs that, as the Associated Press put it, rejected "the American democratic ideal that all should have equality under the law regardless of creed, gender, ethnic origin or race." The second was a recognition that social media was the best means to transform that conviction into reality.

Stoking outrage and seeking attention, Bannon declared Breitbart "the platform for the alt-right." Its editors even invited the movements' leaders to edit their own glowing profile articles. These sorts of arrangements helped restructure the media marketplace. In contrast to how the network of liberal media was distributed across multiple hubs, thousands of smaller, far-right platforms clung to Breitbart in a tight orbit. They happily sent hyperlinks and advertising profit to each other, but almost never anywhere outside of their closed network, which tilted the balance. The change wasn't just that conservatives were abandoning mainstream media en masse, but that the marketplace of information within their community had changed as well. When judged by key measures like Facebook and Twitter "shares," Breitbart had eclipsed even the likes of Fox News, more than doubling "share" rates among Trump supporters.

In this new media universe, not just money, journalism, and political activism mixed, but also truth and hateful disinformation. News reports of actual events were presented alongside false ones, making it hard for readers to differentiate between them. A series of articles on illegal immigration, for instance, might mix stories about real illegal immigrants with false reports of Al Qaeda–linked terrorists sneaking in via Mexico. In some cases, this situation entered the realm of the bizarre, such as when Breitbart quoted a Twitter account *parodying* Trump, instead of his actual feed, in order to make him sound more presidential than he did in reality.

What the stratagem revealed was that on social networks driven by homophily, the goal was to validate, not inform. Internet reporter John Herrman had observed as much in a prescient 2014 essay. "Content-marketed identity media speaks louder and more clearly than content-marketed *journalism,* which is handicapped by everything that ostensibly makes it journalistic — tone, notions of fairness, purported allegiance to facts, and context over conclusions," he wrote. "These posts are not so much stories as sets of political premises stripped of context and asserted via Facebook share — they scan like analysis but contain only conclusions; after the headline, they never argue, only reveal." This was just as well. In 2016, researchers were stunned to discover that 59 percent of all links posted on social media *had never been clicked on* by the person who shared them.

Simply sharing crazy, salacious stories became a form of political activism. As with the dopamine-fueled cycle of "shares" and "likes," it also had a druglike effect on internet partisans. Each new "hit" of real (or fake) news broadcast on social media might be just enough to help their chosen candidate win.

There was also a sort of raw entertainment to it — a no-holds-barred battle in which actual positions on policy no longer mattered. This, too, was infectious. Now taking their lead from what was trending online, traditional media outlets followed suit. Across the board, just one-tenth of professional media coverage focused on the 2016 presidential candidates' actual policy positions. From the start of the year to the last week before the vote, the nightly news broadcasts of the "big three" networks (ABC, CBS, and NBC) devoted a total of just *thirty-two minutes* to examining the actual policy issues to be decided in the 2016 election!

Yet for all its noise and spectacle, the specter of online misinformation didn't begin with the 2016 U.S. presidential race, nor did it fade once the votes were cast. Disappointed Clinton donors vowed to create a "Breitbart of the left," while a new generation of liberal rumor mills and fabulists purported to show why every Republican politician teetered on the brink of resignation and how every conservative commentator was in the secret employ of the Kremlin. Meanwhile, the misinformation economy powered onward. It would pop up in the 2017 French presidential election and roil the politics of Germany, Spain, and Italy soon thereafter. By 2019, viral misinformation had become an inescapable part of the electoral process, affecting contests from Canada to India to Indonesia — anywhere that voting took place.

Nor has the problem been limited to elections. Perhaps the most worrisome example occurred on Christmas Eve 2016, when Pakistani defense minister Khawaja Asif read a false online report that Israel was threatening to attack his country if it intervened in Syria. "We will destroy them with a nuclear attack," the report had quoted a retired Israeli defense minister as saying. Asif responded with a *real* threat of his own, tweeting about Pakistan's willingness to retaliate with nuclear weapons against Israel. Fortunately, Christmas was saved when the original report was debunked before the crisis could escalate further.

Sadly, not all viral falsehood can be stopped before sparking real-life tragedy. Online misinformation has begun to drive ethnic and religious bloodshed around the world. In India, riots erupted in 2017 over fake stories pushed by the Indian equivalent of Breitbart. These prompted a *new* round of fake stories about the riots and their instigators, which reignited the real cycle of violence. That same year in Myanmar, a surge in Facebook rumormongering helped fuel genocide against the nation's Rohingya Muslim minority. The following year in Sri Lanka, wild (and viral) allegations of a "sterilization" plot led a frenzied Buddhist mob to burn a Muslim man alive. "The germs are ours," a Sri Lankan official explained of his country's religious tensions, "but Facebook is the wind."

The online plague of misinformation has even become a problem for some of the least sympathy-inducing groups in the world. In El Salvador, the MS-13 gang faced an unexpected crisis when false stories spread that it was murdering any woman who had dyed blond hair and wore leggings (the hair and leggings were a trademark look of the rival Los Chirizos gang). "We categorically deny the rumor that has been circulated," read the gang's official statement, itself posted online. The criminals solemnly denounced the stories that "only create alarm and increase fear and anxiety in the poor population that live in the city center."

Even the unrepentantly barbaric Islamic State had to deal with false headlines. When ISIS instituted its repressive, fundamentalist government after the seizure of Mosul, reports circulated that it would force genital mutilation on 4 million Iraqi women and girls. Subsequent news stories were shared tens of thousands of times. ISIS propagandists and supporters were aggrieved. Although they'd happily held public beheadings and reinstituted crucifixion as a form of punishment, female genital mutilation wasn't their policy. An ISIS Twitter account, whose Arabic username translated to "Monster," offered a terse rebuttal denouncing the fake news and demanding that the media retract its claims.

In only a few years, online misinformation has evolved from a tabloid-style curiosity to a global epidemic. Ninety percent of Americans believe that these made-up news stories have made it harder to know what's true and what's not. Nearly one-quarter of Americans admit to having shared a fake story themselves. At the end of 2015, the *Washington Post* quietly ended a weekly feature devoted to debunking internet hoaxes, admitting there were simply too many of them. "[This] represents a very weird moment in internet discourse," mused columnist Caitlin Dewey. "At which point does society become utterly irrational? Is it the point at which we start segmenting off into alternate realities?"

The answer is that the whirlwind of confirmation bias and online gratification can swiftly mobilize millions. It can also produce what the World Economic Forum has called "digital wildfires," fast-moving bursts of information that devastate markets, upend elections, or push nations to the brink of war. While these fires may be set by super-spreaders with a specific agenda, as they advance they can cleave huge rifts across society. And if someone's online network has helped fuel a particular fire, he or she is more likely to believe it — and even more likely to help spread the next one.

The human brain was never equipped to operate in an information environment that moves at the speed of light. Even those who've grown up in this world have found it difficult to adjust. Studies show that more than half of U.S. middle schoolers — who spend an average of 7.5 hours online each day outside of school hours — cannot discern advertisements from legitimate news, nor distinguish basic fact from fiction online. "If it's going viral, it must be true," one middle schooler patiently explained to a team of Stanford researchers. "My friends wouldn't post something that's not true."

On the internet, virality is inseparable from reality. A fake story shared by millions becomes "real" in its own way. An actual event that fails to catch the eye of attention-tracking algorithms might as well never have happened. Yet nothing says the people *sharing* the story have to be real either.

ALL HAIL OUR BOT OVERLORDS

Angee Dixson was mad as hell, and she wasn't going to take it anymore.

A "Christian first," as her profile declared, the photogenic brunette had one item on her agenda. "I want my country back. MAGA." Joining Twitter in August 2017, Dixson took to the platform immediately, tweeting some ninety times a day. She made good on her profile's pledge, leaping to President Trump's defense against Democrats, the FBI, late-night comedians, and everyone in between.

Three days after Dixson hopped online, a coalition of alt-right groups descended on Charlottesville, Virginia, for what they dubbed the #UniteTheRight rally. As counterprotesters poured into the streets to oppose what became a vivid expression of hate and white nationalism, a far-right terrorist drove his car into the crowd, killing one young woman and wounding three others. When public sentiment turned against President Trump (who claimed "both sides" were to blame for the violence), Dixson furiously leapt to his defense. "Dems and Media Continue to IGNORE BLM [Black Lives Matter] and Antifa [anti-fascist] Violence in Charlottesville," she tweeted, including an image of demonstrators with the caption "DEMOCRAT TERROR." In the days that followed, her tweets grew even more strident, publicizing supposed cases of left-wing terrorism around the country.

But none of the cases were real — and Dixson wasn't either. As Ben Nimmo, a fellow with the Digital Forensic Research Lab at the Atlantic Council, discovered, "Angee Dixson" was actually a bot — a sophisticated computer program masquerading as a person. One clue to her identity was her frequent use of URL "shorteners," shortcuts that bots use to push out links. (A machine's efficiency can often spill the beans, as lazy humans tend to use the old-fashioned copy-and-paste method.) Another telltale sign was her machinelike language pattern, sometimes lifted from RT and Sputnik. Despite her avowed American focus, Dixson also couldn't help but slip in the occasional attack on Ukraine. And finally, the true giveaway: Dixson's profile picture was actually a photograph of Lorena Rae, a German model who was dating Leonardo DiCaprio at the time.

Dixson was one of at least 60,000 Russian accounts in a single "botnet" (a networked army of bots) that infested Twitter like a cancer, warping and twisting the U.S. political dialogue. This botnet, in turn, belonged to a vast galaxy of fake and automated accounts that lurk in the shadows of Twitter, Facebook, Instagram, and numerous other services. These machine voices exist because they have power — because the nature of social media platforms *gives* them power.

On Twitter, popularity is a function of followers, "likes," and retweets. Attract lots of attention in a short period of time and you'll soon find yourself, and any views you push, going viral. On Google, popularity is a function of hyperlinks and keywords; the better trafficked and more relevant a particular website, the higher it ranks in Google search results. On Facebook, popularity is determined by "likes" from friends and the particular updates that you choose to share. The intent is to keep users emotionally grafted to the network. Bombard your friends with silly, salacious news stories and you'll find yourself receiving less and less attention; describe a big personal moment (a wedding engagement or professional milestone) and you may dominate your local social network for days.

Every social media platform is regulated by such an algorithm. It represents the beating heart of the platform's business, its most closely guarded treasure. But as the world has come to be ruled by the whims of virality and the attention economy, plenty of people seek to cheat their way to fame and influence. Plenty more happily sell them the tools to do so.

The most common form of this cheating is also the simplest. Fake followers and "likes" are easy to produce — all they require is a dummy email address and social media account — and they enjoy essentially unlimited demand. Politicians, celebrities, media outlets, and wannabe "influencers" of all stripes have come to rely on these services. The result is a decade-old black market worth at least $1 billion.

Often, these fake followers are easy to track. In 2016, internet users had a collective chuckle when People's Daily, the main Chinese propaganda outlet, launched a Facebook page that swiftly attracted 18 million "likes," despite Facebook being banned in China. This included more than a million "fans" in Myanmar (out of the then 7 million Facebook users in that country), who instantly decided to "like" China. Likewise, when Trump announced his nationalistically themed presidential campaign in 2015, 58 percent of his Facebook followers, oddly, hailed from outside the United States. Despite his anti-immigrant rhetoric and repeated calls for a border wall, 4 percent supposedly lived in Mexico.

In the nations of Southeast Asia, the demand for fake followers has given rise to a "click farm" industry that resembles the assembly lines of generations past. Amid the slums of places like Dhaka in Bangladesh or Lapu-Lapu in the Philippines, workers crowd into dark rooms crammed with banks of monitors. Some employees follow rigid scripts intended to replicate the activity of real accounts. Others focus on creating the accounts themselves, the factories equipping their workers with hundreds of interchangeable SIM cards to beat the internet companies' spam protection measures.

Yet as with every other industry, automation has begun to steal people's jobs. The most useful form of fakery doesn't come through click farms, but through the aforementioned bots. Describing software that runs a series of automated scripts, the term "bot" is taken from "robot" — in turn, derived from a Czech word meaning "slave" or "servitude." Today, social media bots spread a message; as often as not, it's human beings who become the slaves to it.

Like actual robots, bots vary significantly in their complexity. They can be remarkably convincing "chatbots," conducting conversations using natural language and selecting from millions of preprogrammed responses. Or the bots can be devilishly simple, pushing out the same hashtag again and again, which may get them caught but still accomplishes their mission, be it to make a hashtag go viral or to bury an opponent under countermessages.

For example, the day after Angee Dixson was outed in an analysis by the nonprofit organization ProPublica, a new account spun to life named "Lizynia Zikur." She immediately decried ProPublica as an "alt-left #HateGroup and #FakeNews Site." Zikur was clearly another fake — but one with plenty of friends. The bot's message was almost instantly retweeted 24,000 times, exceeding the reach of ProPublica's original analysis. In terms of virality, the fake voices far surpassed the reports of their fakeness.

This episode shows the power of botnets to steer the course of online conversation. Their scale can range from hundreds to hundreds of thousands. The "Star Wars" botnet, for example, is made up of over 350,000 accounts that pose as real people, detectable by their predilection for spouting lines from the franchise.

If a few thousand or even hundred of these digital voices shift to discussing the same topic or using the same hashtag all at the same time, that action can fool even the most advanced social media algorithm, which will mark it as a trend. This "trend" will then draw in real users, who have nothing to do with the botnet, but who may be interested in the news, which itself is now defined by what is trending online. These users then share the conversation with their own networks. The manufactured idea takes hold and spreads, courting ever more attention and unleashing a cascade of related conversations, and usually arguments. Most who become part of this cycle will have no clue that they're actually the playthings of machines.

As businesses whose fortunes rise or fall depending on the size of their user base, social media firms are reluctant to delete accounts — even fake ones. On Twitter, for instance, roughly 15 percent of its user base is thought to be fake. For a company under pressure to demonstrate user growth with each quarterly report, this is a valuable boost.

Moreover, it's not always easy to determine whether an account is a bot or not. As the case of Angee Dixson shows, multiple factors, such as time of activity, links, network connections, and even speech patterns, must be evaluated. Researchers then take all of these clues and marry them up to connect the dots, much as in the Sherlock Holmes–style investigations the Bellingcat team pursues to chronicle war crimes.

Although botnets have been used to market everything from dish soap to albums, they're most common in the political arena. For authoritarian regimes around the world, botnets are powerful tools in their censorship and disinformation strategies. When Syria began to disintegrate into civil war in 2011, the Assad regime used Twitter bots to flood its opponents' hashtags with random soccer statistics. Those searching for vital information to fight the regime were instead greeted with a wall of nonsense. At the same time, the #Syria news hashtag was flooded with beautiful landscape images. A year later, when international attention turned to the plight of Chinese-occupied Tibet, the Chinese government did the same. Thousands of bots hijacked hashtags like #Free-Tibet, overpowering activists with random photographs and snippets of text.

Botnets have proven just as appealing to the politicians and governments of democratic nations. Among the first documented uses was in 2010, when Massachusetts held a special election to fill the seat vacated by the late Senator Ted Kennedy. At the beginning of the race, there was little notable social media activity in this traditionally Democratic stronghold. But then came a shock: an outlying poll from Suffolk University showed Republican Scott Brown might have a chance. After that came a social media blitz, masterminded by two out-of-state conservative advocacy groups. One was funded by the Koch brothers and the other by the group that had organized the "Swift Boat" negative advertising campaign that had sunk the 2004 presidential bid of Democratic candidate John Kerry.

Suddenly, bots popped up everywhere, all fighting for Brown. Fake accounts across Facebook and Twitter trumpeted Brown's name as often as possible, seeking to manipulate search results. Most novel was what was then called a "Twitterbomb." Twitter users interested in the election began to receive automated replies supporting Brown. Importantly, these solicitations hit users beyond Massachusetts, greatly enriching Brown's coffers. When Brown became the first Republican to win a Massachusetts Senate seat since 1952, political analysts were both floored and fascinated. Bots had enabled an election to be influenced from afar. They had also shown how one could create the appearance of grassroots support and turn it into reality, a tactic that became known as "astroturfing."

Botnets would become a part of every major election thereafter. When Newt Gingrich's promise to build a moon base didn't excite voters in the 2012 U.S. presidential primaries, his campaign reportedly bought more than a million fake followers, to try to create the sense of national support. In Italy, a comedian turned populist skyrocketed to prominence with the help of bot followers. The next year, a scandal hit South Korea when it was revealed that a massive botnet — operated by military cyberwarfare specialists — had transmitted nearly 25 million messages intended to keep the ruling party in power.

Often, botnets can play the role of political mercenaries, readily throwing their support from one cause to the next. During Brexit, Britain's contentious 2016 referendum to leave the European Union, researchers watched as automated Twitter accounts that had long championed Palestinian independence abruptly shifted their attention to British politics. Nor was it an even fight: the pro-Brexit bots outnumbered the robotic champions of "Remain" by a ratio of five to one. The botnets (many since linked to Russia) were also prodigious. In the final days before the referendum, less than 1 percent of Twitter users accounted for one-third of all the conversation surrounding the issue. Political scientists were left to wonder what might have happened in a world without the machines.

The 2016 U.S. presidential race, however, stands unrivaled in the extent of algorithmic manipulation. On Twitter alone, researchers discovered roughly 400,000 bot accounts that fought to sway the outcome of the race — two-thirds of them in favor of Donald Trump. Sometimes, these bots were content simply to chirp positive messages about their chosen candidate. Other times, they went on the offensive. Like the suppressive tactics of the Syrian regime, anti-Clinton botnets actively sought out and "colonized" pro-Clinton hashtags, flooding them with virulent political attacks. As Election Day approached, pro-Trump bots swelled in intensity and volume, overpowering pro-Clinton voices by (in another echo of Brexit) a five-to-one ratio.

To an untrained eye, Trump's bots could blend in seamlessly with real supporters. This included the eye of Trump himself. In just the first three months of 2016, the future president used his Twitter account to quote 150 bots extolling his cause — a practice he would continue in the White House.

Behind this massive bot army lay a bizarre mix of campaign operatives, true believers, and some who just wanted to watch the world burn. The most infamous went by the online handle "MicroChip." A freelance software developer, MicroChip claimed to have become a believer in the alt-right after the 2015 Paris terrorist attacks. With his tech background, he realized he could manipulate Twitter's programming applications, initially testing such "anti-PC" hashtags as #Raperefugees to see what he could drive viral. By the time of the 2016 election, he labored twelve hours at a time, popping Adderall to stay focused as he pumped out pro-Trump propaganda.

Described by a Republican strategist as the "Trumpbot overlord," MicroChip specialized in using bots to launch hashtags (#TrumpTrain, #cruzsexscandal, #hillarygropedme) that could redirect and dominate political conversation across Twitter. When his machine was firing on all cylinders, MicroChip could produce more than 30,000 retweets in a single day, each of which could reach orders of magnitude more users. He took particular joy in using his army of fake accounts to disseminate lies, including #Pizzagate. "I can make whatever claims I want to make," he bragged. "That's how this game works."

MicroChip lived in Utah. Where bots became truly weaponized, though, was in how they expanded the work of Russian sockpuppets prosecuting their "information war" from afar. In 2017, growing public and congressional pressure forced the social media firms to begin to reveal the Russian campaign that had unfolded on their platforms during the 2016 election. The numbers, once begrudgingly disclosed, were astounding.

The bot accounts were putting the disinformation campaign on ste-roids, allowing it to reach a scale impossible with just humans at work. Twitter's analysis found that bots under the control of the Internet Re-search Agency (that lovely building in St. Petersburg where our philoso-phy major worked) generated 2.2 million "election-related tweets" in just the final three months of the election. In the final month and a half be-fore the election, Twitter concluded that Russian-generated propaganda had been delivered to users 454.7 million times. (Though enormous, these company-provided numbers are likely low, as Twitter identified only accounts definitively proven to belong to the Internet Research Agency's portion of the larger Russian network. The analysis also cov-ered only a limited period of time, not the whole election, and especially not the crucial nomination process.) The same army of human sockpup-pets, using automation tools to extend their reach, would also ripple out onto other sites, like Facebook and its subsidiary Instagram. Overall, Facebook's internal analysis estimated that 126 million users saw Rus-sian disinformation on its platform during the 2016 campaign.

The automated messaging was overwhelmingly pro-Trump. For ex-ample, known Russian bots directly retweeted @realDonaldTrump 469,537 times. But the botnet was most effective in amplifying false re-ports planted by the fake voices of the Russian sockpuppets — and en-suring that stories detrimental to Trump's foes received greater viral-ity. They were particularly consumed with making sure attention was paid to the release of hacked emails stolen from Democratic organiza-tions. (The collective U.S. intelligence community and five different cy-bersecurity companies would attribute the hack to the Russian govern-ment.) Indeed, as Twitter's data showed, when these emails were first made public, botnets contributed between 48 percent and 73 percent of the retweets that spread them.

After the news of Russia's role in the hacks was revealed, these same accounts pivoted to the defensive. An army of Russian bots became an army of supposed Americans arguing against the idea that Russia was involved. A typical, and ironic, message blasted out by one botnet read, "The news media blames Russia for trying to influence this election. Only a fool would not believe that it's the media behind this."

As Samuel Woolley, a researcher at Oxford University who studied the phenomenon, has written, "The goal here is not to hack computational systems but to hack free speech and to hack public opinion." The effects of this industrial-scale manipulation continue to ripple across the American political system. Its success has also spawned a legion of copycat efforts in elections from France to Mexico, where one study found over a quarter of the posts on Facebook and Twitter about the 2018 Mexican election were created by bots and trolls.

And yet this may not be the most unsettling part. These artificial voices managed to steer not just the topics of conversation, but also the human language *within* it, even changing the bounds of what ideas were considered acceptable.

After the 2016 election, data scientists Jonathon Morgan and Kris Schaffer analyzed hundreds of thousands of messages spread across conservative Twitter accounts, Facebook pages, and the Breitbart comments section, charting the frequency of the 500,000 most-used words. They cut out common words like "the" and "as," in order to identify the top terms that were "novel" to each online community. The idea was to discover the particular language and culture of the three spaces. How did conservatives speak on Facebook, for instance, as compared with Twitter? They were shocked to find something sinister at work.

Originally, the three spaces were completely different. In January, February, and March 2016, for example, there was not much of a pattern to be found in the noise of online debate on Twitter as compared with Breitbart or Facebook. In the kind of homophily by then familiar, people in the three spaces were often talking about the same things, since they were all conservatives. But they did so with divergent language. Different words and sentence constructions were used at different frequencies in different communities. This was expected, reflecting both how the particular platforms shaped what could be posted (Twitter's allowance of only 140 characters at the time versus Facebook's lengthier space for full paragraphs) and the different kinds of people who gravitated to each network.

But, as the researchers wrote, in April 2016 "the discussion in conservative Twitter communities, the Trump campaign's Facebook page, and Breitbart's comment section suddenly and simultaneously changed."

Within these communities, new patterns abruptly appeared, with repeated sentences and word choices. Swaths of repetitive language began to proliferate, as if penned by a single author or a group of authors working from a shared playbook. It wasn't that all or even many of the users on Twitter, Facebook, and in the comments section of Breitbart during the run-up to the 2016 election were fake. Rather, the data showed that a coordinated group of voices had entered these communities, and that these voices could be sifted out from the noise by their repeated word use. As Morgan and Schaffer wrote, "Tens of thousands of bots and hundreds of human-operated, fake accounts acted in concert to push a pro-Trump, nativist agenda across all three platforms in the spring of 2016." When the researchers explored what else these accounts were pushing beyond pro-Trump or anti-Clinton messaging, their origin became clearer. The accounts that exhibited these repetitive language patterns were four times as likely to mention Russia, always in a defensive or complimentary tone.

The analysis uncovered an even more disturbing pattern. April 2016 also saw a discernible spike in anti-Semitic language across all three platforms. For example, the word "Jewish" began to be used not only more frequently but also in ways easily identifiable as epithets or conspiracy theories, such as being associated with words like "media."

While the initial blast of repeated words and phrases had shown the use of a common script driven by machines from afar, the language soon spread like a virus. In a sense, it was a warped reflection of the "spiral of silence" effect seen in the previous chapter. The sockpuppets and bots had created the appearance of a popular consensus to which others began to adjust, altering what ideas were now viewed as acceptable to express. The repeated words and phrases soon spread beyond the fake accounts that had initially seeded them, becoming more frequent across the human users on each platform. The hateful fakes were mimicking real people, but then real people began to mimic the hateful fakes.

This discovery carries implications that transcend any particular case or country. The way the internet affects its human users makes it hard enough for them to distinguish truth from falsehood. Yet these 4 billion flesh-and-blood netizens have now been joined by a vast number of digital beings, designed to distort and amplify, to confuse and distract. The attention economy may have been built by humans, but it is now ruled by algorithms — some with agendas all their own.

Today, the ideas that shape battles, votes, and even our views of reality itself are propelled to prominence by this whirring combination of filter bubbles and homophily — an endless tide of misinformation and mysterious designs of bots. To master this system, one must understand *how* it works. But one must also understand *why* certain ideas take hold. The ensuing answers to these questions reveal the foundations of what may seem to be a bizarre new online world, but is actually an inescapable kind of war.

6

Win the Net, Win the Day

The New Wars for Attention . . . and Power

Media weapons [can] actually be more potent than atomic bombs.

— PROPAGANDA HANDBOOK OF THE ISLAMIC STATE

"YOU CAN SIT AT HOME AND play Call of Duty or you can come here and respond to the real call of duty . . . the choice is yours."

It would be an unusual slogan for any army, much less the fanatical forces of the Islamic State. But Junaid Hussain was an unusual recruiter. As a stocky Pakistani boy raised in Britain, he was what one would call a nerd. But in the underground world of hackers, he was cool. "He had hacker cred," one of his old acquaintances recalled. "He had swagger. He had fangirls." But Hussain was also reckless — and he got caught. In 2012, at the age of 18, he was jailed for breaking into the emails of an assistant to former British prime minister Tony Blair.

In prison, Hussain was transformed into a holy warrior. He became consumed with radical beliefs, and when his sentence was up, he fled to Syria, becoming an early volunteer for the jihadist group that would eventually become ISIS. He also took a new online handle, "Abu Hussain al-Britani," and posted a new profile picture of himself cradling an AK-47.

But the rifle was only a prop. The weapons that were far more valuable to ISIS were his good English, his swagger, and his easy familiarity with the internet. He helped organize the Islamic State's nascent "Cyber Caliphate" hacking division, and he scoured Twitter for potential ISIS recruits.

Hussain's online persona was infused with charm, pop culture, and righteous indignation. He persuaded hardened radicals and gullible teenagers alike to travel to Syria. It was a striking contrast to how Al Qaeda, the predecessor of ISIS, had bolstered its ranks. The original members of Al Qaeda had been personally known and vetted by bin Laden and his lieutenants. Indeed, the name "Al Qaeda," translated as "the base," had been taken from the name for the Afghan mountain camps where they'd all trained together. By contrast, some 30,000 recruits, urged on by Junaid Hussain and his team of recruiters, would travel from around the world to join a group that they'd never met in person.

Hussain also reached out to people who pledged allegiance to the Islamic State but never left home. He recruited at least nine ISIS converts in the United States who would later be killed or arrested there. From thousands of miles away, Hussain served as a bizarre mix of leader, recruiter, and life coach. In one case, he directly organized a shooting at a Texas community center by two self-proclaimed "soldiers of the Caliphate." "The knives have been sharpened," Hussain bragged on Twitter scarcely an hour before the attack began. "Soon we will come to your streets with death and slaughter!"

Becoming, in effect, a super-spreader of the terror virus, Hussain achieved celebrity status. He even took a wife — a British punk rock musician in her early 40s, whom he met online. However, his growing fame also made him infamous in U.S. military circles. By 2015, the 21-year-old Hussain had risen to become the third most important name on the Pentagon's "kill list" of ISIS leaders, ranking only behind the group's self-declared caliph and top battlefield commander.

Ironically, it was Hussain's nonstop internet use that enabled his execution. The hacker formerly known as "TriCk" was reportedly tricked into clicking a link that had been compromised by British intelligence. His web use allowed him to be geolocated and dispatched by a Hellfire missile fired by a drone. Working at an internet café late at night, Hussain had thought it safe to leave his stepson — whom he frequently used as a human shield — at home.

SEVENTH-CENTURY CYBER-REVOLUTIONARIES

In the case of Junaid Hussain can be seen the wider paradox of the Islamic State. When ISIS first seized global attention with its 2014 invasion of Mosul, many observers were flummoxed. The word of the day became "slick." Indeed, terrorism analysts Jessica Stern and J. M. Berger found "slick" was used more than 5 million times online to describe the Islamic State's well-doctored images and videos. How could a group of jihadists from a war-torn corner of the world be so adept at using all the tricks of modern viral marketing?

The answer was grounded in demography — and one made almost inevitable by social media's wildfire spread. On the one hand, ISIS was a religious cult that subscribed to a medieval, apocalyptic interpretation of the Quran. It was led by a scholar with a PhD in Islamic theology, its units commanded by men who had been jihadists since the 1980s. But on the other hand, ISIS was largely composed of young millennials. Its tens of thousands of eager recruits, most drawn from Syria, Iraq, and Tunisia, had grown up with smartphones and Facebook. The result was a terrorist group with a seventh-century view of the world that, nonetheless, could only be understood as a creature of the new internet.

"Terrorism is theater," declared RAND Corporation analyst Brian Jenkins in a 1974 report that became one of terrorism's foundational studies. Command enough attention and it didn't matter how weak or strong you were: you could bend populations to your will and cow the most powerful adversaries into submission. This simple principle has guided terrorists for millennia. Whether in ancient town squares, in colonial wars, or via ISIS's carefully edited beheadings, the goal has always been the same: to send a message.

If there was any great difference between the effectiveness of the Islamic State and that of terror groups past, it wasn't in the brains of ISIS fighters; it was in the medium they were using. Mobile internet access could be found even in the remote deserts of Syria; smartphones were available in any bazaar. Advanced video and image editing tools were just one illegal download away, and an entire generation was well acquainted with their use. For those who weren't, they could easily find free online classes offered by a group called Jihadi Design. It promised to take ISIS supporters "from zero to professionalism" in just a few sessions.

Distributing a global message, meanwhile, was as easy as pressing "send," with the dispersal facilitated by a network of super-spreaders beyond any one state's control. This was the most dramatic change from terrorism past. Aboud Al-Zomor was one of the founders of the Egyptian Islamic Jihad terror group and a mastermind of the 1981 assassination of Egyptian president Anwar Sadat. Thirty years later, he wondered if — had social media had been around at the time — the entire plot might have been unnecessary. "With the old methods," the aged killer explained, "it was difficult to gather so many people with so much force." Back then, it took a dramatic, high-profile death to seize public attention. Now all you needed was YouTube.

Viral marketing thus became the Islamic State's greatest weapon. A ghastly example could be seen in August 2014, when the American journalist James Foley was murdered on camera as he knelt in the Syrian sand. The moment was carefully choreographed to maximize its distribution. Foley was clothed in an orange Guantanamo Bay–style jumpsuit, the symbolism clear to all. His black-clad killer spoke English, to ensure his message was understood beyond the Middle East. Unlike the videos of killings done by earlier groups like Al Qaeda, the clip was edited so that the image faded to black right as the knife was pulled across Foley's throat. It ripped across the web, propelled by some 60,000 social media accounts that ISIS had carefully prepared in advance, and American public opinion about the wisdom of becoming involved in a third war in the Middle East in a single generation shifted almost overnight. In short order, the U.S. air campaign against ISIS intensified and crossed over into the conflict raging inside Syria. For ISIS, the clip stood among the cheapest, most effective declarations of war in history.

Following the video's release, there was initial puzzlement as to why the brutal ISIS militants hadn't made it even more gruesome. Why had the clip faded to black right as the execution began? Some news outlets unwittingly provided the answer when they linked to the full video. Others filled their stories with dramatic screengrabs of Foley's final seconds, each piece ricocheting onward with more "shares" and comments. The imagery was disturbing, but not too disturbing to post. Even as terrorism experts and Foley's own family members urged social media users, "Don't share it. That's not how life should be," images of Foley in his orange jumpsuit blanketed the web. One aspiring politician, running for a U.S. House seat in Arizona, even incorporated the clip into her own campaign ads. ISIS was using the same tactics as Russia's information warriors: Why shoulder all the hard work of spreading your message when you could count on others to do it for you?

Whenever the attention of global audiences ebbed (as it did when ISIS began to run out of Western hostages), the self-declared caliphate turned to ever crueler displays — akin to how online celebrities continually raise the stakes by feeding their followers a diet of surprises. There were videos of prisoners executed by exploding collars or locked in burning cars. One set of prisoners was trapped in a cage and submerged in a pool, their drowning captured by underwater cameras. The Islamic State also used social media to encourage audience engagement. "Suggest a Way to Kill the Jordanian Pilot Pig," ISIS-linked accounts asked of supporters following their capture of a Jordanian fighter pilot. He was burned alive.

Like any savvy marketer (and Russian sockpuppets), ISIS sought to hijack trending hashtags and inject itself into unrelated stories. "This is our football, it's made of skin #World Cup," bragged one ISIS supporter's tweet, whose accompanying image was exactly what you'd expect. ISIS soon shouldered its way into trending topics as disparate as an earthquake in California (#napaearthquake), and a question-and-answer session with a young YouTube star (#ASKRICKY).

But the Islamic State didn't simply use the internet as a tool; it also lived there. In the words of Jared Cohen, director of Google's internal think tank, ISIS was the "first terrorist group to hold both physical and digital territory." This was where all the accumulated ISIS propaganda resided; where ISIS fighters and fans could mingle; the perch from which it could track and manipulate global opinion; and the locale from which the group could fight on even after it lost its physical turf.

By networking its propaganda, ISIS pushed out a staggering volume of online messaging. In 2016, terrorism analyst Charlie Winter counted nearly fifty different ISIS media hubs, each based in different regions with different target audiences, but all threaded through the internet. These hubs were able to generate over a thousand "official" ISIS releases, ranging from statements to online videos, in just a one-month period. Each then cascaded outward through tens of thousands of ISIS-linked accounts on more than a dozen social media platforms. Such "official" voices were then echoed and supplemented by the personal accounts of thousands of ISIS fighters, who, in turn, were echoed by their tens of thousands of "fans" and "friends" online, both humans and bots.

The price of this online presence was real — and deadly. In Iraq, at least 30,000 civilians would be killed by the group; in Syria, the deaths were literally incalculable in the chaos of the civil war. Beyond the self-declared caliphate, a new recruiting pool of lonely and disenchanted people (a third of whom lived with their parents) fell into the subterranean world of ISIS propaganda, steered toward murdering their own countrymen. Some did so with the help of ISIS taskmasters ("remote-control" attacks), while others did so entirely on their own ("lone wolves"). In the United States, 29-year-old Omar Mateen called 911 to pledge his allegiance to ISIS in the midst of slaughtering forty-nine people in an Orlando nightclub. As he waited to kill himself, he periodically checked his phone to see if his attack had gone viral.

In the West, ISIS's mix of eye-catching propaganda and calculated attacks was designed with the target's media environment in mind. Each new attack garnered unstinting attention, particularly from partisan outlets like Breitbart, which thrived on reporting all the most lurid details of ISIS claims, thus stoking outrage and raking in subsequent advertising dollars. Similarly, the militants' insistence that their actions were in accordance with Islamic scripture — a stance opposed by virtually every actual scholar of Islam — was parroted by this same subsection of far-right media and politicians, who saw it as a way to bolster their own nationalistic, anti-Islamic platforms.

ISIS militants had internalized another important lesson of the social media age: reality is no match for perception. As long as most observers believed that ISIS was winning, it *was* winning. On the battlefields of Libya and Iraq, it concealed its losses and greatly exaggerated its gains. Far from the battlefields of the Middle East, it could take credit for killings that it had nothing to do with — such as the 2017 Las Vegas shootings in the United States and a mass murder in the Philippines — simply by issuing a claim after the fact.

Soon ISIS had so penetrated the popular imagination that *any* seemingly random act of violence across Europe or the United States brought the group immediately to mind. Daniel Benjamin, a former U.S. counterterrorism official, noted that mental health had ceased to factor into discussions of Muslims who committed violent crimes. "If there is a mass killing and there is a Muslim involved," he concluded, "all of a sudden it is, by definition, terrorism."

By successfully translating its seventh-century ideology into social media feeds, ISIS proved its finesse in what its supporters described as the "information jihad," a battle for hearts and minds as critical as any waged over territory. It did so through a clear, consistent message and a global network of recruiters. It also did so through a steady rain of what it called media "projectiles," online content intended to "shatter the morale of the enemy" (or sometimes simply to anger its critics). In the process, ISIS did more than establish a physical state; it also built an unassailable brand. "They have managed to make terrorism sexy," declared a corporate branding expert, who likened ISIS to a modern-day Don Draper, the Kennedy-era adman of the TV series *Mad Men*.

ISIS's legacy will live on long after the group has lost all its physical territory, because it was one of the first conflict actors to fuse warfare with the foundations of attention in the social media age. It mastered the key elements of *narrative, emotion, authenticity, community,* and *inundation,* each of which we'll explore in turn. Importantly, none of these elements are unique to terrorism or the Middle East. Indeed, anyone — digital marketers, conspiracy theorists, internet celebrities, politicians, and national militaries — can employ them.

Whatever or wherever the conflict, these are the weapons that win LikeWar.

NARRATIVE: SPIN A TALE

Spencer Pratt is Southern California personified: blond-haired and blue-eyed, a bro who speaks in bromides. But beneath his surfer-dude appearance, Spencer is also a keen student of people: how they act, how they think, and how to keep their attention. "I always wanted to work for the CIA growing up," he explained. "I'd be a CIA operative in Hollywood that made movies to manipulate the masses.

"But then," he added with a laugh, "I became a reality star."

By his freshman year at USC, he'd figured out how to make $50,000 for a photo he'd taken of Mary-Kate Olsen. But what fascinated him back then in the early 2000s was the bizarre, emerging landscape of reality television. "I saw *The Osbournes* on MTV," Spencer recalls. "I saw that they were getting 60 million viewers to watch — with due respect to Ozzy — a British guy mumbling and his wife yelling, cleaning up dog poop. I was like, 'This is what reality television is? I could make one of these shows.'"

And so he did. Pratt became the creator and producer of *The Princes of Malibu*, an early reality show on Fox that followed two rich brothers who were notable only because of their celebrity father, Bruce (now Caitlyn) Jenner. The show fizzled after a few episodes, but not before unleashing the brothers' stepfamily, the Kardashians, upon the world.

As he faced the prospect of going back to college, Pratt had a better idea. He was telegenic, charming, and shameless. Why not try to be *in* one of those shows instead?

The year was 2006, and MTV was in the midst of launching another reality-television saga, *The Hills*, about four young women trying to make it big in Beverly Hills. So Pratt sought out the venues where *The Hills* was filming. At a nightclub called Privilege, the intrepid hustler sat himself in a booth, surrounded by *Playboy* Playmates. This tableau caught the eye of Heidi Montag, *The Hills* blonde costar; she stole him away from the Playmates for a dance. They hit it off, and Spencer Pratt and Heidi Montag soon became "Speidi."

Pratt had gotten on TV, but now he had to figure out how to stay there. So he gave the supposed reality show what it had lacked: a villain. In short order, *The Hills'* story line shifted to a seemingly psychopathic boyfriend and a woman who kept coming back to him. Each episode brought new shocks and new lows. He flirted with other women in front of Montag and gleefully scorned her family. He stoked rumors about a costar and a supposed sex tape, which was roundly condemned by the entertainment press, but which generated a season's worth of fireworks as friendships exploded.

Of course, the vast majority of it was fake, as most of the "reality" show was staged. Still, it worked and ratings soared. But to gain further fame and fortune, Spencer realized he needed to do something more. "I got into manipulating the media," he told us.

Between seasons, the couple worked to stay in the news by releasing a steady stream of scandalous photographs and shocking interview quotes. "What took us to the next level was working with the paparazzi," Pratt recalled. At a time when most celebrities eschewed the paparazzi, the *Hills* villains embraced them. "I just figured that we could come up with these stories, and work with the magazines, and give them that juicy, gossipy stuff that they usually have to make up," Pratt explained. "Why not help make it up with them . . . and get paid for it?"

By learning how to give the media and audiences what they wanted, Speidi soon ranked among reality television's highest-paid and most visible stars. They were also the most despised. Pratt was twice nominated as the Teen Choice Awards' "Best Villain," an award typically reserved for fictional characters (among his competitors was Superman nemesis Lex Luthor). This was the price of fame: people couldn't look away, but they also hated him for it. Sounding a little remorseful, Pratt described filming a fake pregnancy scare with Montag that ended (on camera) with him stomping on the gas after throwing her out of the car. "We shot that scene like twelve times," he said. "I didn't think anything of it. I should have, because every woman on the planet was like, 'Oh my God, he's the worst person on the earth!' Really, my wife and I just drove away and went to dinner. But the audience sees this guy leaving his girl on the side of the street in tears."

Pratt and Montag had constructed a story line — a remorseless, psychopathic man and a manipulated, unhappy woman — that clung to them long after *The Hills* had ended. They had captivated millions of people and grown famous in the process. But they had also created a kind of cultural gravity they couldn't escape. As he explained, "I was getting paid so much money to be just an awful asshole. You start doing your interviews like that, and next thing is, you've got to stay in character. You're getting paid so much money to not care, it's like, 'Whatever.' But then you forget, like, 'Wait. No. Middle America doesn't get this is all fake.'"

In other words, they'd built a "narrative." Narratives are the building blocks that explain both how humans see the world and how they exist in large groups. They provide the lens through which we perceive ourselves, others, and the environment around us. They are the stories that bind the small to the large, connecting personal experience to some bigger notion of how the world works. The stronger a narrative is, the more likely it is to be retained and remembered.

The power of a narrative depends on a confluence of factors, but the most important is consistency — the way that one event links logically to the next. Speidi wasn't merely insufferable once or twice; they were reliably insufferable, assembling a years-long narrative that kept viewers furious and engaged. As narratives generate attention and interest, they necessarily abandon some of their complexity. The story of Spencer Pratt, the vain villain, was simpler — and more engaging — than the story of the conflicted self-promoter who pretended to leave his girlfriend on the curb in order to pump the ratings.

Human minds are wired to seek and create narrative. Every moment of the day, our brains are analyzing new events — a kind word from our boss, a horrible tweet from a faraway war — and binding them into thousands of different narratives already stowed in our memories. This process is subconscious and unavoidable. In a pioneering 1944 study, psychologists Fritz Heider and Marianne Simmel produced a short film that showed three geometric figures (two triangles and a circle) bouncing off each other at random. They screened the film to a group of test subjects, asking them to interpret the shapes' actions. All but one of the subjects described these abstract objects as living beings; most saw them as representations of humans. In the shapes' random movements, they expressed motives, emotions, and complex personal histories. The circle was "worried," one triangle was "innocent," and the other was "blinded by rage." Even in crude animation, most observers saw a story of high drama, while the one who didn't was an oddity.

By simplifying complex realities, good narratives can slot into other people's preexisting comprehension. If a dozen bad things happen to you on your way to work, you simply say you're having a "bad day," and most people will understand intuitively what you mean. The most effective narratives can thus be shared among entire communities, peoples, or nations, because they tap into our most elemental notions.

Following World War II, for instance, some U.S. statesmen advocated the massive aid package known as the Marshall Plan because of its "psychological political by-products." They saw that the true value of the

$13 billion program was the narrative it would build about the United States as a nation that was both wealthy and generous. This single story line was valuable in multiple ways. It not only countered Soviet narratives about whose economic system was best, but it also cast America as a great benefactor, which linked the U.S.-European relationship to other narrative themes of charity, gratitude, and debt. In the slightly less grand case of Speidi and *The Hills*, Spencer was no less strategic in using his villainy. By playing a familiar role, he created outrage against the couple that was shared almost universally, but that gave him a path to fame and fortune.

Today, Spencer and Heidi are a little wiser, older, and poorer. Pioneers in the world of self-made celebrity, they've watched the development of modern social media with fascination. They described to us how much the game has changed in just a few years. Montag marveled that, in a world of smartphones, "everyone's an editor," tweaking each word and image until it conforms to an idealized sense of self. "Now everyone is a reality star," Pratt added. "And they're all as fake as we were."

The challenge now is thus more how to build an effective narrative in a world of billions of wannabe celebrities. The first rule is *simplicity*. In 2000, the average attention span of an internet user was measured at twelve seconds. By 2015, it had shrunk to eight seconds — slightly less than the average attention span of a goldfish. An effective digital narrative, therefore, is one that can be absorbed almost instantly.

This is where the simple, direct hip-hop vernacular of Junaid Hussain proved so effective in reaching out to millennial youth, compared with the book-length treatises of earlier jihadist recruiters. Donald Trump also capitalized on the premium that social media places on simplicity. During the 2016 election, Carnegie Mellon University researchers studied and ranked the complexity of the candidates' language (giving it what is known as a Flesch-Kincaid score). They found that Trump's vocabulary measured at the lowest level of all the candidates, comprehensible to someone with a fifth-grade education.

This phenomenon might seem unprecedented, but it is consistent with a larger historic pattern. Starting with George Washington's first inaugural address, which measured as one of the most complex overall, American presidents communicated at a college level only when newspapers dominated mass communication. But each time a new technology took hold, the complexity score dipped. It started with the advent of radio in the 1920s, and again with the entry of television in the 1950s, and now once more with social media. To put it another way: the more accessible the technology, the simpler a winning voice becomes. It may be Sad! But it is True!

This explains why so many modern narratives exist at least partially in images. Pictures are not just worth the proverbial thousand words; they deliver the point quickly. Consider one popular photograph, of a shark swimming down a flooded street, supposedly taken from a car window. For years, the (fake) picture has popped up during every major hurricane, captivating social media users and infuriating the biologists who keep having to debunk it. Yet, its longevity makes a lot of sense. For people inundated with news about the latest storm's severity and "record-breaking" rainfall, the image — a shark swimming where it clearly doesn't belong — instantly tells a story with scary consequences. It is fast, evocative, and (most important) easily shared. It is also influential, helping inspire the *Sharknado* franchise.

The second rule of narrative is *resonance*. Nearly all effective narratives conform to what social scientists call "frames," products of particular language and culture that feel instantly and deeply familiar. In the American experience, think of plotlines like "rebel without a cause" or "small-town kid trying to make it in the big city." Some frames are so common and enduring that they might well be hardwired into our brains. In his book *The Hero with a Thousand Faces*, mythologist Joseph Campbell famously argued that one frame in particular — "the Hero's Journey" — has existed in the myths of cultures around the globe. Quite often, these frames can merge with the real-life narratives our brains construct to explain ourselves and the world around us. A resonant narrative is one that fits neatly into our preexisting story lines by allowing us to see ourselves clearly in solidarity with — or opposition to — its actors. Social media can prove irresistible in this process by allowing us to join in the narrative, with the world watching.

Spencer and Heidi achieved resonance by being what every hero or heroine needs — a villainous foe — which they played to caricature. Among its opponents, ISIS achieved resonance by being similarly cartoonishly evil. Among its supporters, it achieved resonance by promising the mystery, adventure, or lofty purpose they'd been hoping for their entire lives. Even for members of Congress, there is a powerful correlation between their level of online celebrity and a narrative of ideological extremism. According to a study by the Pew Research Center, the more unyieldingly hyperpartisan a member of Congress is — best fitting our concept of the characters in a partisan play — the more Twitter followers he or she draws.

This also explains why conspiracy theories have found new life on the internet. It's innately human to want to feel as if you're at the center of a sweeping plotline in which you are simultaneously the aggrieved victim (such as of the vast global cabal that oversees the "deep state") and the unlikely hero, who will bring the whole thing crashing down by bravely speaking the truth. The more an article claims that it contains information that governments or doctors "don't want us to know," the more likely we are to click on it.

The third and final rule of narrative is *novelty*. Just as narrative frames help build resonance, they also serve to make things predictable. Too much predictability, though, can be boring, especially in an age of microscopic attention spans and unlimited entertainment. The most effective storytellers tweak, subvert, or "break" a frame, playing with an audience's expectations to command new levels of attention. At the speed of the internet, novelty doesn't have time to be subtle. Content that can be readily perceived as quirky or contradictory will gain a disproportionate amount of attention. A single image of an ISIS fighter posing with a jar of Nutella, for instance, was enough to launch dozens of copycat news articles.

These three traits — simplicity, resonance, and novelty — determine which narratives stick and which fall flat. It's no coincidence that everyone from far-right political leaders to women's rights activists to the Kardashian clan speaks constantly of "controlling the narrative." To control the narrative is to dictate to an audience who the heroes and villains are; what is right and what is wrong; what's real and what's not. As jihadist Omar Hammami, a leader of the Somali-based terror group Al-Shabaab, put it, "The war of narratives has become even more important than the war of navies, napalm, and knives."

The big losers in this narrative battle are those people or institutions that are too big, too slow, or too hesitant to weave such stories. These are not the kinds of battles that a plodding, uninventive bureaucracy can win. As a U.S. Army officer lamented to us about what happens when the military deploys to fight this generation's web-enabled insurgents and terrorists, "Today we go in with the assumption that we'll *lose* the battle of the narrative."

And yet, as we'll see, narrative isn't the only factor that drives virality, nor are narratives forever fixed in place. Speidi may have been boxed in by their self-created villainy, but they're now writing a new story line. They've rebranded themselves as wizened experts on fame, melded with one of the oldest narratives of all: loving parents. Soon after we spoke with them, the couple proudly announced Montag's pregnancy — for real this time.

But they've not forgotten their old lessons. They chose their new son's name (Gunner Stone) based, in part, on which social media handles were available at the time.

EMOTION: PULL THE HEARTSTRINGS, FEED THE FURY

"When we do not know, or when we do not know enough, we tend always to substitute emotions for thoughts."

The writer T. S. Eliot was despairing over the death of literary criticism thanks to the nineteenth century's "vast accumulations of knowledge." Yet his words are even more applicable in the twenty-first century. What captures the most attention on social media isn't content that makes a profound argument or expands viewers' intellectual horizons. Instead, it is content that stirs emotions. Amusement, shock, and outrage determine how quickly and how far a given piece of information will spread through a social network. Or, in simpler terms, content that can be labeled "LOL," "OMG," or "WTF."

These are the sorts of feelings that create *arousal*, not the sexual kind (at least not usually), but the kind in which the heart beats faster and the body surges with fresh energy. Arousal can be positive or negative. A baby dancing, a politician standing up for what she believes in, the story of a disabled man being robbed and beaten, and an awful flight delay are things that people will likely consume and forward to others in their network. A decade's worth of psychology and marketing studies, conducted across hundreds of thousands of social media users, have arrived at the same simple conclusion: the stronger the emotions involved, the likelier something is to go viral.

But the findings go further. In 2013, Chinese data scientists conducted an exhaustive study of conversations on the social media platform Weibo. Analyzing 70 million messages spread across 200,000 users, they discovered that *anger* was the emotion that traveled fastest and farthest through the social network — and the competition wasn't even close. "Anger is more influential than other emotions like joy," the researchers bluntly concluded. Because social media users were linked to so many others who thought and felt as they did, a single instance of outrage could tear through an online community like wildfire. "The angry mood delivered through social ties could boost the spread of the corresponding news and speed up the formation of public opinion and collective behavior," the researchers wrote. People who hadn't been angry before, seeing so much anger around them, feel inclined to ramp up their language and join in the fury.

A year later, an even larger and more insidious study confirmed the power of anger. Partnering with Facebook, data scientists manipulated the newsfeeds of nearly 700,000 users over the course of a week, without the knowledge of the "participants." For some, the researchers increased the number of positive stories to which they were exposed. For others, they increased the number of negative stories. In each case, Facebook users altered their own behavior to match their new apparent reality, becoming cheerier or angrier in the process. But the effect was most pronounced among those whose newsfeeds had turned negative. The scientists dubbed this an "emotional contagion," the spread of emotions through social networks that resembled nothing so much as the transmission of a virus. "Emotional contagion occurs without direct interaction between people," the scientists concluded, "and in the complete absence of nonverbal cues." Just seeing repeated messages of joy or outrage was enough to make people feel those emotions themselves.

Anger remains the most potent emotion, in part because it is the most interactive. As social media users find ways to express (or exploit) anger, they generate new pieces of content that are propelled through the same system, setting off additional cascades of fury. When an issue has two sides — as it almost always does — it can resemble a perpetual-motion machine of outrage. The graphic online propaganda of ISIS, for instance, served a dual purpose. Not only did it elicit waves of shock and outrage in the West; it also drove a violent anti-Islamic backlash, which ISIS could use to fuel renewed anger and resolve among its own recruits.

Anger is not necessarily bad. After all, nearly every political movement that has risen to prominence in the social media age has done so by harnessing the power of outrage. Sometimes, activists fight for better government policy, using a single, viral moment as their rallying cry: a deadly 2011 train derailment in Zhejiang, China; a massive 2017 apartment building fire in London; or a 2018 school shooting in Parkland, Florida. Other times, the cause is social or racial justice. In 2013, Alicia Garza posted a passionate message about police shootings of African Americans on her Facebook page. She closed it with a simple note: "Black people. I love you. I love us. Our lives matter." A friend then reposted the resonant message on his page, adding the hashtag #BlackLivesMatter. It quickly went viral, fueling a new type of civil rights movement that united 1960s activism with twenty-first-century media platforms. In a matter of days, #BlackLivesMatter would evolve from a mere hashtag to nationwide protests, online organizing, and successfully lobbying for scores of local- and state-level police reforms.

But the bigger picture is grim. If attention is the thing that matters most online — and as we saw in the last chapter, it is — brazen self-promoters will go to any lengths to achieve it. Because anger is so effective at building and sustaining an audience, those who seek viral fame and power have every reason to court controversy and adopt the most extreme positions possible, gaining rewards by provoking fury in others. "Anger leads to hate; hate leads to suffering," observed the wise Master Yoda. And that suffering leads to the Dark Side: what is better known on the internet as trolling.

Although the word "troll" conjures images of beasts lurking under bridges and dates back to Scandinavian folklore, its modern internet use actually has its roots in the Vietnam War. American F-4 Phantom fighter jets would linger near North Vietnamese strongholds, taunting them. If eager, inexperienced enemy pilots took the bait and moved to attack, the Americans' superior engines would suddenly roar into action, and the aces would turn to shoot down their foes. American pilots called this deception "trolling for MiGs."

Early online discussion boards copied both the term and the technique. "Trolling for newbies" became a sport in which experienced users would post shamelessly provocative questions designed to spark the ire of new (and unwitting) users. The newbies would then waste time trying to argue a point that was simply designed to make them argue. An article from a digital magazine of the time succinctly described the appeal of trolling: "If you don't fall for the joke, you get to be in on it."

While early trolling was characterized by wink-and-nudge humor, as more and more people (and real-life problems) penetrated digital sanctuaries, the good-natured joking part soon died. Today, we know trolls as those internet users who post messages that are less about sharing information than spreading anger. Their specific goal is to provoke a furious response. The substance of their messages varies so widely as to be essentially irrelevant. Trolls do everything from slinging incendiary lies about political foes to posing as cancer patients. The only consistency is their use of emotional manipulation. Indeed, the words that best capture this trolling ethos were, appropriately enough, laid down in 1946 by the French philosopher Jean-Paul Sartre in describing the tactics of anti-Semites:

> They know that their remarks are frivolous, open to challenge. But they are amusing themselves, for it is their adversary who is obliged to use words responsibly, since he believes in words ... They delight in acting in bad faith, since they seek not to persuade by sound argument but to intimidate and disconcert.

The modern version is perhaps best expressed by one of the internet's better-known trolls, "Ironghazi," who explained, "The key to being a good troll is being just stupid enough to be believable, keeping in mind that the ultimate goal is making people mad online."

In many ways, the Russian sockpuppets masquerading as American voters and venting fog into the geopolitical system in 2016 were trolls — just paid ones. Most trolling behavior, however, bears little resemblance to that of trained, professional provocateurs. Although a small number of trolls are pathological (exhibiting actual psychopathy and sadism), the vast majority are everyday people giving in to their rage. In a report titled "Anyone Can Become a Troll," a team of researchers found that mounting anger turns users toward trolling behavior. And just like conspiracy theories, the more the anger spreads, the more internet users are made susceptible to it.

After people have trolled once, they're twice as likely to engage in trolling behavior than those who've never done it. And as non-trolls engage with trolls, many embrace trolling tactics themselves. "Such behavior can . . . spread from person to person in discussions and persist across them to spread further in the community," the team wrote. "Our findings suggest that trolling . . . can be contagious, and that ordinary people, given the right conditions, can act like trolls."

There's no doubt that trolling makes the internet a worse place. Trolling targets livelihoods and ruins lives. It silences voices and drives people into hiding, reserving special cruelty for women and racial minorities. Even those who escape the trolls' ire must still contend with a digital environment that amplifies outrage and effectively mutes everything else. The power of trolls — which really represents the power of anger — transforms the internet into a caustic, toxic swamp.

But the worst online trolling doesn't necessarily *stay* online. Think back to the online battles of American street gangs — their "cybertagging" and "Facebook drilling" — or the deliberate antagonism of people from one government or ethnic group against another. These angry flame wars are trolling by another name, intended to grab attention and stir outrage. Such trolling too often ends in real-life violence and tragedy. Or it can yield political power.

Whether the case is swaggering street-fighters or the everyday people who revel in harassing someone after a tweet falls flat, anger is the force that binds them together. Anger is exciting. Anger is addictive. Indeed, in a digital environment suffused with liars and fakes, anger feels raw and *real* in a way that so many other things never do. This authenticity carries an additional power of its own.

AUTHENTICITY: THE POWER OF BEING REAL

Taylor Swift's Instagram comments fell with the power of precision air strikes.

"You have the prettiest, wildest, most child like eyes," the superstar wrote to a young fan dealing with boy troubles. "Feel good about being the kind of person who loves selflessly. I think someday you'll find someone who loves you in that exact way."

And to another, a 16-year-old fan who'd just gotten her driver's license: "YES! You passed!!!!!!!! So stoked for you. 'Don't text and drive' is an obvious piece of advice but people usually forget to tell you 1) don't eat and drive 2) don't apply mascara and drive 3) never let a small animal such as a cat roam free in your car. I'm not saying any of this from personal experience. I repeat. None of that happened to me."

Comments like these felt real because they *were* real. It really was Taylor Swift scrolling through her Instagram feed, learning about the lives of her fans, and tapping out thoughtful comments. She even coined her own hashtag to describe this practice: #Taylurking.

It also was a strategy designed around Swift's intuitive grasp of how social media had changed the cultural landscape. Reflecting on her first record-label meetings, Swift explained how she'd wowed the stodgy music executives by "explaining to them that I had been communicating directly with my fans on this new site called Myspace." She added, "In the future, artists will get record deals because they have fans — not the other way around."

By recognizing this change, Swift transformed from a young millennial with a smartphone and a great voice into the ruler of a billion-dollar music empire, empowered by millions of "Swifties," her army of fervent online fans (a name she strategically copyrighted). She sold 40 million albums, shattered digital streaming records, and, at 26 years old, was named the youngest of *Forbes* magazine's wealthiest self-made women.

Was her virtual authenticity all an act? It was certainly true that Swift penned her Instagram missives with the knowledge that anyone could read them. All those "candid" shots of her celebrity-stuffed parties weren't very candid at all. And whenever Swift fell into a feud that stirred anger online, it was cleverly folded into the marketing for her next album. "Asking whether or not Taylor Swift is genuine is like asking if Kylie Jenner's had plastic surgery, or if Calvin Harris is a real musician," mused entertainment reporter Amy Zimmerman. "There's no simple answer out there — just a whole lot of conflicting opinions."

Yet Swift's online success also showed that question didn't matter. "Authenticity" was becoming as dual in meaning as "fact" or "reality." It really was her dour white cat featured on her Instagram account; it really was her dropping in on a World War II veteran (and Swift superfan) for an impromptu concert or sending out random Christmas gifts with sweet, handwritten notes. But it was also true that each of these actions fed and expanded her juggernaut brand. Swift had married her fame to a sense of intimacy and openness, to a cascade of endless surprises. As she explained, "I think forming a bond with fans in the future will come in the form of constantly providing them with the element of surprise. No, I did not say 'shock'; I said 'surprise.' I believe couples can stay in love for decades if they just continue to surprise each other, so why can't this love affair exist between an artist and their fans?"

Swift hadn't built a fake life; she'd built a performative one. She could approach her fans on their level, and fit the perception of her life into theirs, by uploading a post that spotlighted what made her most relatable: fun with friends, thoughts on the nature of love, and lots of cat pictures. In so doing, Swift harnessed the power of online authenticity and cemented her fame. She also cleared a path toward viral success that today's enterprising marketers — celebrities, corporations, politicians, livestreamers, and terrorists — all seek to follow.

Achieving a sense of authenticity has become an important milestone for any online operation. In bland corporate jargon, this is called "brand engagement" — extending an organization's reach by building a facsimile of a relationship between an impersonal brand and its followers. The Islamic State, for instance, expanded its influence not just through propagandists like Junaid Hussain, but through a general sense of authenticity — a feeling that the terrorist group was somehow more "real" than its rival militant organizations. ISIS fighters proved this by living their lives online, posting images not just of their battles but also of their birthday parties and (naturally) their cats. Like Taylor Swift's clever marketing, ISIS's professionally choreographed videos were complemented by chaotic, seemingly candid footage — albeit taken from Syrian battlefields instead of celebrity-studded Fourth of July parties. And like Swift's strategy, this mix of carefully curated media promotion and surprisingly roughshod moments eventually merged, becoming part of the same identity.

These qualities lay at the heart of ISIS's success in online recruiting. Its fighters would talk up the glory of the caliphate but also muse about their sadness over the death of the actor Robin Williams and their childhood love of his character in the movie *Jumanji*. This authenticity won and inspired followers in a way that government press releases could not. Plenty of radicalized Westerners, pulled back from the brink of recruitment, described online relationships that unspooled over weeks or months. In time, the jihadists living on the other side of the world seemed less like recruiters than friends.

Where this internet-age authenticity has proven most crucial, however, is in electoral politics. Since their very invention in ancient Greece, democracies have been guided by a special class of people discussed in Aristotle's *Politika*: politicians, people who seek to rise above their fellow citizens and to lead them. But this created an enduring paradox of democracy. To gain power over their peers, politicians have often had to make themselves seem like their peers. In the United States especially — a nation whose aversion to a noble class is written into its Constitution — the politician who seems most down-to-earth has long carried the day.

The irony, of course, is that most people who run for political office aren't very relatable at all. They're quite often rich, elitist, and sheltered from voters' daily problems. As a result, American politics has long been a tug-of-war over who *seems* most authentic. In the nineteenth century, even the wealthiest candidates published newspaper biographies that played up their humble farmer's roots. The twentieth century saw the birth of "photo ops" — first painfully staged photographs, then even more painfully staged televised campaign stops, taking place in a seemingly limitless number of Iowa diners.

With the rise of social media, however, the fight to be real turned to what it meant to be real *online*. When Trump first stormed into the 2016 U.S. presidential race, few political analysts took his run seriously. He broke all the cardinal rules of American politics: he didn't try to be an "everyman"; he bragged about being rich; he violated every social taboo he could find; he made outlandish statements; and he never, ever apologized. As "expert" analysts shook their heads in disgust, however, millions of American voters perked up and paid attention. This was a politician who was well and truly *authentic*.

At the heart of Trump's authenticity was his Twitter account. Clearly his own creature, it was unpredictable and hyperbolic and full of id.

Even Trump's most ardent critics found something captivating about a presidential candidate staying up late into the night, tapping out stream-of-consciousness tweets in his bedclothes. "It's a reason why Trump's Twitter feed is so effective," observed reporter Maggie Haberman. "People feel like he's talking to them." This was in stark contrast to his opponent Hillary Clinton, whose tweets were sometimes crafted by a team of eleven staffers. And it was a platform Trump came to love. "My use of social media is not Presidential," he tweeted in response to negative headlines about his continuing Twitter obsession, "it's MODERN DAY PRESIDENTIAL."

It was both a real sentiment and a planned-out strategy that Taylor Swift and Junaid Hussain alike would have immediately recognized.

COMMUNITY: THE POWER OF OTHERS

Internet-age authenticity doesn't just empower an idea or person. It also draws us into contact with others who think and act as we do.

"In the end, what people want is to be united in something bigger than them . . . a sense of belonging," explained a 43-year-old Canadian postal worker when asked why he'd joined a close-knit, 50,000-person Facebook group called La Meute. After all, Facebook's very mission statement is to "bring the world closer together."

But this meeting of the minds illustrated a larger problem: La Meute (The Pack) was an ultra-right-wing extremist group based in Canada and dedicated to fighting Islam and immigrants via paramilitary tactics and hate speech. It was exactly the sort of "interactive community" once prophesied by Licklider and Taylor back in 1968 — except that it was one bonded by hate.

The term "community" connotes a group with shared interests and identities that, importantly, make them distinct from the wider world. In the past, a community resided in a specific location. Now it can be created online, including (and perhaps especially) among those who find a common sense of fellowship in the worst kinds of shared identities that exclude others.

As it has with so many other movements, social media has revolutionized white nationalist, white supremacist, and neo-Nazi groups, spiking their membership and allowing their views to move back into mainstream discourse. In the United States, the number of Twitter followers of such groups ballooned 600 percent between 2012 and 2016, and the Southern Poverty Law Center now tracks some 1,600 far-right extremist groups. Through the web, these groups can link up globally, American neo-Nazis connecting with Hungarian anti-Semites and British fascists.

As these extremists have banded together, they have carved out online spaces where they are encouraged and empowered to "be themselves." They have found warmth and joy in each other's company, even as they advocate for the forced deportation of those whose skin color or religion is different from their own. Beyond hatred of immigrants and Muslims, they have few consistent positions. But hate is enough to draw these communities together and propel some of their members toward lethal violence. Between 2011 and 2017, there were nearly 350 far-right extremist terror attacks across Europe, North America, and Australia. In 2018 alone, fifty Americans were killed by angry young white men fueled by a toxic mix of alt-right ideology and white nationalist social media. The following year saw even more, like a massacre at a Christchurch, New Zealand, mosque that killed fifty worshippers. Before he set out to kill, the shooter posted a manifesto steeped in internet culture that spread like wildfire among the web's most deplorable trolls, who saw him as one of their own. He then livestreamed his attack on Facebook for "fans" to watch. The video of the massacre was shared at least 1.5 million times, by accounts that ranged from anonymous Twitter bots to an Instagram celebrity with over 1.6 million followers.

Ironically, in their aggressive recruiting, inspiration of lone wolf killers, and effective use of authenticity to build a community, these far-right extremists resemble nothing so much as the Islamic State. In northern Europe, the mothers of children who ran away to join the Islamic State recalled how their sons and daughters — reckoning with the social isolation that faces the offspring of many Middle Eastern migrants — looked to ISIS to fill the void. A lonely girl in Washington State — a volunteer Sunday school teacher and part-time babysitter — described how ISIS recruiters gave her the attentive friends she'd always craved. (Only a sharp-eyed grandmother stopped her from boarding a plane to Syria.) ISIS promised adventure and a sense of belonging. "It's a closed community — almost a clique," explained terrorism analyst Seamus Hughes. "They share memes and inside jokes, terms and phrases you'd only know if you were a follower."

In each case, recruits to extremist causes are lured by a warmth and camaraderie that seems lacking in their own lonely lives. In each case, such recruits build communities that attract people from across the world but that show almost no diversity of thought. "Isolation may be the beginning of terror," political theorist Hannah Arendt wrote in a 1953 essay about the origins of totalitarianism. "It certainly is its most fertile ground." If people come to believe that their radical notions are unassailably true — and if they believe that only other people who share the same opinions are "real" or worth protecting — they open the door to violence and bloodshed.

Not by coincidence, the field of study that seeks to counter this process of radicalization, known as countering violent extremism (CVE), also focuses on the powers of community-building. Farah Pandith is a pioneer of this field. Born in the restive Kashmir region of India, Pandith moved to Massachusetts as a young girl. Two moments changed the trajectory of her life. One took place at Smith College, where as a student in 1989 Pandith gave a speech attended by school alumnae, including Barbara Bush. The First Lady was impressed and soon became her pen pal. The other occurred a few years later, back in her birthplace of Srinagar, Kashmir. One family member, who was working to bring peace to the region, was assassinated by extremists. Then, the very same day, another died in violence that broke out during the funeral procession. Pandith's life became guided by a simple question: How could she prevent such tragedy from happening to others?

With the help of her new friend in the White House, Pandith joined the U.S. government. Over the next two decades, she served in various roles in both Republican and Democratic administrations, eventually being appointed the first-ever U.S. special representative to Muslim communities. In this position, established to engage in the post-9/11 "battle of ideas," Pandith traveled to eighty countries and met with thousands of young, disaffected Muslims in places ranging from the slums of Düsseldorf to the mosques of Mali. She foresaw a crisis of identity that would soon sweep the Middle East, culminating in the rise of ISIS. But she also saw something else. "Only peer-to-peer relations can change minds," she concluded. The only way to prevent radicalization was to assemble a crowd of authentic voices to fight back.

Pandith determined that social media would be the key battleground in this fight. She became one of the first high-level U.S. officials to use Facebook in her work. She learned that it was not just a megaphone but also a means to keep her connected to the youth she met around the world, and, even more important, to connect them with each other. "Because I was fully focusing on millennials, I needed to be able to show them in real time what I was hearing from others," she explained. "I wanted to connect the kid I met in Germany with a kid in Australia. The conversation I was having in Mauritania with the cool thing in the Pamir mountains [of Tajikistan] that they were doing." Each could become an ally to the other — and part of a broad collective to push back against the specter of extremism.

Frustrated by a bureaucracy that couldn't get out of its own way and realizing that a teenager's heart and mind are places where "no government is credible," Pandith has since left government. But she hasn't quit the fight. Instead, she has worked to assemble groups around the world into a CVE version of what she has dubbed a "Dumbledore's Army."

The name is taken from the *Harry Potter* series, in which a group of teens mobilize to fight evil. In recent years, a number of these sorts of CVE organizations have arisen. There's the Online Civil Courage Initiative, which links more than a hundred anti-hate organizations across Europe, and Gen Next, which seeks to "deprogram" former jihadists. There's even Creative Minds for Social Good, which has enlisted Middle East YouTube and Instagram stars to visit mosques and churches, sharing interfaith exchanges with their millions of followers.

As Pandith explained, the community is seeking to empower those who know best how to speak to youth: their peers. They can "swarm the content of the extremists online with credible voices that will diminish their standing and showcase a whole host of alternative narratives." For instance, if a 16-year-old girl "is getting more and more interested in what's happening with the 'superhero' guy who's fighting for [ISIS], in real time she'll see her peers push back, 'That's dumb. That's stupid. That doesn't make sense.'"

This community-building has hardly erased the specter of terrorism. But it represents a far more personal and effective approach than staid government broadcasts and press releases. It is also just one example of a new kind of conflict fought largely with bite-sized social media broadcasts, what communications scholar Haroon Ullah has described as "digital world war."

Whether it is politicians or pop stars, hate groups or those that tell haters to "shake it off," the new winners are those who have mastered the power of narrative and primed their audiences with emotion, who have fostered a sense of authenticity and engaged in the community-building that goes with it. But they have another trick up their sleeve. Not only do they do it all on a massive scale — they do it again and again and again at the most personal level.

INUNDATION: DROWN THE WEB, RUN THE WORLD

It was the biggest surprise in internet history. One data scientist found that in the twenty-four hours that followed Donald Trump's election night win of November 8, 2016, the word "fuck" appeared nearly 8 million times on Twitter.

Trump's victory was just as much of a shock to the political system. As the writer Jason Pargin observed, "Trump ran against the most well-funded, well-organized political machine in the history of national politics . . . All of the systems that are supposed to make sure one side wins failed. He smashed a billion-dollar political machine to pieces."

And yet in retrospect, perhaps it shouldn't have been all that surprising, for it was evident at the time that Trump had put to better use the new machine that had already smashed communications and the economy. Indeed, by almost any social media measure, Trump didn't just have more online power than both his Republican and Democratic opponents; he was a literal superpower. He had by far the most social media followers, effectively as many as all his Republican rivals for the GOP nomination combined. He deployed this network to scale, pushing out the most messages, on the most platforms, to the most people. Importantly, Trump's larger follower pool was made up of not just real-world voters, but — as we've discussed previously — a cavalcade of bots and sockpuppet accounts from around the world that amplified his every message and consequently expanded his base of support.

With his Twitter loudspeaker, Trump could drive the national conversation at a pace and volume that left both journalists and his opponents scrambling to keep up. It allowed him not just to dominate the web-borne portion of the 2016 election, but to dominate all other forms of media *through* it, thus capturing $5 billion worth of "free" media coverage (nearly twice that of Clinton). As Republican communications strategist Kevin Madden explained, "Trump understands one important dynamic: In a world where there is a wealth of information, there is always a poverty of attention, and he has this ability to generate four or five story lines a day . . . He is always in control."

In an interview shortly after the election, Trump reflected on how he had won. "I think that social media has more power than the money they spent, and I think . . . I proved that."

But Trump's power lay not just in @realDonaldTrump but in the wider online army mobilized behind it. In his quest for the White House, Trump attracted the regular coalition of evangelical conservatives and traditional Republican partisans. But his crucial, deciding force was a new group: a cohort of mostly tech-savvy angry, young, white men who inhabited the deepest bowels of internet culture.

While many had gotten their start on 4chan, a notorious image board where anonymous users fight an endless battle of profane one-upmanship, the group is better understood through what is known as "Poe's Law." This is an internet adage that emerged from troll-infested arguments on the website Christian Forums. The law states, "Without a winking smiley or other blatant display of humor, it is utterly impossible to parody a [fundamentalist] in such a way that someone won't mistake it for the genuine article." In other words, there is a point at which the most sincere profession of faith becomes indistinguishable from a parody; where a simple, stupid statement might actually be considered an act of profound meta-irony. Taken to its logical conclusion, Poe's Law could lead to a place of profound nihilism, where nothing matters and everything is a joke. And this was exactly where many of these internet denizens took it.

From the beginning, many of these lifelong trolls found something to admire in Trump. Part of the reason was cultural; they felt marginalized by national conversations about race and gender ("identity politics") and saw Trump as the cure. Part of it was economic; although hardly coal miners, they bought into Trump's economic populism and his vow to "Make America Great Again." But most of all, they liked Trump because, in the fast-talking, foulmouthed, combative billionaire, they saw someone just like them — a troll.

Trump's digital force organized in many dark corners of the internet, but their main roost was Reddit. The discussion board /r/The_Donald was launched a week after Trump's June 2015 presidential campaign announcement. What had started with a few dozen tongue-in-cheek supporters grew to 100,000 by the time he clinched the nomination in April 2016 and then to 270,000 by November 2016. (After the election, its size would double again as it became a willful propaganda arm of the administration.) On /r/The_Donald, supporters obsessed over Trump's every utterance and launched endless, crowdsourced attacks against his foes. They were soon consumed with the narrative of Trump, standing strong against the forces of "globalism," aligning with many fervent conspiracy theories. Their meta-irony turned to white-hot anger at what they perceived as increasingly one-sided attacks by the "mainstream media." And serving in the trenches of a seemingly endless internet war, they also found camaraderie and friendship.

Although they labored tirelessly for Trump, the participants in this online collective were not formal members of his campaign. This provided Trump the best of both worlds. Whenever his online army launched attacks that were clearly profane or bigoted, Trump could deny any association. Yet when the activists struck gold, their work could be incorporated into official campaign messages by the Trump aides who regularly monitored their efforts. Sometimes, the work of these anonymous "shitposters" would even find its way into the Twitter feed of Trump himself — a pattern that continued after he won the presidency. These supporters and aggressive proxies (figures like Jack Posobiec quickly joined the bandwagon) came to echo their "Dear Leader," eschewing all notions of defense in order to attack, attack, attack. "The pro-Trump media do not appear to ever stop or take days off," writer

Charlie Warzel concluded. "They are endlessly available and are always producing. Always." In their frantic mania, they set a tempo that no traditionally organized campaign could match.

The collective efforts of Trump's troll army helped steer the online trends that shaped the election. They dredged up old controversies, spun wild conspiracy theories that Trump's opponents had to waste valuable political capital fighting off, and ensured that the most impactful attacks continued to fester and never left public attention. Although neither presidential candidate was well liked, an analysis, by the firm Brandwatch, of tens of millions of election-related tweets showed a near-constant decline in the number of messages that spoke positively about Clinton. For Trump, the trend was the reverse. Essentially, the longer the campaign went on, the louder Trump's proxies grew. And because they reveled in building botnets or assuming fake identities, they were omnipresent.

There was no doubt that this effort was viewed through the lens of information warfare, showing the blurring of lines that Clausewitz would recognize. As General Michael Flynn himself exulted just after the election to a crowd of young supporters, "We have an army of digital soldiers . . . 'cause this was an insurgency, folks, 'cause it was run like an insurgency. This was irregular warfare at its finest, in politics."

Trump's new kind of volunteer online army was so effective, though, because it was backed by another organization never before seen, which followed all the new lessons that had begun to fuse politics, marketing, and war. It was an organization that, reflecting social media itself, combined massive scale with personalized micro-targeting.

The effort was overseen by Trump's son-in-law, Jared Kushner, the famously private, immaculately coifed real estate baron who (ironically) avoided social media himself. In a rare postelection interview, Kushner explained how early on, the campaign had realized that the very identity of their unconventional candidate meant they would have to eschew the traditional pathways to victory. Television advertising or field offices would not win this fight for Trump. The campaign would instead pump its strategic efforts into social media, utilizing the new techniques that it allowed, like message tailoring, sentiment manipulation, and machine learning.

The strategist behind the operations was Brad Parscale, a scraggly-bearded former web designer from Texas, who had risen to the top of Trump's online business and then his election campaign. The emphasis was clear from start to finish. Parscale famously blew every cent of his first $2 million on Facebook ads. ("Who controls Facebook controls the 2016 election," Parscale purportedly said.) By Election Day, it was his team — not the Silicon Valley–friendly Democrats — that bought every last bit of advertising space available on YouTube.

The digital effort Parscale oversaw was both fundamental to the Trump campaign and more massive in scale than anything seen before in political history. At the center was Project Alamo, named for its election last-stand location in Texas. A 100-person team, aided by embedded social media company employees, drew upon a database whose size and depth would come to dwarf all other campaign operations that had come before it. Into this database was pumped basic information about all of Trump's donors (including anyone who purchased the ubiquitous red "Make America Great Again" hat). Then there was the data archive of the Republican National Committee, which claimed to have nearly 8 *trillion* pieces of information spread across 200 million American voters. And, finally, came the massive data stores of a controversial company called Cambridge Analytica.

A UK-based firm that Breitbart chairman and Trump campaign CEO Steve Bannon had helped form in 2013, Cambridge Analytica had previously been active in conducting information warfare–style efforts on behalf of clients ranging from corporations to the "Leave" side of Brexit. It would later be reported to have provided to the Trump campaign some 5,000 data points on 220 million Americans. Controversially, a subset came from data collected via various Facebook apps (ranging from a survey to a "sex compass"), which scraped data on not just 87 million users but also their friends — without their consent or knowledge. Included in the dataset was information gleaned not only from public posts, but also from direct messages that users assumed were private.

This data was a "gold mine," according to one cybersecurity researcher who was able to review a small portion of it when it leaked online. Through the clever use of this mountain of information, one could infer much more through "psychometrics," which crosses the insights of psychology with the tools of big data. Teams of psychometric analysts had already shown how patterns of Facebook "likes" could be used to predict characteristics of someone's life, from their sexual orientation to whether their parents had divorced. The researchers had concluded that it took only ten "likes" to know more about someone than a work colleague knew and just seventy to know more than their real-world friends. As a whistle-blower from the Cambridge Analytica part of the project said in 2018, "We exploited Facebook to harvest millions of people's profiles. And built models to exploit what we knew about them and target their inner demons."

By slicing and dicing the data, the Trump team didn't just gain a unique window into the minds of its supporters; it could also use advertising tools like Facebook's Lookalike Audiences to track down users who shared the same political disposition or psychological profile. This tool literally changed the economics of the battle for votes. Suddenly, isolated patches of rural voters — long neglected because of the cost of television advertising — could be selectively targeted. Thanks to Facebook and big data, Parscale marveled, he could reach "fifteen people in the Florida Panhandle that I would never buy a TV commercial for."

Importantly, the wealth of data didn't just allow a new kind of micro-targeting of voters, with exactly the message they cared most about, but it also provided new insights into how to tailor that message to influence them most. As opposed to a TV or print ad that could run in only one form at a time, the campaign would regularly run thousands of variations of an online outreach effort simultaneously. The key was that every single message to win hearts and minds was also an experiment. Messages might differ in the phrasing, the choice of photo, and even tiny changes in color that would influence one person's particular psychological profile more than another. The reason was that social media had turned the conversation into a mass-scaled but two-way street. The targets' feedback (who clicked it, who "liked" it, who shared it) went back into the profiles, not just for that one person, but for all the other people in the dataset who shared similar characteristics. This allowed the campaign to find the "perfect" messages for engaging different groups of voters — all dynamically and all at the same time. Once, the number of variations on a single message approached 200,000. By the end of the campaign, the Trump team had run almost 6 million different versions of online ads. Clinton, by contrast, had run just 66,000.

Plugged into the subconscious of millions of likely voters, Trump's digital team began to guide the candidate's travel, fundraising, rally locations, and even the topics of his speeches. "[The campaign] put so many different pieces together," Parscale said. "And what's funny is the outside world was so obsessed about this little piece or that, they didn't pick up that it was all being orchestrated so well." The importance of Parscale's digital strategy to Trump's 2016 campaign would perhaps be best illustrated by his promotion to lead all campaign operations for Trump's 2020 reelection.

There was little *political* precedent for Trump's strategy. But there *was* precedent. It could be found at an internet giant famous for journalistic classics like "15 Hedgehogs with Things That Look Like Hedgehogs" and "Which Ousted Arab Spring Ruler Are You?"

In 2006, a young MIT postgraduate named Jonah Peretti cofounded a "viral lab." Peretti's intention was to understand what content took off and what didn't. Within a decade, the spinout company called BuzzFeed would grow to become a billion-dollar network with hundreds of employees and offices scattered around the world.

If BuzzFeed had a secret, it was *scale.* It wasn't one person angling for viral hits; it was an army, applying a systematic formula of the same kind of weaponized experimentation, constantly testing to map the depths of the attention economy and then make it their own. BuzzFeed could churn out more than 200 articles, "listicles," and videos each day. It then monitored the performance of each item in real time, tweaking titles and keywords and shifting marketing focus in an algorithmically driven process that was a precursor of the sort of real-time focus-testing conducted by the Trump team. With every viral success, the writers and marketers got a little more experienced, their dataset got a little bigger, and their machines got a little smarter.

Importantly, BuzzFeed's model didn't depend on handcrafting any particular item to go viral; it depended on throwing out dozens of ideas at once and seeing what stuck. For every major viral success, like "12 Extremely Disappointing Facts About Popular Music," there were dozens of duds, like "Leonardo DiCaprio Might Be a Human Puppy." What mattered most was scale and experimentation, inundating an audience with potential choices and seeing what they picked. The lesson for BuzzFeed, and for all aspiring social media warriors, was to make many small bets, knowing that some of them would pay off big.

How BuzzFeed made its money was not all that different from how Brad Parscale helped the Trump campaign to victory. It was also strikingly reminiscent of how Russian propagandists drown their opponents in what RAND researchers describe as a "firehose of falsehood," weaponizing the very same Facebook micro-targeting tools. And it was also a crucial aspect of ISIS's online efforts to overwhelm its opponents with messaging that was both scaled and tailored. Recall that ISIS could generate over a thousand official propaganda releases each month. In each case, this continuous cascade allowed these savvy viral marketers to learn what worked for the next round.

This is the last part of the equation explaining how combatants can conquer social media and penetrate the minds of those who use it. To "win" the internet, one must learn how to fuse these elements of narrative, emotion, authenticity, community, and inundation. And if you can "win" the internet, you can win silly feuds, elections, and deadly serious battles alike. You can even warp how people see themselves and the world around them.

But the fact that these lessons are now available to anyone means that not all online battles will be one-sided blitzkriegs. As more and more users learn them, the results are vast online struggles that challenge our traditional understanding of war.

Donald J. Trump ✓
@realDonaldTrump

〔 Follow 〕 ⌄

Be sure to tune in and watch Donald Trump on Late Night with David Letterman as he presents the Top Ten List tonight!

11:54 AM - 4 May 2009

▲ Donald J. Trump's first foray onto Twitter was part of a promotional campaign for the season finale of the TV show *Celebrity Apprentice*. 2,819 days later, @realDonaldTrump would be inaugurated the forty-fifth president of the United States.

(Donald J. Trump / Twitter)

 Jun 15
ISIS: **O #Baghdad we are coming!** ISIS pic.twitter.com/XbUNQsapu6"

▲ Social media proved integral to the rise of ISIS and its surprising military successes. Here, in a tweet that went viral during its 2014 invasion of Iraq, an ISIS fighter and flag are superimposed on images of Iraq's capital city.

(Twitter via PBS)

▶ On October 29, 1969, computers became communication devices, changing the world as we know it. This BBN Interface Message Processor routed information from a computer at UCLA to one at Stanford, creating ARPANET, the progenitor of the modern internet. Costing $82,200 (about $600,000 today), it had only 12KB memory—1/10,000th of a first-generation iPhone.

(Photo by Steve Jurvetson)

◄Known as the "father of the internet," Vint Cerf poses before the calculations for the protocol he helped establish in 1973, which enabled the early computer networks to knit together into a single system. Online communication could now grow to limitless size and complexity.

(Jose Mercado / Stanford News Service)

► The millennial genera- tion would grow up with the internet and then transform it. Working from his Harvard dorm room, Mark Zuckerberg registered "TheFacebook.com" on January 11, 2004. Within a decade, Facebook's users would number one billion, and soon after double again. This image, posted to Zuckerberg's care- fully curated Facebook feed in 2011, would garner nearly 400,000 "likes."

(Mark Zuckerberg / Facebook)

◄Social media has been used to both foment conflict and reveal new aspects of it. In 2014, Russia launched a "stealth" invasion of Ukraine, sowing false stories online to help spark what became the most significant European conflict in twenty years. Yet the Russian government's strident denials were undermined by some of the very citizens it sought to "liberate," who happily snapped photos with Russian commandos (whom they called "little green men") and posted them to social media.

(Via Instagram)

▶ ▼ Social media platforms have become inextricably linked with the performative violence of gangs and drug cartels. José Rodrigo Aréchiga Gamboa, a hitman and high-ranking member of the Sinaloa Cartel, posted on Instagram everything from poses with his gold-plated guns to a "photo bomb" of fellow online celebrity Paris Hilton during a trip to Vegas.

(Via Instagram)

Felina
@Miut3

SE QUE NO SOY FÁCIL PERO, DEFINITIVAMENTE VALGO LA PENA | SDR´s y más | youtu.be/u-PyTLPKAUA

📍 Ubicación: Lado Oscuro

🔗 miut3.com

🕐 Se unió en enero de 2010

▶ Social media has empowered new voices, operating in some of the world's most dangerous places. "Felina" reported tirelessly from the Mexican state of Tamaulipas, filling the gap left by media that had fled the escalating drug war. Over half a million people would read the Catwoman's reports before cartel killers captured María del Rosario Fuentes Rubio, the real hero behind the account, and broadcast her murder.

(Via Twitter)

General Flynn ✓
@GenFlynn

U decide - NYPD Blows Whistle on New Hillary Emails: Money Laundering, Sex Crimes w Children, etc...MUST READ!
truepundit.com/breaking-bombs…

10:18 PM - 2 Nov 2016

↩ ⟳ 9,154 ♥ 8,126

▲ Once considered among the nation's leading military officers, Lieutenant General Michael Flynn foresaw how social media would transform the world of war and politics. But after being fired from the Defense Intelligence Agency, he morphed into an adamant internet conspiracy theorist, falling victim to the very same forces he had warned against.

(Michael Flynn / Twitter via Time)

▲ As part of its disinformation campaign during the U.S. presidential election of 2016, Russia flooded American social media with tens of thousands of fake accounts. Among the most successful was "TEN_GOP," which posed as a hub for Tennessee Republicans. On Election Day, it was the seventh-most-retweeted account across all of Twitter.

(Via @Ushadrons and @Toolmarks)

◄ Russian operatives also built popular Facebook groups, buying ads to extend their reach. The overall campaign reached 147 million Americans. Leveraging Facebook's sophisticated audience engagement tools, this particular ad targeted users whom the algorithms had identified as interested in "Christianity, Jesus, Ron Paul, and Bill O'Reilly."

(Via Facebook)

◀ The network of Russian influence ops often bootstrapped each other, amplifying their reach. Here, a fake Russian account on Twitter promotes another fake Russian account, urging Americans to show up for a real event.

(Via @Ushadrons and @Toolmarks)

Trump is my Leader! (i.imgur.com)
submitted 1 year ago by rubinjer
4 comments share save hide report

FOR GOD AND COUNTRY!

IT IS OUR HOLY DUTY TO
GUARD AGAINST THE
FOREIGN HORDES

▶ No platform was spared in this wide-ranging Russian disinformation campaign. On Reddit, the fake user "rubinjer" became a popular presence on pro-Trump discussion boards. "Rubinjer" garnered 99,493 "upvotes" from fellow users before being unmasked as a Russian operative.

(Via Reddit)

PIZZAGATE

◀ Illustrating the growing political influence of viral conspiracy theories, "Pizzagate" alleged that Hillary Clinton's campaign was involved in satanic worship and underage sex trafficking, operating out of a Washington, DC, pizza parlor. It attracted a large online following, which only grew after the election. It also provided fodder to conspiracy-minded graphic artists, who produced dozens of fake images of children in distress.

(Via @media_nc / Twitter)

▶ On December 4, 2016, 28-year-old part-time firefighter Edgar Welch burst into Comet Pizza with an assault rifle, seeking to rescue the children he believed to be imprisoned in an underground sex dungeon. There were no children, nor even a basement; Welch would be sentenced to four years in prison.

(Edgar Welch / via Facebook)

► Super-spreaders, and the marketplace that incentivizes them, play a key role in deciding what goes viral. Former U.S. Navy intelligence officer Jack Posobiec sought online fame and influence through far-right trolling and conspiracy theories like Pizzagate. A few months later, Posobiec would receive press access to the Trump White House, followed by the ultimate reward: a retweet from the president.

(Carlos Barria / Reuters)

◄ "You can sit at home and play Call of Duty or you can come here and respond to the real call of duty . . . the choice is yours." Junaid Hussain was a British-Pakistani hacker turned online propagandist for ISIS. From thousands of miles away, "Abu Hussain al-Britani" served as a bizarre mix of leader, recruiter, and life coach for would-be terrorists. He would ultimately rise to number three on the Pentagon's "kill list," dying in a 2015 drone strike.

(Junaid Hussain / via Twitter)

▲ As extremists learned to use social media to greater effect, the self-decided defenders of the web began to fight back. This 2014 video was produced by a member of the Anonymous hacktivist collective, one of numerous declarations of war against the digital forces of the Islamic State. #OpISIS mixed Twitter trolling, content policing, and forum board infiltration.

(TheAnonJournal / via YouTube)

◄ Mastering the new tools of social media, singer-songwriter Taylor Swift rose to become not just a superstar, but, in her ability to engage and mobilize armies of online fans, a role model for businesses, politicians, and extremist groups alike. Here, she performs in a 2012 concert in Sydney, Australia.

(Photo by Eva Rinaldi)

► Pepe the Frog began life as a simple comic book character, described by its creator, Matt Furie, as "drinkin', stinkin', and never thinkin.'" It soon transformed into one of the internet's most enduring and politically charged memes. Everyone from weightlifters to neo-Nazis repurposed the image for their own purposes.

(Via Reddit)

▲ In modern wars, the online fight for attention and influence has become just as important as the physical conflict, sometimes even more so. Militaries have begun to change their organizations and tactics to reflect this new front. Here, the Israeli Defense Forces encourages youths to fulfill their military enlistment requirement by serving as social media proxies and graphic designers.

(IDF recruitment website)

◀In contrast to the centralized approach of the IDF, Hamas propaganda is produced by a global network of proxies and supporters. This grisly image—an anthropomorphized knife saying "good morning" and wrapped in the colors of the Palestinian flag—was released amid a spate of stabbing attacks targeting Israeli civilians in 2015.

(Via Facebook)

▶ This widely shared meme, released at the height of Operation Pillar of Defense in 2012, was designed by the IDF to engage Western audiences, transposing the threat of Hamas-launched missiles and mortars to their own national capitals.

(IDF)

▼AI can emulate speech, masquerade as human, or even stitch together image and video from nothing but its own "imagination." Here, the "FaceApp" uses neural networks to alter the gender of a person in an online image. In the battles of the future, the line between truth and reality will only grow more blurred.

(TudorTulok)

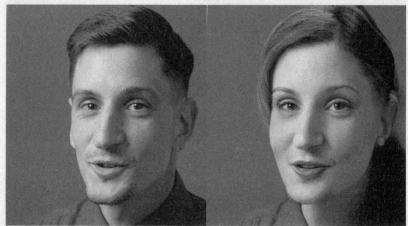

7

LikeWar

The Conflicts That Drive the Web and the World

It was the first non-linear war . . . Now four coalitions collided.
Not two against two, or three against one. No. All against all.

— VLADISLAV SURKOV, "WITHOUT SKY"

WORLD WAR I HAD BEEN RAGING for less than five hours when
the crew of the CS *Alert* undertook one of the most important operations of the entire conflict. Floating in the English Channel, the crew
unspooled dozens of meters of heavy wire into the water, anchored by
an iron hook. Dredging the ocean floor, they waited for a telltale tug before reeling their prize back up to the surface. One by one, all seven of
Germany's transatlantic cables were severed that day. For the rest of the
war, Germany would be forced to use unencrypted radios or sneak messages into public telegraph channels — messages that lay at the mercy of
British code breakers.

But in losing these transatlantic lifelines, Germany was dealt an even
bigger blow. For the rest of the war, it could not communicate directly
with the American public, leaving Britain to control the narrative inside the one major nation yet to join the fight. Stories of German barbarity spread, abetted by British propagandists and unanswerable by a
muted German government. In one case, a purposeful mistranslation of
the German word *kadaver* (translated as "corpse" instead of "animal")
led many newspapers to report that Germany had begun incinerating
battlefield dead for their fat, turning fallen soldiers into candles and lu-
bricants. Such tales of the "monstrous Huns" helped drive Americans

into a fury. When the United States finally entered the war in 1917, it was thanks in part to an information blockade that had stripped Germans of their humanity.

For nearly all of recorded history, this was the way things were. When a citizen of one nation tried to communicate directly with a citizen of another, the government was likely to play a part in the process, whether certifying postage or regulating the traffic of international telegraph lines. If the two states fell into war or a trade dispute or simply didn't like each other, such communication halted. Letters were intercepted. Cables were cut. The flow of information fell to a trickle. Two people, living in two countries, separated by a hostile border, could well have lived on different planets.

You know the punch line already. The internet changed this, fast. Making an international connection soon became as easy as knowing an email address, then just a name. Images, videos, livestreams, and ever-improving online translation make it simple to strike up conversations across the globe. Any of these connections can be observed or joined by audiences from yet other parts of the world.

In 1990, a generation after Licklider and Taylor made their prediction about computer communication, two political scientists with the Pentagon's think tank at the RAND Corporation started to explore the security implications of this emerging internet. Many of John Arquilla and David Ronfeldt's colleagues considered their line of investigation into computers used mostly by geeks a bunch of mumbo jumbo. But an important few realized that it was game-changing. The first time Arquilla and Ronfeldt put their thoughts into a short memo, the Pentagon immediately classified it.

Their findings were made public in a revolutionary 1993 article titled "Cyberwar Is Coming!" At a time when the internet hadn't even yet been opened up to commercial activity, they observed that "information is becoming a strategic resource that may prove as valuable and influential in the post-industrial era as capital and labor have been in the industrial age." Accordingly, future conflicts would be won not by physical forces, they argued, but by the availability and manipulation of information. They warned of "cyberwar," battles in which computer hackers might remotely target economies and disable military capabilities.

But Arquilla and Ronfeldt went further. They also predicted that cyberwar would be accompanied by something else: "netwar." They explained:

> It means trying to disrupt, damage, or modify what a target population "knows" or thinks it knows about itself and the world around it. A netwar may focus on public or elite opinion, or both. It may involve public diplomacy measures, propaganda and psychological campaigns, political and cultural subversion, deception of or interference with the local media . . . In other words, netwar represents a new entry on the spectrum of conflict that spans economic, political, and social as well as military forms of "war."

Netwar entailed more than simply a propaganda campaign launched over the internet. It meant a new way of thinking and a new kind of conflict. It meant understanding that online information itself was a weapon, used to dismantle some realities and to build others in their place. It suggested a future where groups and nations alike might effect massive political change — the kind that at the time usually took years of bloody struggle — without firing a shot.

Like most theories about the early internet, however, the rhetoric ran far ahead of what was happening in the real world. For most national militaries and governments, a few online bulletin boards connected by unreliable dial-up modems hardly looked like the future of warfare. Instead, their attention turned to robots, drones, and precision-guided munitions. By the late 1990s, the "weaponization" of online information was essentially a dead topic.

Instead, early netwar became the province of far-left activists and democratic protesters, beginning with the 1994 Zapatista uprising in Mexico and culminating in the 2011 Arab Spring. In time, terrorists and far-right extremists also began to gravitate toward netwar tactics. Observing these developments with interest, Arquilla and Ronfeldt, still at work tracking conflict trends, began to liken what had happened to the Roman deity Janus, the two-faced god of beginnings and endings (as well as war and peace). "Our hope was at the least, there would be a balance between the two," Arquilla told us, with the benefit of twenty years' reflection. "I think what we've seen is a greater prevalence of the darker side of Janus. I'm troubled to see that."

There was no exact moment when the balance shifted. For disenchanted activists like the Belarusian Evgeny Morozov, it came when dictators learned to use the internet to strengthen their regimes. For us, the moment came when we saw how ISIS militants used the internet not just to sow terror across the globe, but to win its battles in the field. For Putin's government, it came when the Russian military reorganized itself to strike back against what it perceived as a Western information offensive. For many in American politics and Silicon Valley, it came when that same Russian effort poisoned their networks with a flood of disinformation, bots, and hate.

Today, online battles are no longer just the stuff of science fiction or wonky think tank reports, but an integral part of global conflict. As a result, governments around the world have begun to adapt to it. Russia is the most obvious example — a government whose state media, troll factories, and botnets all conspire to wage (in the words of its own military doctrine) "global information warfare." Echoing the language of ISIS propagandists, Russian military strategists describe how a strong information offensive can have a strategic impact on a par with the release of an atomic bomb. They warn of the power of foreign information to "blur the traditional Russian spiritual and moral values," and argue instead for "a system of spiritual and patriotic education" (aka censorship) and the development of "informational . . . measures aiming to pre-empt or reduce the threat of destructive actions from an attacking state." In this line of reasoning, the Russian government doesn't resort to netwar because it wants to. Rather, it sees no other choice. The best defense, after all, is a good offense.

China's Great Firewall, social engineering, and online armies of positivity can be seen in much the same light. But one shouldn't think that there isn't an offensive side. Since 2003, the Chinese military has followed an information policy built on the "three warfares": psychological warfare (manipulation of perception and beliefs), legal warfare (manipulation of treaties and international law), and public opinion warfare (manipulation of both Chinese and foreign populations). Where China is strong, its strengths must be amplified even further in the public imagination; where China is weak, attention must be diverted. China must be seen as a peaceful nation, bullied by powerful adversaries and "reluctantly" responding by building its armies and laying claim to new lands. China's critics must be confused and scattered, even as its own people are forged into a singular, iron will. In 2015, China's official military strategy would put the challenge even more starkly: "War is accelerating its evolution to informatization."

Even the United States, birthplace of the free and open internet, has started to accept netwar as a matter of policy. In 2011, DARPA's research division, which once created the internet itself, launched the new Social Media in Strategic Communications program to study online sentiment analysis and manipulation. Around the same time, the U.S. military's Central Command began overseeing Operation Earnest Voice, a several-hundred-million-dollar effort to fight jihadists across the Middle East by distorting Arabic social media conversations. One part of this initiative was the development of an "online persona management service" — essentially sockpuppet software — "to allow one U.S. serviceman or woman to control up to 10 separate identities based all over the world." And, beginning in 2014, the U.S. State Department poured vast amounts of resources into CVE efforts, building an array of online organizations that sought to counter ISIS by launching information offensives of their own.

The initiatives have begun to spread through governments around the world. In 2015, Britain formed the 1,500-soldier-strong 77th Brigade, intended to be an "agent of change through targeted Information Activity and Outreach." The NATO alliance inaugurated its Strategic Communications Centre of Excellence, focused on "the weaponization of social media." Add to this the impressive digital arm of the Israeli Defense Forces (IDF), Turkey's growing patriotic troll army, the burgeoning botnets of the Mexican government, and the cyber-propaganda initiatives of dozens of other countries.

But there's a second revolution at work — even stranger and more pressing than the one foreseen by Arquilla and Ronfeldt. As national militaries have reoriented themselves to fight global information conflicts, the domestic politics of these countries have also morphed to resemble netwars. And the two spheres have become linked. Just as rival states and conflict actors use the internet to manipulate and deceive, so, too, do political candidates and activists of all stripes. Online, there's little difference in the information tactics required to "win" either a violent conflict or a peaceful campaign. Often, their battles are not just indistinguishable in their form but also directly linked in their activities (such as the alignment of Russian sockpuppets and alt-right activists). The realms of war and politics have begun to merge.

In essence, every LikeWar is a battle for attention with a specific objective in mind (promoting a candidate; gaining concessions; winning a war) and challenged by an opponent (other people, groups, or nations). Victory requires an appreciation of the nature of virality and the whimsical ways of the attention economy, as well as a talent for conveying narrative, emotion, and authenticity, melded with community-building and a ceaseless supply of content (inundation). And because it all takes place on the open internet, each of these conflicts becomes a global tug-of-war with an unknown number of players.

The world of LikeWar can feel inescapable and overwhelming. Yet, in seeking to explain how its battles work, there is perhaps no one better for the task than a big-eyed, big-lipped amphibian, who looks like he's stepped right out of the hellscape of Microsoft Paint.

MEMES AND MEMETIC WARFARE

To his critics, he was a blazing symbol of hatred and bigotry. To his supporters, he was both a joke and a badge of honor. To the artist who created him, he was just a "chilled-out [dude] who likes to eat snacks and talk on the phone [and] smoke weed." He wore everything from a blue shirt to a baggy suit to hot-pink lingerie. He was thin or fat; sad or smug or angry. Sometimes, he looked like Donald Trump; other times, Vladimir Putin, the rapper Nicki Minaj, or even Adolf Hitler. But three things about him were always the same:

1. He was green.
2. His name was Pepe the Frog.
3. He was a dumb internet meme.

If you spent any time on the internet, you couldn't *not* see him. Soon enough, you wanted to *unsee* him. In 2015, Pepe was adopted as the banner of Trump's vociferous online army. By 2016, he'd also become the symbol of a resurgent tide of white nationalism, declared a hate symbol by the Anti-Defamation League. Trump tweeted a picture of *himself* as an anthropomorphized Pepe. By 2017, Pepe was ascendant. Trump supporters launched a crowdfunding campaign to erect a Pepe billboard "somewhere in the American Midwest." On Twitter, Russia's UK embassy used a smug Pepe to taunt the British government in the midst of a diplomatic spat ("No trust in Britain's best friend and ally?").

But . . . *why?* It made little sense that a cartoon frog had become a standard-bearer for a white ethno-state, or a campaign symbol or tool of international diplomacy.

But it was never about the frog. Instead, Pepe was the product of an evolutionary cycle that moved at digital warp speed on the internet, piling meaning atop meaning until everyone lost track. Pepe was also the product of a conflict of reinvention and appropriation that twisted him in directions that no one might have expected. In understanding Pepe, one can understand memes, and through them the life cycle of ideas on the internet.

Pepe the Frog was born in 2005 to the San Francisco–based artist Matt Furie. One of four teenage monsters in Furie's Boy's Club comic series, Pepe was just a cartoon layabout who spent his days "drinkin', stinkin', and never thinkin'." In 2008, an anonymous 4chan user shared a panel in which Pepe pulled his pants down to his ankles to urinate in a public bathroom, along with his shameless explanation for doing so: "Feels good man." Pepe's simplicity and irreverence perfectly captured the spirit of the freewheeling, profane 4chan community. He went viral among internet forum users and — more impressively — *surpassed* virality to become part of the underlying culture of the internet. When the original meme seemed exhausted, users mined Furie's comics for more drawings. When the comics ran out, users began to make their own. In a sense, Pepe became the ideal online phenomenon — popular and endlessly adaptable, while remaining too weird and unattractive to ever go fully mainstream.

With these attributes, it was not much of a shock when Pepe became the unofficial mascot of 4chan's /pol/ politics board. When these trolls went to battle for Trump, they took Pepe with them. What began as a mockery of political activism soon became, for many of these users, a serious effort to help Trump win. At the same time, clusters of traditional Trump supporters began to adopt the same mannerisms and tactics as actual trolls. As a result, Pepe underwent another transformation. The meme was still dumb and irreverent, but now he was suffused with *political* meaning as well. Pepe had entered the real world, with real consequences.

Amidst the election excitement, there was another, darker battle raging for Pepe's soul. It was led (or so they wanted you to believe) by a group of thirty "shitposters" and casual bodybuilders, who feared that their meme was being co-opted by "normies," people with no appreciation of the internet's subterranean culture. Their solution was to turn the cartoon frog into a literal Nazi, flooding social media with Pepe memes laced with swastikas, Hitler quotes, and Third Reich iconography. In a tactic that would become commonplace, they also targeted unwitting reporters, bombarding them with Pepe-related hate speech in order to convince them that the meme was a white supremacist and anti-Semitic. The gambit worked. Pepe became inextricably linked to white nationalism, denounced by most journalists and the American left and embraced unironically by *actual* neo-Nazis, who finally had a "hip" symbol to call their own. The notorious white supremacist leader Richard Spencer even took to wearing a lapel pin of Pepe in public, and in an instantly viral video tried to explain its symbolic value to his cause — until a passerby punched him in the face.

Trump's troll army wielded Pepe like a weapon, poking and prodding at mainstream journalists and Clinton supporters, trying to elicit a backlash. The moment they were called out as racist or white supremacist, they replied with smug outrage, asking *how in the world* a cartoon could be construed as anything more than a silly, dumb frog.

Pepe formed an ideological bridge between trolling and the next-generation white nationalist, alt-right movement that had lined up behind Trump. Third Reich phrases like "blood and soil," filtered through Pepe memes, fit surprisingly well with Trump's America First, anti-immigrant, anti-Islamic campaign platform. The wink and nod of a cartoon frog allowed a rich, but easily deniable, symbolism.

When Trump won, Pepe transformed again. The green frog became representative of a successful, hard-fought campaign — one that now controlled all the levers of government. On Inauguration Day in Washington, DC, buttons and printouts of Pepe were visible in the crowd. Online vendors began selling a hat printed in the same style as those worn by military veterans of Vietnam, Korea, and World War II. It proudly pronounced its wearer as a "Meme War Veteran."

In the months that followed, Pepe continued his evolution, popping up at the events of the far-right, where grizzled, camo-clad militiamen marched alongside gangly white teenagers. When a white nationalist terrorist drove his car into a crowd in Charlottesville, Virginia, and killed a peaceful demonstrator, it was discovered that his Facebook page was peppered with Pepe memes. In response, anti-fascist demonstrators flooded the streets and the internet with signs of their own. Again, Pepe featured prominently: this time as a corpse, a Ku Klux Klan mask torn from his green face.

Had Pepe really been racist? The answer is yes. Had Pepe been an innocent, silly joke? Also, yes. In truth, Pepe was a prism, a symbol continually reinterpreted and repurposed by internet pranksters, Trump supporters, liberal activists, ultranationalists, and everyone who just happened to glimpse him in passing. Pepe was a "meme," an empty vessel, like the chromatin that shields DNA; a protective layer over a rich, ever-multiplying strand of ideas. As it was with Pepe, so it is with all memes. They are the vessels by which culture is transmitted — and a crucial instrument by which LikeWar is fought.

Yet the *concept* of memes has nothing to do with the internet. In the late 1960s, biologists had begun to unravel the basic nature of the genetic code, discovering how cellular instructions are passed from one generation to the next. Flush with excitement, they pondered whether their work could be applied more broadly. If the rules of genetics could explain life, could they not explain many other things — even the nature of information? After all, just as biological life had to ceaselessly copy itself in order to survive, ideas had to do so, too. In his 1976 book *The Selfish Gene,* evolutionary biologist Richard Dawkins put a name to these bits of organic, self-multiplying information: "memes."

"The computers in which memes live are human brains," Dawkins wrote. Memes are born from human culture and shaped and transmitted by language. Over time, a meme might become increasingly self-referential and complex, spawning clusters of *new* memes. A meme is "alive" only so long as it exists in the human mind. For a meme to be forgotten means that it goes extinct, the same as a species that can no longer pass on its genetic code.

Religion, for instance, can be viewed as a series of memes both broad and narrow: a general belief in a higher power to the more specific catechisms of the Christian faith to even the warping of a religion to push bigotry against members of another faith. For example, one of history's cruelest and most enduring memes is the web of conspiracy theories that infuse anti-Semitism. The belief in a secret Jewish conspiracy aiming to run the world built upon itself from the Middle Ages to a fake pamphlet created by Russian secret police in 1903 (*The Protocols of the Meetings of the Learned Elders of Zion*) to the pamphlet's mass printing and distribution in America by Henry Ford and its subsequent use in Nazi propaganda in Germany.

The arrival of the internet sped up this memetic evolution. Dawkins, who had picked up programming as a hobby, observed as much in *The Selfish Gene*'s 1989 edition. "It was obviously predictable that manufactured electronic computers, too, would eventually play host to self-replicating patterns of information — memes," he wrote. ". . . It is a perfect milieu for self-replicating programs to flourish and spread."

Through the 1990s, memes proliferated across the chaotic, emerging network of websites and forum boards. Longstanding memes (like anti-Semitic conspiracy theories) found fresh, receptive audiences. Meanwhile, entirely new memes could seize popular attention. "The internet is a first class ecology for memes," Dawkins observed in 2014. By this point, Dawkins himself had become something of a meme — as an uber-rational, dyspeptic, atheistic defender of Reason, he managed to transform himself into a toxic Twitter troll.

As the web became more image- and social media–friendly, what we know as the "internet meme" was born. These were images or short GIFs, often overlaid with text and easily shareable, that relayed ideas fast. Grasping their full meaning, however, required understanding not only the content at hand but also its previous iterations. For instance, the LOLCats phenomenon, comprised of tens of thousands of cat pictures with misspelled captions, becomes funnier (to a point) only if you're familiar with the context to which it refers — the pervasiveness of cat images on the internet. Indeed, the most effective memes often build not merely on themselves but on other memes as well. One reason for Pepe the Frog's enduring popularity was the way he could be used to mock or replicate other viral content — to essentially "meme" memes.

Importantly, it takes only one event, group, or person to alter a meme's meaning for everyone who might use it. "Digital content can travel further, be decontextualized more quickly and accessed instantaneously — without the original creator's consent or even awareness — by millions of people," write internet ethicists Whitney Phillips and Ryan Milner. Pepe the Frog's creator went so far as to sue to try to stop the misuse of his creation. But it was to no avail. By either hijacking or chance, a meme can come to contain vastly different ideas than those that inspired it, even as it retains all its old reach and influence. And once a meme has been so redefined, it becomes nearly impossible to reclaim. Making something go viral is hard; co-opting or poisoning something that's already viral can be remarkably easy.

The study of internet "memetics" has steadily merged with studies of online warfare, attracting strange bedfellows. Both psychological warfare professionals and shitposting trolls began to explore how to co-opt old memes and spin off new ones. For U.S. defense analysts, one of the first examinations of the subject came in 2006, when U.S. Marine Corps Major Michael Prosser published a far-reaching thesis: "Memetics — a Growth Industry in US Military Operations." In an echo of the earlier netwar writings of Arquilla and Ronfeldt, Prosser argued that armed conflict would increasingly be decided by dueling ideologies on a "nonlinear battlefield." Accordingly, militaries would need to track the memes promulgated by their adversaries, counter them, and respond with memes of their own.

In the United States, Prosser's work would kick off a tiny, DARPA-funded industry devoted to "military memetics" — the analysis and weaponization of memes to gain advantage in an invisible, all-consuming information war. As of 2018, the work was still going strong. A perfect illustration came when the Center for Naval Analyses, a U.S. military–funded think tank, released a report titled "Exploring the Utility of Memes for U.S. Government Influence Campaigns." Naturally, its cover was a meme of *Toy Story*'s Buzz Lightyear.

Of course, the U.S. government has hardly been alone in this quest. Over the past decade, shadowy groups in the weirdest corners of the internet also began to write more explicitly about transforming memes into weapons. One now mostly defunct organization is the Bureau of Memetic Warfare, part of 8chan (a board for users too extreme for

4chan), whose tagline brags to visitors, "He who controls the memes, controls the world." The conversations were a horrifying mix of unrepentant neo-Nazism, plots to hijack or undermine popular online movements, and fairly nuanced discussions about social engineering and the nature of ideas.

One user grandly summarized the promise of memetic warfare for ultranationalist agitators: "Now as never before we have the ability to reach out, learn, and spread truth as we know it to be ... We are presented with a state of affairs unique to history, an age of ideological memetic warfare in which the controlling principles of mankind are loosed to spread with no physical barriers." Naturally, the user's profile was a photograph of Joseph Goebbels.

These two worlds — the leading edge of military theory and the dark kingdom of internet trolls — unsurprisingly found each other online. The union came in the form of Jeff Giesea, a tech consultant who worked as an early and avid organizer for Trump. He was one of the cofounders of MAGA3X, a meme-generating hub for the Trump online army, which described itself as "Freedom's Secret Weapon." Giesea felt that the election's relentless creation and co-option of memes echoed a larger shift in global affairs — one that had caught the United States and most democracies off guard. So he put his thoughts to paper in an article titled "It's Time to Embrace Memetic Warfare." The document wasn't published on a Trump fan site, however, but in the journal of NATO's Strategic Communications Centre of Excellence. Giesea drew parallels between the messaging of pro-Trump forces and the equally effective influence strategies of Russian propagandists and the Islamic State. "It's time to drive towards a more expansive view of Strategic Communications on the social media battlefield," he wrote. "It's time to adopt a more aggressive, proactive, and agile mindset and approach. It's time to embrace memetic warfare."

The memetic warfare envisioned by Prosser, Giesea, DARPA researchers, and internet anti-Semites alike turns on essentially the same principle. It recognizes the power of virality — the need to produce and propel viral content through the online system. But it also recognizes that the content that *goes* viral — the meme — can be quite easily hijacked. And whoever does that best determines what reality looks like: whether Pepe the Frog draws chuckles or revulsion; whether a terrorist group sows fear and inspires real-life attacks or simply drowns in online derision.

In the larger picture, memes are akin to the skirmishes of LikeWar, the micro-battles that shape and determine the outcome of the global tug-of-war. Win enough of these skirmishes, and, for a time, victory is yours for the taking. "Own the moment, own the hour," writes defense analyst August Cole. "Own the hour, own the country."

Yet, while these skirmishes start online, they don't necessarily stay there. In a world of LikeWar, internet conflicts now merge seamlessly with those of flesh and blood.

WAR IN THE OPEN

To many Palestinians who live in Gaza City, Ahmed al-Jabari was a hero: the commander of Hamas, the militant wing of resistance to the Israeli occupation. To Israelis, he was a villain: a terrorist who exploded bombs on packed school buses and rained mortar shells down on cities. But most of all, Jabari was a survivor. He'd lived through five assassination attempts and boasted that he no longer feared bullets or bombs.

His reckoning came on November 14, 2012, as Jabari and his bodyguard were driving down a residential street in Gaza City. High above them, an Israeli Heron drone loitered. Its high-powered camera zoomed in as Jabari's car sped past a packed minibus and onto a stretch of open road. Then the drone fired a missile.

Jabari never saw the explosion that ended his life, but millions of other people did. Even as his body smoldered, the Israeli military's official Twitter account spun into action. "The IDF has begun a widespread campaign on terror sites & operatives in the #Gaza Strip," declared @IDFSpokesperson. Then came an infographic that listed Jabari's crimes in bullet points, with a big red box reading "ELIMINATED" slapped across his glowering face. After that came the YouTube clip. "In case you missed it — VIDEO — IDF Pinpoint Strike on Ahmed Jabari, Head of #Hamas Military Wing." You could watch Jabari's car trundling down the street before it exploded in a ball of fire. You could watch him die as many times as you wanted (the video has since been viewed nearly 5 million times) and share it with all your friends.

Within a few hours, Israeli aircraft had destroyed dozens of weapons caches hidden across Gaza City. "We recommend that no Hamas operatives, whether low level or senior leaders, show their faces above ground in the days ahead," @IDFSpokesperson taunted. The challenge didn't go unanswered. "Our blessed hands will reach your leaders and soldiers wherever they are," a Hamas spokesperson, @AlqassamBrigade, fired back. "(You Opened Hell Gates On Yourselves.)"

The Israelis called it Operation Pillar of Defense. IDF air strikes perforated the buildings in which suspected Hamas fighters gathered, killing militants and innocent families alike. Hamas fighters responded with hundreds of unguided rockets, eager to kill any Israeli they could. Few reached their targets. Israel had a new, U.S.-provided defenses, the Iron Dome, a missile shield that could intercept the projectiles in midair. The result was an eight-day, one-sided campaign. The IDF hit every intended target; Hamas, almost none. Two IDF soldiers and 4 Israeli civilians were killed, and another 20 Israelis were wounded. On the Palestinian side, roughly 100 militants and 105 civilians were killed, and another thousand wounded.

But this wasn't the only fight that counted. There were now three fronts at work in any conflict, Israel's chief information officer explained. Two were predictable: the "physical" fight, which Israel easily dominated, and the "cyber" fight, in which the IDF just as easily beat back the efforts of Palestinian hackers. But there was a third front, he said, "the world of social networks." This front would prove more troublesome, and impossible to contain, soon seeping into every corner of the internet. A comparatively tiny physical conflict, fought in an area the size of Portland, Oregon, became a global engagement, prompting the exchange of more than 10 million heated messages on Twitter alone.

The IDF deployed a Twitter account, Facebook pages in multiple languages, Tumblr blog pages, and even a Pinterest page. There were slick infographics and a stream of videos and statistics. Maximizing follower engagement, the official IDF blog offered small digital rewards for repeat users. Visiting the blog ten times got you a "Consistent" badge; searching the website got you recognized as a "Research Officer." Memes were fired off in volleys and tested for engagement, the best ones deployed extensively. The IDF's most widely shared image showed Hamas rockets bearing down on cartoon versions of Sydney, New York, London, and Paris. "What Would You Do?" the caption asked in bold red letters.

By contrast, the propaganda efforts of Hamas's militants were less structured. Most of its social media response came from millions of unaffiliated observers around the world, who watched the plight of Palestinian civilians with horror and joined the fray. The Twitter hashtag #GazaUnderFire became an unending stream of atrocities: images of bombed-out buildings, dead children, crying fathers.

The scourge of war left nothing untouched — including video games and fast-food chains. The IDF hijacked the hashtags for the World Cup, a new James Bond movie, and even the same Call of Duty franchise that Junaid Hussain of ISIS would weave into his own recruiting ("Playing war games on Call of Duty last night? Over a million Israelis are still under REAL fire#BlackOps2"). Meanwhile, pro-Hamas hackers took over the Israeli Facebook page of Domino's Pizza, using the opportunity to threaten a merciless reprisal of "more than 2000 rockets" against Israeli cities. When Domino's regained control of the account, it had a message of its own: "You cannot defeat . . . the Israeli hunger for pizza!"

Even as the missiles flew, the IDF and Hamas continued to narrate the conflict, each posting alerts, updates, and a steady string of taunts. "Warning to reporters in Gaza," wrote @IDFSpokesperson. "Stay away from Hamas operatives and facilities. Hamas, a terrorist group, will use you as human shields." @AlqassamBrigade couldn't let this stand. "Stay away from Israeli IDF," the Hamas spokesperson mimicked. "We are just targeting Israeli soldiers, fighter jets, tanks and bases." It was a remarkably juvenile exchange. But these taunts couldn't be dismissed as easily as the ones you might hear in a kindergarten classroom. After all, they were salvos lobbed by two combatants in a real, shooting war.

There was a temptation, after the sides had settled into an uneasy cease-fire, to dismiss this weird internet flame war as a bunch of digital noise. The angriest tweets were still just *tweets* — literally, "short bursts of inconsequential information." But that would have been a mistake. Years after Operation Pillar of Defense had slipped from the public mind, American University professor Thomas Zeitzoff conducted a painstaking study of hundreds of thousands of tweets, which he then mapped across each hour of the physical side of the eight-day conflict.

What he found was shocking. In the case of Israel, a sudden spike in online sympathy for Hamas *more than halved* the pace of Israeli air strikes and resulted in a similarly sized leap in Israel's own propaganda efforts. If you charted the sentiment (pro-Israel or pro-Palestine) of these tweets on a timeline, not only could you infer what was happening on the ground, but you also could predict what Israel would do next. Israeli politicians and IDF commanders hadn't just been poring over battlefield maps. They'd also been keeping an eye on their Twitter feeds — the battlefield of the social network war.

Taking place in 2012, Operation Pillar of Defense offered a glimpse of an emerging way of warfare. It was a conflict in which each side had organized to taunt and troll each other online, even as they engaged in a life-and-death struggle in the real world. Their battles drew in millions more international fighters. Some were passionate supporters of one side; others had just stumbled upon the war while looking for video game news or pizza. They shaped the fight all the same, strengthening the voice of one faction or another — and, by tiny degrees, altering the course of events on the ground. The lesson was clear: Not only did modern war require a well-planned military campaign. It required a viral marketing campaign as well.

Moving forward, both Israelis and Palestinians would apply this lesson, albeit in very different ways. Their approaches are broadly representative of the two strategies — loosely networked or centrally organized — that characterize how combatants approach the information battles of LikeWar today.

The next major outbreak of both web and "real" war came in 2014, as Israel and Hamas descended into another, bloodier, even more lopsided conflict, culminating in Operation Protective Edge, the IDF's ground invasion of Gaza City. Sixty-seven IDF soldiers and three Israeli citizens would perish, as would hundreds of Hamas militants and more than a thousand Palestinian civilians. Hamas actively solicited images of the victims of Israeli air strikes (children were best) and posted them online as soon as possible. "There is nothing wrong with publishing images of the injured," one web video urged. Real images of the devastation were soon blended with an ocean of fakes and forgeries. One 16-year-old Twitter user, confronted with proof that their recently posted viral image was fake, was as dismissive as a White House spokesperson. "People don't need to take it as a literal account," the teen said. "If you think of bombs going off, that's pretty much what it looks like."

Although the IDF won every battle, as casualties mounted, social media chatter grew unrelenting in its criticism of the Israeli military. In a single month, the hashtag #GazaUnderAttack was used more than 4 million times — twenty times that of the one the IDF was pushing, #IsraelUnderFire. When Operation Protective Edge ended seven weeks after it began, many Israelis were furious. They felt that their government had crumpled under international pressure; nine out of ten believed the military had failed to achieve its objectives.

Yet as the fighting ebbed, the images and videos didn't. By 2015, more than one-third of Palestinians had a Facebook account; even more had smartphones and access to the internet. And they began to use them, even — and especially — the very young.

Janna Jihad was a little girl with long brown hair and fierce hazel eyes. Like the young reporter from Pennsylvania we met in chapter 3, she plunged into citizen reporting when she was just 7 years old. Yet Janna was driven not by opportunity but by loss: the killing of a local boy in her village and the death of a cousin and uncle at the hands of Israeli security forces. She took to traveling the occupied territories, drifting from one protest and checkpoint showdown to the next, her mom's iPhone in hand. In time, she launched Facebook, Twitter, YouTube, and Snapchat feeds, amassing hundreds of thousands of followers.

To watch her dispatches was to feel torn in two. On the one hand, Janna was undoubtedly courageous, and her reports were heartbreaking. She charged into protests with a green, red, and white bandana (the colors of the Palestinian flag), braving tear gas and shouting down heavily armed soldiers. "They're killing us!" she screamed into her iPhone during one such encounter, as a young IDF trooper tried to corral the crowd. "We don't need any war, we don't need any blood. That's enough! We just want to live in peace and freedom. Without war, without killing, without hitting!"

On the other hand, Janna's goal wasn't journalism — to capture the context and explain the events. Instead, her intent was to spark the outrage that often drives virality. She was a child soldier in a new kind of war. As she explained simply, "My camera is my gun."

It was much the same for the protesters on whom she reported. When asked why they were marching into battle against Israeli soldiers, one young Palestinian explained that it was to "upload the pictures to the internet so that the world sees how the children are being fired at with bullets and gas." For these children, the most important thing wasn't launching a protest or hurting Israeli soldiers; it was getting footage of *themselves* being hurt. Even elementary-school-age children were very aware of the cross of optics and marketing into modern war.

This online effort added new fuel to age-old conflicts. Spying opportunity, Hamas and other militant groups launched their own digital campaigns, urging Palestinians to spill Israeli blood in lone wolf attacks. There were YouTube songs (sample lyrics: "I will attack you, tear you apart, and stab you"), "how-to" videos showing the human body's most vulnerable arteries, and even a viral fad in which Palestinian parents posed their toddlers with knives. In solidarity, ISIS started a parallel hashtag campaign: #killajew. Speaking to reporters, one older Palestinian man was blunt about the consequences for youth, echoing almost exactly what has been said about the gang problems in Chicago: "I have a job to save my son. From his friends, from social media . . . from himself."

Although it would be wrong to suggest that children like Janna were responsible for this cycle of violence, it is also true that their dispatches formed part of this terrible, self-perpetuating system that drove people to kill and be killed. And online, the fighting never stopped.

Yet this was a fight that Israelis were uniquely prepared to face, for they'd been gearing up for it, too. Their young nation had spent its infancy beset by powerful enemies and dependent on allies for support and survival. Even as Israel grew stronger, its government remained keenly attuned to the tides of global opinion. So the country invested heavily in overseas lobbying efforts, relying on the wider Jewish diaspora to get its message across. This effort was called *hasbara* — Hebrew for "explaining."

In 2000, activists took hasbara online, with the founding of the Jewish Internet Defense Force. By the mid-2000s, Jewish organizations were offering hasbara fellowships to students to rewrite Wikipedia articles in a more positive light. In 2007, Israel established a formal "information directorate," and in 2009 it became the first democratic nation to fund an "internet warfare squad" to respond to hostile blog comments. Nonetheless, even Israel was caught off guard by the magnitude of the social media revolution. The critical moment came in 2010, when human rights activists organized the heavily publicized Freedom Flotilla to bring supplies to Gaza, then under Israeli blockade. The IDF intercepted the ships, but only after a bloody boarding that left ten activists dead and sparked a furious global backlash.

"They tried to block the cell signal around the flotilla, but that didn't work," remembers Gilad Lotan, a data scientist and former IDF intelligence officer. "Videos were coming out and people were getting really angry around the world . . . They couldn't just neglect them and say, 'We're doing what we need to do, we'll explain later.' No, that doesn't work. You lose legitimacy, and then you pay for that over the years."

Afterward, Israel shifted course. Where before, Israeli propaganda efforts had been tinged with fatalism, akin to U.S. soldiers speaking of narrative ("In the media we are going to lose the war . . . It doesn't matter what we do," one IDF spokesperson said), now the government drastically increased investment in hasbara online outreach. Just a year earlier, the IDF YouTube channel had been launched by a few soldiers in their twenties as a fun side project. By 2012, those same twenty-somethings held some of the most important positions in the Israeli armed forces. The @IDFSpokesperson Twitter account, the IDF's official voice to the world, was the work of one energetic 26-year-old. "We turn it over to the kids, and they translate [our messages] into the new language of social media," an IDF general marveled. "I say it's magic."

The effort was as much out in the open as the battles themselves. The IDF's recruiting page invited young people to "join the international social media desk"; its web ads showed a handsome Israeli man in an olive-drab uniform, smiling as he gazed at an iMac. The form explained how young Israeli citizens could fulfill their two-year draft requirement by serving as content creators, graphic artists, and photographers. There was even a catchy tagline that seemed torn from a Silicon Valley startup's recruiting manual: "Create. Engage. Influence."

But the IDF's official military efforts constituted only one part of the new apparatus of online battle. "Hasbara war rooms" were organized at Israeli universities to build out online armies. Students could hang out and socialize in these designated computer labs as they fought for their nation on social media. During the 2014 conflict, one college boasted of 400 volunteers, their "operations" encompassing 31 different languages.

As the years passed, the war rooms became permanent campus fixtures. "This war doesn't end with the last rocket," a young Israeli activist explained. On the internet, it didn't end at all.

By 2017, Israel's hasbara offensive had its first smartphone app, trumpeted by its creators as the "Iron Dome of Truth." The web ad for the app showed two scantily clad young women crooning in a man's ear, "You are going to tell the whole world the real truth about Israel!" The app worked by pairing users with different online "missions" and rewarding them with points and badges. In one case, the app urged users to write positive things on comedian Conan O'Brien's Instagram page during his visit to Israel. In another, it prompted them to "report" a Facebook image that had superimposed the Israeli flag over a picture of a cockroach. It offered a glimpse of war's future: organized but crowdsourced, directed but distributed.

Of course, Israel had another, more direct way of gaining an advantage in online warfare: its flesh-and-blood police and soldiers. As Palestinians grew more vocal in their social media use, Israeli courts broadened the definition of speech they considered "incitement" to violence. Police units monitored social networks for particular keywords and messages; they studied suspicious users, checking to see if they gave vocal support to any militant causes. In the occupied West Bank, IDF soldiers in armored Humvees prowled the neighborhoods of suspected social media "inciters." Between 2011 and 2015, more than 400 Arabs were arrested for social media–related crimes. Their fate was easy to predict. Israeli military courts had a conviction rate of 99 percent.

Today, the battles between Israelis and Palestinians continue, both in the occupied territories and online. Yet they are only one tiny front in a world of wars. The social media accounts of every military organization, diplomatic envoy, world leader, soldier, and civilian exist in the same digital milieu. For the most part, they're accessible to the same global audience, who either use English as the internet's lingua franca or rely on increasingly sophisticated language-translation software. And when the sides come into conflict — as they often do — their supporters, critics, and trolls are all thrown into battles of their own. Every tweet or public statement, in other words, is a new front of LikeWar waiting to happen.

Sometimes, the spark is as simple as a disagreement about numbers. A puffed-up propaganda statement about enemy casualties — an act of exaggeration as old as war itself — draws a great deal more scrutiny when both sides are on Twitter and can fact-check each other. In Afghanistan, Taliban spokesmen complain often about not getting "credit" for their attacks. Such exchanges also lead to weird moments of impropriety. When, in the midst of an argument over air strikes, the Taliban feed made reference to a NATO commander's mistress, it left observers aghast. The online consensus was that the Taliban, a group that murders women for the crime of learning to read, had "gone too far."

In places where the physical battle lines are less clear, the migration of conflict to the internet becomes similarly perplexing. Imed Lamloum, a Libyan correspondent for Agence France-Presse, described in 2017 how the war-torn country had splintered into two rival social media–savvy governments. "Each government has its own press agency," he explained. "Both are named *Lana*. Each *Lana* puts out statements by its respective government and also seeks to discredit the other." Libya's dozens of armed militias each had their own Facebook account, where they did everything from negotiating cease-fires to posting a mix of bluster and factual "updates" that were nearly impossible to verify, just like the gangs and presidents. "Sometimes I think that the situation in the country could improve if internet access was cut," Lamloum wrote. "Then people would no longer have access to rumors, which represent roughly 90 percent of information that's out there."

When two nations with similar languages and shared borders fall to bloodshed, the result is an even more surreal mess of public disparagements and digital proxy battles. Since Russia's invasion and occupation of eastern Ukraine in 2014, the two countries have remained locked in a simmering conflict that has stolen tens of thousands of lives and that reverberates each day through Ukrainian-, Russian-, and English-language social media. As homophily and group polarization have kicked into high gear (wars have a way of accelerating the process), supporters of each side have assembled an inexhaustible list of real or fabricated crimes, which they throw at each other like knives. Each public posting across Facebook, Twitter, or VK presents a new opportunity to argue and troll. When a famous Russian-backed separatist commander was killed in a Ukrainian rocket attack, for instance, his funeral was livestreamed on Facebook — and promptly discovered by Ukrainian nationalists. The livestream became an extension of the physical battlefield, as laughing and crying emojis both jockeyed for supremacy.

This mix of deadly serious conflict and weird digital levity reverberates through the highest levels of international diplomacy. In one instance, the official Russian Twitter account (@Russia) warned, "Whoever comes to us with #sanctions will perish." The boast was paired with a heavy-metal music video of Russians parading with medieval swords and armor. The response from the official Ukrainian account (@Ukraine) was swift: "If you'd respected international law, you would've avoided sanctions & would've been sending missions to Mars now, not running with sticks." Ukraine then underscored its taunt with a meme, a GIF of two characters from the television series *South Park* hitting each other with sticks. Russia responded the same day in a different way, with shelling that killed a Ukrainian soldier. Indeed, armed confrontations have become inextricably linked to internet trolling. In 2019, a suicide bomber with ties to a Pakistan-linked terrorist group killed 40 Indian soldiers. The two nuclear powers sprang into red alert, edging as close as they had in a generation to all-out war. Yet the most prominent battlefield turned out not to be their tense border region — which saw a few inconsequential skirmishes — but rather WhatsApp, Twitter, and Facebook. Government mouthpieces spouted propaganda about fantastic (and often nonexistent) victories, amplified in turn by armies of digital activists. Hashtags like #IndiaWantsRevenge, #PakistanStrikesBack, or #SayYesToWar became new fronts in a furious tug-of-war between rival partisans. Even the action that ultimately defused the crisis, Pakistan's return of a captured Indian fighter pilot, went through this churn. Long before the pilot landed safely again in India, he'd become an internet meme.

Whether the conflict is a civil war echoing across YouTube, a dispute over missile tests that culminates in one leader tweeting threats at another, a swaggering Facebook argument between gangs, or just a celebrity flame war, all of these socially mediated conflicts are overtly theatrical. It is about bolstering friends and dissuading foes, just like the kind of macho showmanship that precedes a bar fight. It is about persuading someone to back off before the first punch is thrown. Failing that, it's about weakening and embarrassing them, sapping their supporters while energizing your own.

But there's also another kind of internet conflict — subtle, often invisible — that simmers beneath such public disputes. These hidden battles of influence comprise the side of LikeWar that doesn't *want* to be noticed, but it may be the part that has most decisively reshaped the modern world.

THE WAR YOU CANNOT SEE

To start the process of a "pro-Russian drift" of Crimea and eastern Ukrainian territories, [certain] events should be created beforehand that can support this process with political legitimacy and moral justification; also a PR-strategy should be built that draws attention to the forced, reactive nature of the actions of Russia and the pro-Russian political elites of southern and eastern Ukraine.

In early 2014, a policy paper began circulating in the Kremlin, outlining what steps Russia should take if President Viktor Yanukovych, the pro-Russian autocrat who controlled Ukraine, was toppled from power. Russia had to be ready, the memorandum's author urged, to create a new set of political conditions on the ground — to manipulate the "centrifugal aspirations" of ethnic Russians, pushing them to declare independence from Ukraine. In essence, if their guy was ever forced from power, Russia had to be ready to start a war.

Just two weeks later, amidst mounting protests, the unpopular Yanukovych fled the country in what was known as the Euromaidan. As proof of the emerging power of social media, the name of this Ukrainian revolt was taken from a Twitter hashtag (it combined "Europe," for the demonstrators' desire to partner with Europe instead of Russia, and "Maidan Nezalezhnosti," the square in Kiev where they gathered). But just as the revolutionaries had used the new form of the internet to unite and topple their foe, so Russia now used it to tear Ukraine apart.

The strategic vision behind the operation was subsequently explained by Dmitry Peskov, Putin's long-serving media advisor, in a 2017 interview. Particularly notable was not just how candid Peskov was, but whom he drew upon for inspiration. He spoke of "a new clash of interests" between Russia and the world, brought about by the social media revolution. But he also marveled at the incredible influence wielded by new online powers, citing Kim Kardashian. "Let's imagine that one day she says, 'My supporters — do this,'" he said. "This will be a signal that will be accepted by millions and millions of people." But importantly, Peskov noted, Kardashian had "no intelligence, no interior ministry, no defense ministry, no KGB."

The implication was clear. Russia *did* have these things — and unlike Kim Kardashian, it would use them for more important fights than a feud with Taylor Swift. "The new reality creates a perfect opportunity for mass disturbances or for initiating mass support or mass disapproval," Peskov said. This was the grounds for "an informational disaster — an informational war."

Ukraine proved a critical test case. Negative Russian-language news articles about Ukraine doubled, then tripled, in number. Ethnic Russians inside Ukraine, already on edge, soon boiled with resentment toward the activists who had overthrown the government they supported. Meanwhile, Russian commandos filtered into Crimea, then eastern Ukraine, recruiting and arming cells of pro-Russian separatists. There were waves of protests, then violence, then mounting tragedy.

The tipping point came in the city of Odessa, where dozens of pro-Russian protesters — many armed — retreated to a large, Soviet-era trade union building that soon caught fire amid the hail of bullets and Molotov cocktails. At least thirty-one people died.

Spying opportunity, the Russian "PR-strategy" kicked into high gear. Russia deftly seized on the tragedy, orchestrating a media campaign that Ukraine couldn't hope to match. RT reveled in publishing gory details that were impossible to verify: pro-Ukrainians had sprinted through the flames to "strangle" pro-Russians; "17-year-old hooligans were finishing people with bats." Legions of trolls then seeded the stories through fringe outlets around the world. Conspiracy theory memes were deployed. "US Media Covers Up Mass Murder in Odessa," blazed one Infowars headline. Meanwhile, the Russian government fed off the very headlines it had ordered written. The Russian foreign minister declared that, given the news of atrocities, it was now Russia's solemn duty "not to allow fascism to spread throughout Europe and the world at large."

As time passed, the supposed atrocities grew even more troubling. Russian state media described how Ukrainian soldiers had stripped a

3-year-old to his underwear, then crucified him "just like Jesus" — right before strapping the mother to a tank and dragging her around the town square. There was nothing in the way of proof, but there didn't have to be. The point wasn't to be truthful, but to justify an invasion.

Soon enough, thousands of Russian troops were streaming into Ukraine. Although these troops did what they could to conceal their identities, as we saw in chapter 3, they couldn't escape the spotlight of the same social media platforms they'd deftly manipulated. Some ethnic Russians in Crimea, enthusiastic about the prospect of reunification, took selfies with the occupiers and posted them on Instagram ("Sweetest guys," one caption read). In a bit of internet wordplay, they took to calling these heavily armed soldiers in unmarked uniforms "little green men." The Russians themselves couldn't keep from talking. On his VK account, one artilleryman bragged, "We pounded Ukraine all night." This chatter fed a new flurry of OSINT analysis. The investigative team at Bellingcat found the Russian military had awarded more than 10,000 medals for "combat operations" at a time when Russia wasn't *officially* fighting anywhere.

Across the world, no one knew what to make of it. The United States and its European allies imposed sanctions and went on their highest military alert since the Cold War, all for something that *officially* wasn't happening. It was an invasion that wasn't, a major conflict that one side flatly refused to acknowledge it was fighting. Russia had used social media not only to stoke the fires of conflict, but also to create something akin to a "Schrödinger's war": a perception-warping, reality-bending conflict that existed in two simultaneous states. "This, in short, was no traditional military invasion; it was hybrid warfare in which goals were accomplished even before the adversary understood what was going on," former U.S. ambassador to NATO Ivo Daalder explained. His military counterpart, General Philip Breedlove, then supreme allied commander of NATO, called it nothing less than "the most amazing information blitzkrieg we have seen in the history of information warfare."

But it was something even more than that. On the ground in separatist-occupied Donetsk, journalist David Patrikarakos scrolled through frenzied Ukrainian social media feeds even as he listened to shells falling on the city outskirts. He realized these two worlds were inextricably linked. "I began to understand that I was caught up in two wars: one fought on the ground with tanks and artillery, and an information war fought . . . through social media," he wrote. "And, perhaps counterintuitively, it mattered more who won the war of words and narratives than who had the most potent weaponry." The result was a violent, confusing, paralyzing mess — precisely as Russia intended.

Following the success of the Ukrainian offensive, these information campaigns grew in number, intensity, and daring. Against the Baltic states of Latvia, Lithuania, and Estonia — newer members of the NATO alliance with large ethnic-Russian minorities — the strategy was much the same as it was in Ukraine. There were the wildfire rumors, like a fake viral story that NATO soldiers had raped a 15-year-old Lithuanian girl. There were the troll brigades that consistently swarmed Latvian news portals, decrying the Latvian government and the West while praising Russia. There was even a glittering new media service, Baltica, launched as part of the aforementioned RT network of government media, and funded by the Russian government through a series of shell companies. The Estonian version of the site marketed itself as "an experimental publication with the goal to show readers a positive outlook on life." In actuality, it spread strange and scary rumors, such as how American soldiers in Estonia for NATO defense exercises were actually planning to confiscate Estonian cars.

Here, the main goal was to till the soil, planting seeds of future operations if they were ever needed. Lithuanian officials described how a stealthy social media campaign was essentially rewriting history, spreading rumors that large swaths of Lithuanian territory had been — and would always be — Russian. Secession causes were seeded. Some pro-Russian Lithuanian activists began to create Facebook pages that called for independent, ethnic-Russian enclaves similar to the puppet states Russia had carved out of eastern Ukraine. "If we lose the information war today," a Lithuanian military information specialist warned, "tomorrow we may be fighting with weapons."

Farther afield, the purpose of these Russian information offensives was to add a fifth "D" to the classic disinformation goals we explored in chapter 4. In addition to dismiss, distort, distract, and dismay, these messages were intended to *divide*. An obvious wedge was the issue of Syrian refugees, who had fled their homeland by the millions to seek asylum in Europe, leading to sharp disagreements between the members of the European Union. Germany — bedrock nation of both NATO and the EU — had made waves when it announced that there would be no limit on the number of migrants it admitted. And so it was to Germany that Russian information warriors turned their gaze.

The story, when it broke, was horrifying. A 13-year-old Russian-German girl had been kidnapped, beaten, and raped by three Arab migrants. Moreover, the police were refusing to investigate! This sparked a tiny protest — covered only by RT — that was followed by a much bigger one as the news rocketed through the far-right German media. The truth was, as the weary German government explained again and again, the story was a hoax, made up by a girl who was embarrassed after running away from home. But their explanation was ignored. Soon the Russian foreign minister himself waded into the controversy, citing Russian reporting on rumors that had been spread by Russian proxies. "It is clear that [she] did not exactly decide voluntarily to disappear for 30 hours," he said with a smirk. "I truly hope that these migration problems will not lead to attempts to 'gloss over' reality for political motives — that would be just be [*sic*] wrong." With this statement, the Russian minister was himself glossing over reality for political motivation. But that was the point.

The hoax launched a legion of copycat stories, stirring anti-refugee sentiments in Germany and across Europe. Russian media covered every one, and they spiraled out through the network. The meme of the savage, woman-defiling, brown-skinned migrant (#Rapefugees) proved an immense aid to the resurgence of Germany's stridently far-right nationalist political party. For the first time in nearly sixty years, it won enough votes to see its candidates seated in the German parliament.

Wherever internal division festered, Russian propagandists supported it from afar. In 2014, they weighed in on the Scottish independence referendum. In 2016, they promoted the UK's Brexit proposal even more aggressively, and subsequently labored to guide the outcome of the U.S. presidential election. In 2017, when the Spanish region of Catalonia teetered on the brink of secession, Russian media and proxies emerged as the dominant pro-independence voices, their every message amplified by an army of bots. When tiny Montenegro moved to become the twenty-ninth member of NATO, online operatives worked doggedly to tear the nation apart. This army of bots was later revealed to be laying the foundation for a plot by Russian-supported extremists to assassinate the prime minister and overthrow the government. The plot was narrowly averted when local police discovered machine guns, sniper rifles, and an RPG hidden near the prime minister's home.

Eventually, even the typically somnolent legislature of the European Union had had enough. We are "seriously concerned by the rapid expansion of Kremlin-inspired activities in Europe," the European Parliament declared. "Hostile propaganda against the EU and its member states seeks to distort the truth, provoke doubt . . . and incite fear and uncertainty among EU citizens."

Putin answered with disdain. "We are observing a certain, quite obvious, degradation . . . of how democracy is understood in Western society," he said. He expressed mock surprise that after "trying to teach us democracy," the EU was silencing alternative opinions. In his response, the Russian president seemed like nothing so much as an internet troll made flesh.

For all of Russia's efforts, though, there is nothing inherently Russian about this new, potent kind of information conflict that has infested war and politics alike. Instead, it is merely emblematic of larger truths in the social media age. Bots, trolls, and sockpuppets can invent new "facts" out of thin air; homophily and confirmation bias ensure that at least a few people will believe them. On its own, this is grim enough, leading to a polarized society and a culture of mistrust. But clever groups and governments can twist this phenomenon to their own ends, using virality and perception to drag their goals closer within reach. Call it disinformation or simple psychological manipulation. The result is the same, summarized best by the tagline of the notorious conspiracy website Infowars: "There's a war on . . . for your mind!"

These offensives abide by two basic principles. The first is *believability*. Engineered falsehoods work best when they carry what seems like a grain of truth. They play on existing prejudices, seeking to add just one more layer to a narrative that already exists in the target's mind. Recall the KGB's Operation INFEKTION, the Cold War claim that the U.S. military invented AIDS. During that era, Ladislav Bittman served alongside the KGB in its allied Czechoslovakian intelligence service's disinformation department. In a 1985 book, he explained how "every disinformation message must at least partially correspond to reality or generally accepted views." The AIDS disinformation campaign, for example, didn't invent a new threat; instead, it leveraged people's fears about a well-known but then mysterious disease.

The internet, and especially its memetic elements, enables this further. For instance, as bizarre as the #Pizzagate hoax may have seemed, it effectively leveraged a canon of Clinton-linked controversies that stretched across every social media platform for over a decade. So, too, the stories told to prime the Ukrainian war — of babies being crucified — built upon the real atrocities of past wars. And social media's very form lowers the credibility threshold even more: whatever the news, if it comes from friends and family, it is inherently more believable.

The second principle of these stealthy information campaigns is *extension*. The most devastating falsehoods are those that extend across vast numbers of people as well as across time. They spread by how they linger, formed in such a way that the very act of denial breathes new life into the headline, helping it burrow deeper into the collective consciousness. Such engineered stories work like barbed arrows, spreading more infection and rot even as the victim struggles to remove them. The more salacious the accusation, the better. There is a famous political legend that Lyndon Baines Johnson, falling behind in one of his first local elections, ordered his campaign manager to spread the rumor that his opponent was a "pig-fucker." The manager protested, saying there was no proof. "I know," Johnson replied. "But let's make the sonofabitch deny it."

The internet makes it even easier to strike and then prolong the agony. Social media algorithms work by drawing attention to content that trends on their networks, even (and especially) when people are outraged by it. The result is the virtual equivalent of a grease fire, where widespread condemnation of something ensures that new groups of users see it and condemn it in turn. Because virality is incompatible with complexity, as content trends, any context and details are quickly stripped away. All that remains is the controversy itself, spread unwittingly by people who feel the need to "weigh in" on how fake or nonsensical it sounds. Even as they complain about how big it has gotten, they make it bigger.

Shortly after the election, Jack Posobiec, the infamous #Pizzagate architect and urinalysis officer, provided a perfect illustration of this dynamic at an anti-Trump protest in Washington, DC. Posobiec would later be captured in online reports infiltrating the crowd gathered outside the Trump International Hotel and holding aloft a shocking sign — "Rape Melania" — just long enough for one of his coconspirators to snap a picture of it.

Thanks to Posobiec's network of alt-right proxies and promotion on Russia's RT, the image quickly went viral. Soon #rapemelania became one of Twitter's top-trending hashtags, with tens of thousands of Twitter users all screaming at the same nonexistent adversary. Even the very trend itself became a point of outrage, spurred on by Breitbart's smug headline: "Twitter Allows 'Rape Melania' to Trend After Site Explodes with Trump Assassination Threats." The controversy then launched a series of angry follow-on conversations about everything from the hypocrisy of Trump fans crowing about a sexual assault that didn't happen to the dangers of the perceived "violent left" that would seemingly endorse such a crime. (In text messages later published online, Posobiec also admitted that he started an "assassinate Trump" chant, hoping others would join in.) Then came the realization for some that it had all been a hoax. As the "Rape Melania" sign swept social media, a dedicated group of activists, reporters, and Wikipedia editors worked to tie Posobiec's name to the stunt. This effort started a fresh battle of accusations and arguments, as Posobiec denied the published text messages showing him and his collaborators brainstorming the plot ("Fuck Melania" was decided to be "too subtle"), while pro-Posobiec forces resorted to the classic troll tactic of playing the victim, angrily accusing other Wikipedia users of "character assassination."

But one story was lost amid the commotion: the motivations of any of the hundreds of demonstrators who had shown up in the first place. With a single act of misdirection, their purpose and message essentially disappeared.

Importantly, these veiled information operations are effective because they're part of a broader strategy that combines the unseen and seen in a vicious one-two punch. Posobiec's "Rape Melania" gambit worked because he could count on a vast network of pro-Trump social media accounts to make it go viral, a situation that was then amplified by the combination of a network of media outlets and Twitter's own trending algorithms. The stratagem was little different from one that ISIS or Russia or any other information warrior would use.

It might seem that these battles never truly end (indeed, in discussing the incident, we now become part of its back-and-forth, as will you if you post online what you just read), but they do have clear winners and losers. The point comes — often quickly — where the main objective has been achieved. With a few seconds of work, "Rape Melania" is now forever associated with a chilly November 2016 protest. With a few weeks of online foment, Ukraine was plunged into war. And as the targets try to untangle themselves and fight back, their opponents are busy plotting the *next* offensive. Just as with ISIS's prodigious propaganda output or BuzzFeed's relentless grinding for viral hits, all of these battles are also experiments, both for the actors in them and for the wider world, teaching all what works for the next fight.

And in this sort of war, Western democracies find themselves at a distinct disadvantage. Shaped by the Enlightenment, they seek to be logical and consistent. Built upon notions of transparency, they seek to be accountable and responsible. These are the qualities that made them so successful, the form of government that won both world wars and the contest of superpowers in the last century. Unfortunately, they are not the values of a good troll, be it a urinalysis officer turned provocateur, a celebrity turned politician, or a nation applying them as a global strategy for information war.

When Ukraine announced the creation of a volunteer "Internet army," Russian propagandists turned it into a joke. When Germany launched the Center of Defense Against Disinformation to combat fake rumors and teach citizens to be critical of Russian sources, an RT broadcaster observed (not incorrectly) that it carried echoes of a "ministry of truth." When U.S. intelligence services documented evidence of Russian interference in the 2016 presidential election in order to increase public awareness of information war and generate support for U.S. countermeasures, it was quickly twisted into an alleged plot by the "deep state." The openness of these nations has become just another wedge to be exploited.

But all is not so dark. For that very openness, both of democracies and the internet that only an open democracy could create, also widens the fight. The more confused and nonlinear war gets, the more participatory it becomes.

WAR BETWEEN EVERYONE

The shrouded figure gazes into the camera. It wears a white mask with cartoonishly arching eyebrows, a wide mustache turned up at both ends, and a thin, pointed beard. The mask's mouth carries an oversized, sinister smile. Its computerized voice rings out, condemning the crimes being committed by the self-declared Islamic State and also promising vengeance. "We are held by a code of honor to protect those who are defenseless, both in the cyber world and the real world."

The figure folds its hands, black suit contrasting with the sepia-tinted globe behind it. The video ends with a motto intended to send chills down the spine of any internet user.

We are Anonymous.
We are Legion.
We do not forgive.
We do not forget.

And finally, as the screen fades to black:

Expect us.

When ISIS rocketed to digital prominence, it drew the attention of the world, but also an unexpected adversary: members of the hacktivist group Anonymous. The Islamic State's viral propaganda struck a nerve. Viewing themselves as the guardians of the internet, Anonymous decided to fight back.

The counterattack began in stumbling, jumbled confusion, a mob of random trolling and harassment. Digital vigilantes surfed the massive, public ecosystems of platforms like Twitter and YouTube, relentlessly taunting ISIS militants about their sex lives (you quickly felt sorry for the goats) and flagging accounts for deletion. ISIS websites were overloaded with message traffic and knocked offline. Japanese hacktivists, spurred by the brutal beheading of two Japanese hostages, targeted one of the Islamic State's most precious assets: its Google ranking. They wrote programs to bombard ISIS-related search terms, replacing messages of jihad with a black-robed, green-haired anime character (named ISIS-chan) with a cute smile and a strange obsession with melons.

Although the Anonymous effort inspired plenty of thrill-seekers who jumped into anti-ISIS operations for only a few days (spending most of that time editing their own epic YouTube declarations of war), the churn and madness gradually forged a cadre of hacktivists cut from a different cloth. For these men and women, throwing their lives into battling the "Cyber Caliphate," hatred of the Islamic State was visceral — and often intensely personal. There was the U.S. military veteran DigitaShadow, based in the backwoods of the American South, who hunted the dark web for ISIS sanctuaries; a former Miss Jordan beauty pageant queen, Lara Abdallat, who pushed back against ISIS propaganda on Twitter; and the European transient Mikro, a product of foster homes and juvenile prison, who harnessed his restless energy to locate, flag, and eliminate ISIS social media accounts across the web.

Over time, these dissidents pooled their efforts into a weird, new online organization that was equal parts hacking collective, troll factory, and amateur spy ring. ISIS websites were systematically targeted and destroyed with military efficiency. Teams of volunteers maintained a rotating list of tens of thousands of ISIS Twitter accounts, crafting algorithms to concentrate their fire and banish ISIS mouthpieces as quickly as they were created. Other hacktivists posed as would-be recruits, penetrating deep into ISIS networks to harvest snippets of personally identifying information and pass them on to governments. In a surreal turn of events, one faction even founded an internet security group, promising "cyber terrain vigilance" and competing for contracts with other cybersecurity firms that had themselves been created to defend against Anonymous.

As this weird internet war rumbled onward, it sometimes blurred with the "real" one. In Tunisia, eagle-eyed members of the collective spotted evidence of an impending attack on social media, leading to the arrest of eleven suspected militants. ISIS recruiters in Indonesia, careless in masking their IP address, awoke to a police raid, their identities leaked by hacktivists halfway around the world.

Devoted as the hacktivists were, the accumulated impact of their campaign was modest. They couldn't stop the flow of foreign recruits to ISIS. Their Twitter trolling could hardly liberate Syria and Iraq.

But that wasn't the goal. The point of the anti-ISIS campaign, expressed repeatedly by those who fought it, was simply to push back; to mount a resistance where before there had been none. Most netizens, much as they despised ISIS, weren't going to march into a military recruiting station and undergo a year's worth of training in order to one day possibly — *possibly* — get to fight ISIS on the ground. But almost anyone could join this fight. They could spend their lunch breaks scanning Twitter for new ISIS accounts, feeding them into a community spreadsheet to ensure they were reported and deleted as quickly as possible. This was a tiny but perceptible way to chip away at the Islamic State's propaganda apparatus, open to all comers with a basic internet connection.

This fight was just one of the many wars that have broken out online, in which an army of volunteers have rushed into the battle. Wikipedia editors fought to preserve the truth of the "Rape Melania" hoax, while unpaid Bellingcat analysts investigated the shootdown of MH17. Indeed, when both the U.S. government and social media firms delayed taking action against Russian bots and sockpuppets in 2016, it was a group of think tank wonks who started to track them down.

Yet the combatants extend beyond these cyber-minutemen consciously volunteering for the cause. As modern warfare turns increasingly on the power of internet operations, it renders everyone a potential online combatant. Any particular message, image, video, or status update has the chance of achieving virality. Such content can work to the benefit of one side in a conflict and the detriment of another. As two or more online adversaries fight over the fate of that content — your individual choice whether to amplify, expand, suppress, or distort it — even a single "like" or retweet becomes a meaningful action in an ever-evolving information war.

We've seen how this dynamic has divided and destroyed, spinning out lies and empowering some of the worst possible people and movements. But this very same force can work to the benefit of the weakest and most downtrodden, giving them a voice where before they had none. That is illustrated by another young girl, in this case motivated not by opportunity or loss, but by a desire for peace.

Bana Alabed was just 7 years old when her Twitter account spun to life in October 2016, broadcasting live from the besieged Syrian city of Aleppo. Her first message was poignant in its simplicity: "I need peace." Over the following days and weeks of the siege, Bana's messages were a surreal mix of harrowing updates ("We are sure the army is capturing us now"), images of bombed-out buildings, and the musings of a little girl trapped by circumstances beyond her control ("I miss school so much").

In many ways, Bana's online diary of thoughts made her a modern-day Anne Frank, except in this case she was exposing the horrors of war in real time. (Frank's written diary wasn't published until six years after her death in the Holocaust.) In just two months, Bana amassed 200,000 Twitter followers, in the process becoming a human face for the hundreds of thousands of civilians caught in a messy civil war. At the same time, she gained powerful fans. Before an audience of millions, she corresponded with *Harry Potter* author J. K. Rowling, receiving free copies of Rowling's books and exposing a massive cross-section of Potter fans to the horrors of Aleppo.

As Bana's account rocketed to global prominence, she sparked new information conflicts of her own. Critics alleged that Bana was actually a sockpuppet manufactured by the Syrian opposition, or that she was part of a top-secret propaganda operation run out of Britain. Rogue accounts were created, trying to trick people interested in Bana to click on them instead. They purported to show a little girl in a hijab with a gun, the insinuation that Bana was actually a jihadist. Then, as Bana weathered night after night of artillery bombardment ("I am very afraid I will die tonight. This bombs [sic] will kill me now"), even the president of Syria personally attacked the child. She was part of a "game of propaganda," Bashar al-Assad said, concocted by "terrorists." The social media account of a 7-year-old girl had become a new flash point in the Syrian civil war's ever-evolving information battle.

The truth (eventually cobbled together by a Bellingcat investigation) was that Bana was indeed real: a young girl who had found a powerful outlet for her hopes and dreams through her instantly viral Twitter account. There was one caveat, though. Bana wasn't doing the tweeting alone; her English-literate mother, trained as a reporter, was helping her. This either proved or disproved her authenticity, depending on which side you took in the fight. Yet everything else — the videos of her standing amidst playground rubble, the plaintive cries for help as the shells fell around her house — was as real as it was heart-stopping. Even after Bana and her family escaped Syria, she would remain a powerful voice for Syria's refugees. It was a voice that couldn't have existed just a decade before.

But in its democratization of conflict and elevation of new voices, social media presents a challenge. Whereas users like Bana employ the internet to call for the *end* of war, others are eager to use it to start a new one. This dilemma is best captured in the Chinese proverb "If you ride a tiger, it's hard to get off." And China, where the internet has been appropriated as a tool to promote unitary nationalism, may be the best illustration of this danger.

As we have seen, through a careful strategy of web and social engineering, the Chinese regime has brought the masses online and woven the online discourse into a strongly patriotic community. But this same massive community can become a hive that erupts in anger at the slightest international provocation. Amidst Taiwan's 2016 general election, for instance, one of the most popular phrases on Weibo translated as "use force to unify Taiwan." In discussions with its neighbors over disputed islands, the common tone on Chinese social networks has been conveyed in messages like this: "Even if China is a graveyard, still need to kill all Japanese. Even if no grass grows in China, still need to recover Diaoyu Islands." Following an international court ruling that rejected China's sweeping territorial claims to the South China Sea, Chinese social media erupted with hundreds of thousands of furious comments, many calling for war. The anger, which had originally been stoked by the government, quickly spooked senior party officials; in response, censors and state media worked overtime to restrain the very forces they had helped unleash.

What has been most alarming for the regime, however, is that the hive no longer roils at foreigners alone, but also at Chinese government actions that fall short of the most stridently patriotic standards. After the transit of a U.S. Navy destroyer through contested waters, for example, the fury of Chinese social media users was directed not just at the United States, but also at its own military — once an unassailable institution. "Stop boasting and fight!" became a common refrain.

For the Chinese regime, dependent above all else upon the illusion of consensus, the spontaneous political movements enabled by the internet represent a potentially existential threat. When the crowd cries for violence, its desires cannot be satisfied — but neither can they be wholly ignored. "Domestic voices calling for a more muscular Chinese foreign policy have created a heated political environment," writes former U.S. State Department official Thomas Christensen. "Gone are the days when Chinese elites could ignore these voices." Even if politicians disregard this fervor, it's unclear if the generals will. The door is being slowly opened to a bizarre but not impossible future where the world's great powers might fall to bloodshed due — in part — to matters getting out of hand online.

In this dynamic, one is reminded of how the First World War began. As war clouds gathered over Europe in 1914, the advisors to both the German kaiser and the Russian tsar came to the same curious conclusion. Confiding in their diaries at the time, they wrote that they feared more the anger of their populace if they *didn't* go to war than the consequences if they did. They had used the new communications technology of the day to stoke the fires of nationalism for their own purposes, but then found that these forces had moved beyond their control. Fearing that *not* going to war would cost them their thrones, the monarchs started the war that . . . cost them their thrones.

The thread that runs through all these strange internet skirmishes is that they take place simultaneously, in the same space. Sometimes, the conflict is between feuding celebrities; other times, nations embroiled in a life-and-death struggle. Sometimes, these battles dominate social media chatter completely; other times, they pass with nary a mention.

There aren't two or ten of these conflicts, but many thousands, all unfolding at once and leaving no one and nothing untouched. By merely giving them our attention, we become a part of them. Like cyberwar, these LikeWars are also about hacking. But they're not targeting computer networks — they're targeting human minds.

There's one more aspect that makes them different from the conflicts of the past. Anyone can fight in them, but all the combatants are equally powerless in one key, new way. For while these warriors of LikeWar each fight their own personal and global wars across the internet, they aren't the ones writing its rules.

8

Masters of the Universe

The New Rules and Rulers of LikeWar

> There's a war out there, old friend. A world war. And it's not
> about who's got the most bullets. It's about who controls the
> information. What we see and hear, how we work, what we
> think . . . it's all about the information!
>
> — COSMO, in *Sneakers*

THEY SAY NECESSITY IS THE MOTHER of invention. For Chad
Hurley, Steve Chen, and Jawed Karim, that necessity was seeing Janet
Jackson's nipple.

During the live television broadcast of the 2004 Super Bowl, super-
stars Janet Jackson and Justin Timberlake teamed up for a duet at halftime, coming onstage soon after a salute to the troops fighting in Iraq. At
the very moment that Timberlake sang "Bet I'll have you naked by the
end of this song," he reached across to tear off a piece of Jackson's top.
For nine-sixteenths of a second, her bare breast stood exposed to the
Houston air — and to 140 million viewers. What followed was dubbed
"Nipplegate," weeks of cultural commentary and hand-wringing about
small children and stolen innocence. The Federal Communications
Commission was flooded with a record-shattering 540,000 complaints.
America Online, which had spent $10 million to sponsor the show, de-
manded its money back.

Yet as much as everyone was talking about what had happened, it was
nearly impossible to find the uncensored evidence. Newspapers weren't
showing it. Networks weren't showing it. And angry, soon-to-be-irrel-

evant America Online *certainly* wasn't showing it. The videos of Jackson's "wardrobe malfunction" were out there; they just needed to be hosted in one searchable, shareable place.

And so YouTube was born.

Yes, the website that would become the video archive of the human race was launched by an errant nip-slip. Yet the strangest part of the story wasn't how unusual it was, but rather how *typical.* Nearly all of today's internet giants can be traced to such beginnings. Facebook was a product of the juvenile, hacked-together Facemash. Twitter was the result of a hasty pivot from a failing podcasting service, originally marketed to the San Francisco rave scene. Even Google started with two Stanford nerds just trying to write a half-decent dissertation.

This is the DNA of the social media ecosystem: nearly universally male, white, drawn from America's upper middle class, and dedicated, at least initially, to attacking narrow problems with equally narrow solutions. Despite their modest, usually geeky origins, these founders now rule digital empires that dictate what happens in politics, war, and society at large. It has been an uneasy reign as they come to grips with what it means to rule their kingdoms.

ACCIDENTAL EMPERORS

There's no historical analogue to the speed and totality with which social media platforms have conquered the planet. The telegraph was pondered for at least two generations, while the internet gestated for decades in U.S. government labs. Beyond science fiction and the grand predictions of a few sociologists, social media was something that simply *wasn't*—until suddenly it was.

The surprise is most obvious among the creators themselves. As their platforms graduated to their first thousand, million, and even billion users, these bright young founders weren't pondering how their systems might be used to fight and win wars; they were mostly struggling to keep the lights on. More users required more servers. More servers required more investors. More investors required a sustainable business or, failing that, even more users. "This is the original sin of Silicon Valley," writes design ethicist Mike Monteiro. "The goal of every venture-backed company is to increase usage by *some* metric end over end until the people who gave you that startup capital get their payday."

Everything about these services was designed with growing the business in mind, engineering toward more users, and drawing them more deeply into the online experience. Consider something as innocuous as the "notification" icon — the red dot that has haunted the Facebook app for a decade. No part of the design is an accident. Red is the color of agitation and psychological arousal, the mere glimpse of which can lead to a spike in heart rate. It feels *good* to make red things go away. Because notifications are purposefully vague until touched, following them can feel like opening a present. (Is the notification a long, heartfelt comment from a close friend, or just another forgotten acquaintance's birthday?) While the notification icon was certainly intended to make Facebook users' lives easier, it was *also* intended to keep users buried in the app — part of what leads the average person to touch their phone 2,617 times each day. In short, that red button provides the good news of the company's value to hand to Facebook shareholders at their annual meeting. Although Facebook engineers were essentially putting a drug in users' pockets, it wasn't their — or anyone's — job to consider the potential side effects.

This single-minded push for growth was best illustrated by an internal memo circulated among Facebook leadership in the summer of 2016, at the very same time Russian propagandists were running rampant on the service and the Trump campaign was poring over tens of millions of ill-gotten Facebook profiles. "We connect people. Period," a senior Facebook vice president wrote. "That's why all the work we do in growth is justified. All the questionable contact importing practices. All the subtle language that helps people stay searchable by friends. All of the work we do to bring more communication in. The work we will likely have to do in China some day. All of it . . . Maybe it costs someone a life by exposing someone to bullies. Maybe someone dies in a terrorist attack coordinated on our tools."

This Silicon Valley myopia would matter less if services like Facebook, YouTube, and Twitter were simply inventions, like the telegraph or radio, which might be repurposed by other tech pioneers while the original creators enjoyed fat royalty checks. But they're not. These companies aren't inventions but *platforms* — services that deliver the most value to users who visit them frequently, often addictively.

A good parallel to the present-day titans isn't Samuel Morse, tinkering away in his workshop to build one of several competing telegraph services. Instead, it's Alexander Graham Bell, whose Bell Telephone Company (later AT&T) would control virtually all telephone wire in the United States, under the motto "One Policy, One System, Universal Service."

Scale allows the most successful Silicon Valley entrepreneurs to rule like absolute sovereigns over their platforms — and, by extension, over everyone who relies on them. If Mark Zuckerberg authorizes a small tweak to Facebook's design — like replacing comment boxes with bubbles — the change will be seen by more than 2 billion users, immediately ranking it among the largest collective experiences in human history. In turn, tiny, imperceptible shifts in the newsfeed algorithm can turn previously niche media outlets into hulking behemoths while wrecking the fortunes of others. It can even, as we've seen, alter the course of American elections and of wars in the Middle East.

In some ways, we're lucky that these mighty figures have chosen to rule their empires like benign and boring figureheads. They're almost universally progressive, pledging themselves to the cause of social justice, but also striving to be militantly inoffensive in their public statements. They've established rules and regulations that seek to mirror or exceed the permissive speech codes of the United States. "In the process of building private communities," writes John Herrman, "these companies [put] on the costumes of liberal democracies." In so doing, however, they've avoided reckoning with the extent of their own burgeoning influence and power.

Part of the reason is an inherent contradiction. For all the talk of "community," these platforms are businesses. Their "boss" isn't the United Nations or even their user base; it's shareholders. At the end of the day, the metrics that matter most aren't the number of violent crimes averted or the number of humans shielded from harm; they are stock price and year-over-year revenue. In turn, for all the transparency these companies have forced upon the world, their most important decisions still originate in closed corporate boardrooms.

There's also the technocratic, optimistic worldview that comes from companies staffed mostly by engineers working hard to build products. "You're so focused on building good stuff," explained Mike Hoefflinger, a former Facebook executive and author of *Becoming Facebook,* "you're not sitting there thinking, 'If we get lucky enough to build this thing and get two and a quarter billion people to use it, then this other bad stuff could happen.'"

Finally, the companies are haunted by a very real, deep cultural tension. Most of the people who build and maintain these politically pivotal platforms don't particularly *like* politics. This ethos was on clear display at Hacker News, a popular Silicon Valley forum board, when "Political Detox Week" was announced shortly after the 2016 U.S. presidential election. As the rest of the country struggled to come to terms with the election results, an administrator declared:

> Political stories are off-topic. Please flag them. Please also flag political threads on non-political stories. For our part, we'll kill such stories and threads when we see them. Then we'll watch together to see what happens.
>
> Why? Political conflicts cause harm here. The values of Hacker News are intellectual curiosity and thoughtful conversation. Those things are lost when political emotions seize control. Our values are fragile — they're like plants that get forgotten, then trampled and scorched in combat. HN is a garden, politics is war by other means, and war and gardening don't mix.

In other words, this was a forum for "intellectual curiosity and thoughtful conversation," and to this community of tech geeks, politics was by definition the opposite of that. Thus, at the very moment the forum members' work had been shown to be remaking the political landscape, they felt they could just tune it out.

This "engineering first" mentality applies to both problems and potential solutions. Whenever these companies have had to reckon with a political, social, or ethical dilemma — ironically spawned by their platforms' very success — they often grasp for another new technology to solve it. As a senior executive at one of these companies put it to us, "If we could use code to solve all the world's problems, we would."

But for all the reasons outlined in this book, the excuses have begun to fall short. It is one thing to run a social media service whose biggest headaches are copyright infringement and naughty pictures; it is quite another when that same service is being used to abet terrorism, stoke racism, and shatter entire political systems. When Mark Zuckerberg entertains pleas from Ukrainian activists to break a Russian "informational blockade" or dispatches Facebook engineers to "ensure the integrity of the German elections," he is no longer simply the custodian of a neutral platform. When he vows that his company will devote itself to "spreading prosperity and freedom" or "promoting peace and understanding," he is no longer simply a tech CEO. He is a new kind of leader, who has begun moving, reluctantly, to claim his position on the world stage.

Ultimately, the greatest challenge that confronts these social media giants has nothing to do with software code. It is a problem of corporate incentives, clashing cultures, and a historic revolution that has left both politics and Silicon Valley reeling. It is a problem of bestowing carefree engineers uninterested in politics with grave, nation-sized political responsibilities.

And although this is a problem with endless dimensions, at its heart it has always revolved around the same three questions: *Should* these companies restrict the information that passes through their servers? *What* should they restrict? And — most important for the future of both social media and the world — *how* should they do it?

Naturally, it's another story that begins with the desire to see someone's breasts.

THE DIRTY ORIGINS OF DIGITAL FREEDOM

"Marketing Pornography on the Information Superhighway: A Survey of 917,410 Images, Descriptions, Short Stories, and Animations Downloaded 8.5 Million Times by Consumers in over 2000 Cities in Forty Countries, Provinces, and Territories."

When it was published in the *Georgetown Law Journal* in June 1995, Marty Rimm's remarkably titled study became an overnight sensation. Rimm concluded that more than four-fifths of the nascent internet consisted of pornography, which the author claimed to have exhaustively cataloged. The finding was reported in every major newspaper, debated across TV and talk radio, and featured on the cover of *Time* magazine with a picture of a shocked toddler looking at a computer screen emblazoned with the new word "CYBERPORN."

In many ways, it's fitting that this instantly viral report was also a total fabrication. Marty Rimm was an attention-seeking Carnegie Mellon University student who'd gotten his work published by dodging peer review. Rimm did have another publication under his belt, *The Pornographer's Handbook: How to Exploit Women, Dupe Men, and Make Lots of Money*, which suggested he wasn't being entirely sincere in his statements about the pornographic menace. As his work came under scrutiny and his claims that the internet was almost completely porn were debunked, Rimm vanished and eventually changed his name.

Nonetheless, the damage had been done. While 1995 was a banner year for the internet, as the U.S. government officially relinquished control and millions of new users jumped online, in Congress this excitement gave way to moral panic. Thanks to one viral story and an audience of mostly tech-illiterate legislators, the internet now meant only one thing: *pornography*.

"The information superhighway should not become a red-light district!" bellowed Senator James Exon, an elderly Democrat from Nebraska. He introduced the Communications Decency Act (CDA), which made it a crime to "send to or display in a manner available to" those under 18 years of age any communication that depicted "sexual or excretory activities . . . regardless of whether the user of such service placed the call or initiated the communication." The punishment was two years in prison and a fine of $100,000. In 1996, the law passed with overwhelming bipartisan support.

The final law had one crucial tweak, however. Two younger U.S. representatives — Chris Cox, a Republican from California, and Ron Wyden, a Democrat from Oregon — realized that unless something was done to protect websites that tried to police their own networks, the entire internet would be paralyzed by fear of lawsuits and prison time. Their consequent amendment became 47 U.S.C. § 230 (1996), known as Section 230. It was, in the words of *Wired* magazine, "the most important law in tech."

Section 230 provided "protection for 'Good Samaritan' blocking and screening of offensive material," essentially ruling that no website could be held accountable for the speech of its users. And no website that made a "good-faith" effort to enforce applicable U.S. regulations could be punished, even if its efforts fell short. It was an amazingly permissive statute buried in one of the strictest "obscenity" laws ever passed by Congress.

It was fortunate, then, that before the ink on the CDA had dried, the Supreme Court struck it down. *Reno v. American Civil Liberties Union* (1997) was the first and most important Supreme Court case to involve the internet. In a unanimous decision, the justices basically laughed the CDA out the door, noting that it massively violated the First Amendment. The only part of the CDA that survived was Section 230. Over the ensuing years, it would be consistently challenged and upheld. With each successful defense, its legal standing grew stronger. Outside of two specific exemptions (federal criminal law and intellectual property), the internet was mostly left to govern itself. As a result, most early corporate censorship — more politely known as "content moderation" — would come not because of government mandate, but to avoid involving government in the first place.

As ever, money was the driver. Over the next decade, questions regarding what constituted permissible online speech centered not on politics or propriety, but on property. Blogger (eventually acquired by Google), for example, was an early self-publishing hub that enabled millions of users to set up websites free of charge. And yet a visitor to Blogger's home page in 1999 would find no list of rules, only a friendly reminder to properly configure your URL so you could be added to the master blog roll. Some blogs might host racist rants and pornography, the thinking went, but so what? This wasn't a problem — it was the whole reason services like Blogger existed: to share the panoply of human expression, emotions, and beliefs.

By contrast, violations of intellectual property rights were covered not by the permissive Section 230, but by the much stricter 1998 Digital Millennium Copyright Act (DMCA). This law imposed a maximum prison sentence of five years or a fine of $500,000 for the *first* offense of posting material for which someone else held a copyright. Fortunately, much like Section 230, the law also included a "safe harbor" provision.

If websites promptly responded to a takedown notice filed by the copyright holder — without pausing to consider the merits of the request — they could avoid being sued or jailed.

Ground zero for the copyright battle was YouTube, whose nature made it an irresistible target for hosting copyrighted songs or videos. In 2006, YouTube placed a ten-minute limit on its videos, reasoning that longer clips were likely to be illegally hosted TV shows or movies. A year later, notably following Google's $1.7 billion acquisition of the company, YouTube introduced its "Content ID" system, which assigned a unique digital fingerprint to tens of millions of copyrighted files. If Content ID spotted a match on YouTube's servers, the file was automatically flagged for removal. This was the first use of sophisticated, wide-scale automation to control user content on a U.S. website. It was a sign of things to come.

In another sign of things to come, the automated system went too far, disabling videos that contained even an incidental glimpse of copyrighted material. Just a few wayward seconds — like Katy Perry's "I Kissed a Girl" playing in the background of a video shot in a crowded bar — was enough to nuke a whole clip. In 2008, Republican presidential candidate John McCain complained that his political ads were being automatically removed because they contained brief clips from broadcast news. Digital rights activists gleefully reminded McCain that he'd voted the DMCA into law a decade earlier.

Happily, a small reprieve from copyright laws would arrive later that year, following an epic legal battle between the artist formerly known as Prince and a 13-month-old baby. The baby had been marked as a "copyright violator" after his mother uploaded a video of him pushing a toy stroller and laughing as Prince's "Let's Go Crazy" played for precisely twenty seconds in the background. The judicial ruling found, essentially, that the whole thing was ridiculous and that internet users had a right to plead "fair use" before their content was eradicated. YouTube's copyright-sniffing algorithm was allowed to relax — but not much.

Even as they strengthened their copyright controls, however, emerging social media titans confronted a far more horrifying problem: child pornography. Under Section 230, websites enjoyed broad legal immunity against charges of child abuse or exploitation. But use of a platform by child porn distributors wasn't just a legal problem; it was a moral imperative, one whose mere mention could sink a company's reputation. In 2009, Microsoft announced a free service called PhotoDNA. Applying a system much like that of Content ID, PhotoDNA compared each posted image and video with a massive database and instantly flagged any matches. Every major social media platform would eventually adopt the tool, essentially eliminating child pornography from their networks. Today, this top-secret, U.S. government–sanctioned database hosts more than a million instances of child pornography.

Other than addressing copyright infringement and child porn, by the mid-2000s Silicon Valley companies still did little to regulate user speech, clinging to the laissez-faire principles of the first generation of internet pioneers. But as the internet's population passed 1 billion, that age had clearly ended. Social media platforms were clogged with eager users, including half of all American teenagers. Flush with hormonal drama and anguish, the vast digital commons increasingly resembled a powder keg. All it needed was a spark.

That spark was provided in 2006, when the handsome, athletic, 16-year-old Josh Evans joined Myspace, the then-dominant social network. He liked the bands Rascal Flatts and Nickelback; among his "turn-ons" were tongue piercings and ear nibbling. He'd lived a hard life, born to a single mother who bounced between jobs. Nonetheless, Josh was upbeat. His goal, he confided, was to "meet a great girl."

Sadly, Josh had one flaw: he wasn't real. He was a hoax — in the words of journalist Lauren Collins, "an online Frankenstein's monster." He was a sockpuppet built to exploit the hopes and vulnerabilities of one teenage girl.

That target was 13-year-old Megan Meier, who — like most teenagers — maintained roller-coaster relationships with her peers. As she entered eighth grade, she fell out with a friend who lived just four houses away. That friend's mother, 47-year-old Lori Drew, created Josh to spy on Meier, to see if the girl was saying mean things about her daughter. Drew solicited help from a 19-year-old employee of her small business and two other teenagers to run the fake account. Josh began a warm, flirtatious online friendship with Meier. The attractive boy always seemed to know exactly what to say to make Meier happy. Indeed, how could he not? His creator knew Meier well; in a happier time, she had even joined the Drew family on vacations.

The ruse soon turned to tragedy. After Meier got into an angry on-line argument with other classmates, Josh abruptly turned on her, tak-ing the other kids' side and peppering her with insults. In shock, Meier fled from the family computer and retreated to her room. When her mother checked on her a short time later, the 13-year-old was dead, hanging from an Old Navy belt. Her bereaved father uncovered the last message sent by Josh: "You're a shitty person, and the world would be a better place without you in it."

The story swelled into a major scandal. Drew and her accomplices were tried for conspiracy, convicted, but then acquitted. Their actions were simply too new to have clearly violated any existing laws. That soon changed. As outrage spread and more reports emerged of online harassment, dozens of states enacted "cyberbullying" laws. For the first time, many Americans were forced to reckon with the potentially deadly offline consequences of online actions.

For social media platforms, the death of Megan Meier was also a wake-up call. Myspace had never been in serious legal jeopardy. Indeed, Myspace was technically a *victim*, listed alongside Megan in the high-profile court case, because Drew had falsely represented herself to the company. But in the court of public opinion, the world's largest social network faced a public relations disaster. If Myspace had done some-thing, anything, might that young girl still be alive?

For Myspace, Megan's death was a setback from which it never re-covered. For Silicon Valley at large, it was a sign that "terms of service" could no longer be a simple check box to placate jumpy investors. These user agreements had to become a new kind of law, penned by private corporations instead of governments, to administer communities of un-precedented scale. To be effective, these terms would have to be regu-larly monitored and updated, pulling engineers ever closer to the task of "regulating" free speech. To do otherwise — to leave hundreds of mil-lions of users to say or do whatever they pleased — risked the govern-ment jumping in and passing ever more stringent legislation. In other words, to allow truly *free* speech would be financially ruinous.

Among the companies destined to rule the new social web — Twitter, Google, and Facebook — a shaky consensus emerged. All three banned personal threats or intimidation, soon expanding to include a more general ban on harassment. All but the free-spirited Twitter banned graphic violence and pornography, which included taking a hard-line stance against nudity. These rules seemed simple. They'd prove to be anything but.

For Twitter, staffed by many of the same members of the idealistic team that had founded the original Blogger, the goal was to create a free-flowing platform, a libertarian ideal. "We do not actively monitor and will not censor user content, except in limited circumstances," declared Twitter's longstanding mission statement. A Twitter executive proudly described the company as "the free speech wing of the free speech party," reminding all that it was the place to launch protests and topple dictators. By design, Twitter accounts were anonymous and essentially untraceable. All accounts could speak to all other accounts, unfiltered.

The free speech haven became a perfect platform for rapid news distribution, but also a troll's paradise. A graphic, personal threat wasn't allowed, but anything short of that — like telling a Jewish user what would *hypothetically* happen to them in a second Holocaust — was fair game. The worst fate that could befall a Twitter user was an account ban, but as one neo-Nazi derisively pointed out to us, it took mere seconds to create a new one. As a result, the free speech haven became, in the words of one former employee, a "honeypot for assholes."

The first case of sustained harassment on Twitter occurred in 2008, as a female tech blogger endured months of insults, threats, and stalking from a network of anonymous accounts. "Twitter is a communication utility, not a mediator of content," one founder coolly replied, as the backlash grew. For years, despite the paltry protection of its terms of service, Twitter would remain a brutally hostile place for women and nonwhite users. It would take until 2013 — amid a massive, sustained harassment campaign against female members of the British Parliament — for Twitter to introduce a way for users to directly report abusive tweets.

A year later came "Gamergate," an absurd scandal that began over the complaints of an obsessive ex-boyfriend and protests over "ethics in gaming journalism." It ended with literally millions of abusive tweets hurled at a handful of female video game developers, the effective birth of the alt-right political movement, and an inquest by the United Nations. "Freedom of expression means little as our underlying philosophy if we continue to allow voices to be silenced because they are afraid to speak up," Twitter's new general counsel concluded. By the end of 2015, the company's promise not to censor user content had vanished from its mission statement.

Ironically, YouTube's more restrictive terms of service ("Don't cross the line," it said in big bold letters) led it almost immediately into thorny political questions. The platform banned "unlawful, obscene [or] threatening" videos. But this content proved difficult to define and regulate. The first challenge came in 2007, when Mexican drug cartels flooded the service with music videos that featured the mutilated bodies of their enemies, intended to boost recruitment. YouTube tried to remove the videos as it discovered them. That seemed easy enough. However, the same year, YouTube also deleted the videos of an Egyptian anti-torture activist, whose work necessarily documented torture in order to combat it. Following an angry backlash from human rights groups, those videos were reinstated. In 2008, YouTube removed video of an air strike on a dozen Hamas fighters, whereupon the Israeli Defense Forces complained about the loss of its "exclusive footage."

But these relatively simple approaches by Twitter (avoid intervention) and YouTube (content bans) hardly registered next to the complex content moderation policies that emerged at Facebook, the company born from a website comparing the hotness of college coeds. From the beginning, Facebook had been gunning for Myspace, so it wanted to avoid as many Myspace-style scandals as possible. Facebook's internal guidebooks soon came to resemble the constitutional law of a mid-sized nation. In 2009, its first attempt to codify its "abuse standards" ran to 15,000 words.

Each new rule required more precise, often absurd clarification. It was one thing for Facebook to ban "incitement of violence"; it was quite another to say what that meant. If a user pleaded for someone to shoot the U.S. president, it was a clear incitement and could be deleted. But if a user urged others to "kick a person with red hair," it was a more general threat and therefore allowable. Indeed, leaked slides of Facebook policy gave a horrifying example of the nuance. The message "Unless you stop bitching I'll have to cut your tongue out" was permissible because the threats were *conditional* instead of *certain*.

Even a seemingly black-and-white rule — like a blanket ban on "nudity and sexual activity" — sowed a minefield of controversy. First came the protests of historians and art critics, who pushed Facebook to allow nudity in photographs of paintings or sculptures but not in *digital* art, which the classicists considered pornography. Then came the protests of new mothers, furious that their images of breastfeeding were being deleted on the grounds of "obscenity." They launched a mommy-sourced lobbying campaign, coining their own hashtag, #freethenipple (which was, unsurprisingly, hijacked by porn distributors). These nipple wars led Facebook into years of heated, internal deliberation. Eventually, Facebook's senior leaders settled on a new policy that permitted portrayals of breastfeeding, but only so long as the nipple was not the *principal* focus of the image.

The engineers who had built the world's largest digital platform — which raked in billions of dollars in revenue and shaped news around the world — had neither expected nor wanted to spend hundreds of hours in corporate boardrooms debating the spectrum of nipple visibility. But they did. With great power came great and increasingly expansive responsibilities.

Then came global politics. Originally, a firm could escape onerous censorship requests by arguing that it was a U.S. company subject to U.S. laws. By the early 2010s, this was no longer a realistic defense. These companies had become multinational giants, grappling with regulations across dozens of national jurisdictions. As the scope of Silicon Valley's ambition became truly international, its commitment to free speech sagged. In 2012, both Blogger (originally marketed as "Push-Button Publishing for the People") and Twitter ("the free speech wing of the free speech party") quietly introduced features that allowed governments to submit censorship requests on a per-country basis.

If there was a moment that signified the end of Silicon Valley as an explicitly *American* institution, it came in 2013, when a young defense contractor named Edward Snowden boarded a Hong Kong–bound plane with tens of thousands of top-secret digitized documents. The "Snowden Files," which would be broadcast through social media, revealed an expansive U.S. spy operation that harvested the metadata of every major social media platform except Twitter. For primarily U.S.-based engineers, this was an extraordinarily invasive breach of trust. As a result, Google, Facebook, and Twitter began publishing "transparency reports" that detailed the number of censorship and surveillance requests from *every* nation, including the United States. "After Snowden," explained Scott Carpenter, director of Google's internal think tank, "[Google] does not think of itself all the time as an American company, but a global company."

From this point on, social media platforms would be governed by no rules but their own: a mix of remarkably permissive (regarding threats and images of graphic violence) and remarkably conservative (regarding nudity). In essence, a handful of Silicon Valley engineers were trying to codify and enforce a single set of standards for every nation in the world, all in an attempt to *avoid* scandal and controversy. As any political scientist could have told them, this effort was doomed to fail.

DOWN THE SLIPPERY SLOPE

"Literally, I woke up in a bad mood and decided someone shouldn't be allowed on the internet. No one should have that power."

It was August 2017 and Matthew Prince, cofounder and CEO of the Cloudflare web hosting service, had just made a decision he'd spent a decade dreading. Cloudflare had been built to protect websites from cyberattacks, the kind that often happened when someone attracted too much negative attention. Thanks to Cloudflare, dissidents around the world were shielded from unfriendly hackers. But so, too, were the internet's most abhorrent voices.

For years, Stormfront, a neo-Nazi forum board, had relied on Cloudflare to keep its servers running. Now, in the aftermath of a deadly white nationalist terror attack in Charlottesville, Virginia, Stormfront users were openly celebrating the murder. As outrage against Stormfront and its media outlet, The Daily Stormer, intensified, the anger also targeted Cloudflare, whose technology was keeping the neo-Nazis online. The company meekly explained that it couldn't revoke Stormfront's accounts without "censoring the internet" — a position that led Stormfront to brag that Cloudflare was on its side. Seeing this, a furious Prince abruptly reversed course and pulled the plug. He explained his change of heart in an email to his staff:

This was my decision. Our terms of service reserve the right for us to terminate users of our network at our sole discretion. My rationale for making this decision was simple: the people behind the Daily Stormer are assholes and I'd had enough.

Let me be clear: this was an arbitrary decision. It was different than what I'd talked with our senior team about yesterday. I woke up this morning in a bad mood and decided to kick them off the internet . . . It's important that what we did today not set a precedent.

Saying that something *shouldn't* set a precedent doesn't stop it from doing so. This was a landmark moment. An ostensibly "content-neutral" company had made a decision to destroy content — a decision that was obviously *not* neutral. And it had happened because a single person at the top had changed his mind. But at least he was being transparent about it.

Prince's decision echoed the dilemma that increasingly fell to social media's ruling class. Faced with vocal campaigns to censor or delete speech, the companies could either ignore their users, risking a PR disaster, or comply and be drawn deeper into the political brush. In essence, in avoiding governance, these companies had become governments unto themselves. And like any government, they now grappled with intractable political problems — the kind always destined to leave a portion of their constituents displeased.

Yet they also had little choice. Nations around the world had gradually awoken to the influence that these U.S. social media giants exerted over domestic politics. Between 2012 and 2017, some fifty countries passed laws that restricted the online speech of their citizens. And these weren't just the wannabe authoritarians discussed in chapter 4; they were also some of the most liberal nations in the world, fearful of terrorism, extremism, or even simply "fake news." Even in the United States, a new generation of tech-savvy politicians hovered, ready to slap these companies with onerous new government regulations if they didn't tighten the rules on their own.

No longer was it enough to police copyright infringements, naughty pictures, and the most obvious forms of harassment. Now Silicon Valley firms would be pushed ever closer to the role of traditional media companies, making editorial decisions about which content they would allow on their platforms. Many engineers argued that this was a "slip-pery slope." But their founders' ingenuity and the internet's exponential growth had placed them in this treacherous terrain. It was now their task to navigate it.

The first and most obvious challenge was terrorism. Very early on, Al Qaeda and its copycats had begun to post their propaganda on You-Tube. This included grisly recordings of snipers killing U.S. soldiers in Iraq. Although YouTube technically prohibited graphic violence, it was slow to remove the clips, while the American public was quick to vent its fury.

But the challenge proved even starker with the first internet-inspired terror attacks. The same year YouTube was created, an American-born Islamic cleric named Anwar al-Awlaki became radicalized and moved to Yemen. Charismatic and English-speaking, he began uploading his Quranic lectures to the platform, accumulating millions of views across a 700-video library. Although there was no explicit violence portrayed in the clips of the soft-spoken, bespectacled al-Awlaki, his words *promoted* violence. And they were incredibly effective, inspiring dozens of deadly attacks around the world, such as the 2009 shooting at Fort Hood, Texas, that claimed thirteen lives.

Moreover, the YouTube algorithm exacerbated the threat by creat-ing a playlist of "related videos" for its viewers. In al-Awlaki's case, this meant the platform was helpfully steering viewers of his videos to vid-eos by other terrorist propagandists.

By 2011, the U.S. government had had enough, and al-Awlaki was sen-tenced to death in absentia by an Obama administration legal memo stating that his online propaganda "posed a continuing and imminent threat of violent attack." Soon after, he was slain by a U.S. drone strike. On YouTube, however, al-Awlaki's archive became something else: a digital shrine to a martyr. In death, al-Awlaki's online voice grew even more popular, and the U.S. intelligence community began noticing an uptick in views of his videos that accompanied spikes in terrorist at-tacks. This illustrated another conundrum. The government had done what it could to silence the "bin Laden of the internet," but it was up to the engineers at YouTube to determine the terrorist's future influ-ence. It would take the company another six years, until 2017, to decide to block the videos.

Yet it was Twitter, not YouTube, that became terrorists' main social media haven. In a horrifying irony, terrorists who wanted to destroy freedom of speech found perfect alignment with Twitter's original commitment to freedom of speech. The only line a terrorist couldn't cross was personal harassment. You could tweet, generally, about how all *"kuffar"* (non-Muslims) deserved a violent death; you just couldn't tell @hockeyfan123 that you were going to cut off his head. Although many voiced frustration that terrorists were allowed on the platform, Twitter brushed off their complaints. If the NATO coalition could tell its side of the story in Afghanistan, the thinking went, why not the Taliban? For aspiring terror groups, Twitter then became not just the space to connect with followers, but also the perfect platform to build brand recognition among both recruits and Western journalists.

But then came headlines Twitter couldn't ignore. In 2013, four gunmen stormed Nairobi's Westgate shopping mall, murdering 67 people and wounding nearly 200 more. The attackers belonged to Al-Shabaab, an East African terror organization whose members had been early and obsessive Twitter adopters. Shabaab applied digital marketing savvy to the attack, pumping out a stream of tweets, press releases, and even exclusive photos (snapped by the gunmen themselves). "#Westgate: a 14-hour standoff relayed in 1400 rounds of bullets and 140 characters of vengeance," summarized one terrorist's post. Soon Shabaab became the main source for international journalists writing about the attack — a position the group used to spread misinformation and confuse the situation on the ground even more. Reeling from bad press, Twitter intervened in a way it had been unwilling to do just a few years earlier. It suspended the terrorists' account. It didn't matter; Shabaab simply registered new ones.

And then, in 2014, the Islamic State roared onto the global stage, seizing hold of the internet's imagination like a vise. At its peak, the ISIS propaganda machine would span at least 70,000 Twitter accounts, a chaotic mix of professional media operatives, fanboys, sockpuppets, and bots. As ISIS propaganda seeped across the platform in more than a dozen languages, Twitter executives were caught flat-footed. Their content moderation team simply wasn't equipped to deal with the wholesale weaponization of their service. This wasn't just for lack of interest, but also for lack of resources. Every employee hour spent policing the network was an hour not spent growing the network and demonstrating investor value. Was the purpose of the company fighting *against* propaganda or *for* profitability?

Meanwhile, public outrage mounted. In 2015, Congress edged as close as it had in a decade to regulating social media companies, drafting a bill that would have required the disclosure of any "terrorist activity" discovered on their platforms (the definition of "terrorist activity" was kept intentionally vague). The same year, then-candidate Donald Trump seemed to endorse the internet censorship and balkanization practiced by authoritarian nations. "We have to talk to [tech CEOs] about, maybe in certain areas, closing that internet up in some ways," he declared. "Somebody will say, 'Oh freedom of speech, freedom of speech.' Those are foolish people."

Twitter tried to act, but ISIS clung to it like a cancer. Militants developed scripts that automatically regenerated their network when a connection was severed. They made use of Twitter blocklists — originally developed to fight harassment by bunching together and blocking notorious trolls — to hide their online activities from users who hunted them. (ISIS media teams soon added us to this list.) Some accounts were destroyed and resurrected literally hundreds of times, often with a version number (e.g., @TurMedia335). When the rather obvious Twitter handle @IslamicState hit its hundredth iteration, it celebrated by posting an image of a birthday cake. Nonetheless, the growing suspensions changed the once-free terrain for ISIS. "Twitter has become a battlefield!" lamented one ISIS account by mid-2015.

Thanks to diligent volunteer efforts, steady improvements to Twitter's systems, and unrelenting public pressure, ISIS's use of the platform gradually declined. In 2017, Twitter announced that its internal systems were detecting 95 percent of "troubling" terrorist accounts on its own and deleting three-quarters of them before they made their first tweet. It was a remarkable achievement — and an extraordinary turnabout from Twitter's laissez-faire approach of just a few years before.

Although Twitter's transformation was the most dramatic, the other Silicon Valley giants charted a similar path. In 2016, Google piloted a program that used the advertising space of certain Google searches (e.g., "How do I join ISIS?") to redirect users to anti-ISIS YouTube videos, carefully curated by a team of Google counter-extremism specialists. It spoke to the seriousness with which Google was starting to treat the problem. Meanwhile, Facebook built a 150-person counterterrorism force to coordinate its response effort, comprised of academics and former intelligence and law enforcement officers.

At the end of 2016, Facebook, Microsoft, Twitter, and Google circled back to where online censorship had begun. Emulating the success of Content ID and PhotoDNA, which had curbed copyright violations and child porn respectively, the companies now applied the same automated technique to terrorist propaganda, jointly launching a database for "violent terrorist imagery." Just a few years before, they had insisted that such a system was impossible, that the definition of "terrorism" was too subjective to ever be defined by a program. This was another sign of how decisively the political landscape had shifted.

Yet no matter how much these social media companies evolved, there were always outside forces pressuring them to do more. In 2015, Facebook was sued for $1 billion by relatives of Americans who had been killed during a spate of lone wolf terror attacks in Gaza. The tech firm was accused of having "knowingly provided material support" to the terrorists, simply by giving them the means to transmit their propaganda. Around the same time, 20,000 Israelis brought suit against Facebook not just for the violence they'd suffered, but for future violence they feared they *might* suffer. "Facebook and Twitter have become more powerful today . . . than the [UN] secretary-general, the prime minister of Israel, and the president of the United States," declared one of the plaintiffs, whose father had been murdered in a Palestinian terror attack. Although the lawsuits were eventually dismissed, each new terror attack prompted further lawsuits by victims. The legal protections granted by Section 230 — originally meant to police pornography — had now been pushed to the limit.

Meanwhile, the precedent set by Silicon Valley's well-publicized purge of ISIS accounts steered it toward other, even more painfully ambiguous political challenges. By 2015, ultranationalists, white supremacists, and anti-immigrant and anti-Islamic bigots had begun to coalesce into the alt-right movement. Feeling emboldened, they increasingly took their hatred into the open.

But they were sly about it. They shrouded their sentiments in memes and coy references; they danced to the very edge of the line without crossing it. The alt-right leader Richard Spencer, for instance, didn't use his popular (and verified) Twitter profile to directly champion the killing of all Jews and blacks; instead, he simply observed how much nicer things would be if America were made white and pure. The extremists toyed with new ways of targeting people with anti-Semitic harassment. As an example, the last name of someone known or thought to be Jewish would be surrounded by triple parentheses, so that "Smith" became "(((Smith)))." Such tactics made it easier for Gamergate-style hordes to find their targets online and bury them with slurs and abuse. If challenged, they claimed that they were "just trolling." If their user accounts were threatened, they'd flip to play the victim, claiming that they were being targeted for practicing "free speech." It represented both a twist on Russian tactics and a deft use of the same language that companies like Twitter had invoked for so many years.

For a time, Google, Facebook, and Twitter essentially threw up their hands and looked the other way. Racism and bigotry were distasteful, the companies readily admitted, but censoring distasteful things wasn't their job. They also lay within the political spectrum of American politics — at the extreme, to be sure, but gradually becoming mainstream, pushed by their technology. Plus, the tactics of these extremists — winking and nudging, dog-whistling and implying — were often too subtle for any terms of service to adequately address.

But as Silicon Valley cranked up the pressure on terrorists and their supporters, it became easier to contemplate the next step: moving to combat a more general kind of "extremism" that evaded labels, but whose victims — women and ethnic and religious minorities — were easy to name. In mid-2016, Twitter fired the first salvo, kicking the Breitbart writer and far-right provocateur Milo Yiannopoulos out of its service. Having won fame with his race-baiting, Yiannopoulos had finally crossed the line when he organized a campaign of online harassment targeting an African American actress for the crime of daring to star in a *Ghostbusters* remake.

While Yiannopoulos would insist that he'd been wrongly smeared as a bigot — that he'd "just been trolling" — the evidence suggested otherwise. A year later, when a trove of Yiannopoulos's files leaked online, it was revealed that he used email passwords like "Kristallnacht" (a November 1938 attack on German Jews in which dozens were murdered) and "LongKnives1290" (a reference to both the Night of the Long Knives, a 1934 Nazi purge that solidified Hitler's rule, and the year in which Jews were banished from medieval England).

Following a spate of more than 700 hate crimes across the country after the election of Donald Trump in November 2016, pressure began to build on the social media giants to do more about the hate that was not just allowed but empowered by their platforms, especially as it spurred violence. The crackdown started with the long-overdue Twitter suspension of white supremacist leader Richard Spencer. He issued a dramatic rebuttal on YouTube titled "The Knight of Long Knives." "I am alive physically," he explained to his followers, "but digitally speaking, there has [*sic*] been execution squads across the alt-right . . . There is a great purge going on."

But this digital purge was actually only a time-out. Confident and mobilized in a way that hate groups had not been since the mass KKK rallies of the 1920s, the alt-right used social media to organize a series of "free speech" events around the nation, culminating in that infamous Charlottesville rally. "As you can see, we're stepping off the internet in a big way," one white supremacist bragged to a reporter as the air was suffused with neo-Nazi chants. "We have been spreading our memes, we've been organizing on the internet, and so now [we're] coming out."

Amid the national outcry that followed, the social media giants moved to expand their definition of "hate speech" and banish the worst offenders from their services. Twitter banned the most virulent white supremacist accounts, while Facebook removed pages that explicitly promoted violent white nationalism. Reddit rewrote its terms of service to effectively outlaw neo-Nazi and alt-right communities. White supremacists even found themselves banned from the room-sharing service Airbnb and the dating site OkCupid. In 2019, Facebook announced that it was applying the same standard to "white nationalism" that it did to the extremism of groups like ISIS (though once again belatedly, changing its policy only after a series of mass killings at sites ranging from a Pittsburgh synagogue to a New Zealand mosque).

This was a massive shift for an industry barely over a decade old. Since their founding, social media companies had stuck by the belief that their services were essentially a "marketplace of ideas," one in which those that came to dominate public discourse would naturally be the most virtuous and rational.

But Silicon Valley had lost the faith. Social media no longer seemed a freewheeling platform where the best ideas rose to the surface. Even naïve engineers had begun to recognize that it was a battlefield, one with real-world consequences and on which only the losers played fair. Their politics-free "garden" had nurtured violence and extremism.

The trouble went deeper than the specters of terrorism and far-right extremism, however. Silicon Valley was beginning to awaken to another, more fundamental challenge. This was a growing realization that all the doomsaying about homophily, filter bubbles, and echo chambers had been accurate. In crucial ways, virality *did* shape reality. And a handful of tech CEOs stood at the controls of this reality-shaping machine — but they hadn't been working those controls properly.

It was the election of Donald Trump that drove this realization home. Most deeply impacted was Facebook, whose mostly young and progressive employees feared that their work had elevated Trump to high office. Indeed, there was strong evidence that it had. Although Twitter had served as Trump's treasured microphone, it was Facebook in which Americans had been at their most politically vulnerable, trapped in networks of people who thought just like them and who accorded hundreds of millions of "shares" to stories that were obvious hoaxes. Indeed, the whispers were already beginning that Facebook had been saturated not just with profit-motivated misinformation and "fake news" spun by Macedonian teenagers, but also with a pro-Trump disinformation campaign orchestrated by the Russian government.

As stupefied liberals searched for someone or something to blame, Mark Zuckerberg could see the tidal wave coming. What followed was essentially a corporate version of psychiatrist Elisabeth Kübler-Ross's five stages of grief: denial, anger, bargaining, depression, and acceptance.

Zuckerberg's first impulse was to deny. It was "a pretty crazy idea," he said a few days after the election, that misinformation on his platform had influenced the outcome of anyone's vote. After his initial denial was met with widespread fury and even a private scolding from President Obama, Zuckerberg shifted gears, penning a series of notes in which he vowed to try harder to counter hoaxes and misinformation on Facebook. At the same time, he tried to reassure users that this was a comparatively small problem. Meanwhile, frustrated Facebook employees began meeting in private to crowdsource solutions of their own. The truth then came out that some at the company *had* been concerned about rampant misinformation taking place on their platforms during the election, but had been prevented from making any changes for fear of violating Facebook's "objectivity," as well as alienating conservative users and legislators.

By mid-2017, Facebook had struck a very different note. In the first report of its kind, Facebook's security team published "Information Operations and Facebook," a document explaining how its platform had fallen prey to "subtle and insidious forms of misuse." In another first, Facebook publicly named its adversary: the government of the Russian Federation. Critics noted, however, that the company had waited a crucial nine months between when its executives knew that a massive campaign of Russian manipulation had occurred on its networks and when it informed its customers and American voters about it.

Reflecting its ability to implement change when so motivated, however, Facebook expanded its cybersecurity efforts beyond regular hacking, turning its focus to the threat of organized disinformation campaigns. Where the company had studiously ignored the effects of disinformation during the 2016 U.S. election, it now cooperated closely with the French and German governments to safeguard their electoral processes, shutting down tens of thousands of suspicious accounts. Facebook even launched a specially built "war room" to monitor the 2018 U.S. midterm elections, staffed with dozens of political specialists. A year after calling the idea of electoral influence "crazy," Zuckerberg apologized for having ever said it. And in a very different speech, delivered via Facebook Live, Zuckerberg addressed his 2 billion constituents. "I don't want anyone to use our tools to undermine democracy," he said. "That's not what we stand for."

This shift was driven in part by a reckoning that their creations had been used and disfigured. Even at the freewheeling Reddit, CEO Steve Huffman spoke of how he realized Russian propaganda had penetrated the site, but removing it would not be enough. "I believe the biggest risk we face as Americans is our own ability to discern reality from nonsense, and this is a burden we all bear."

Yet much of the impetus for change came in the form it always had — mounting legal and political pressures. In 2017, over the strenuous objections of Silicon Valley lobbyists and free speech advocates, German lawmakers passed a bill that levied fines as high as $57 million on companies that failed to delete "illegal, racist, or slanderous" posts within twenty-four hours. The floodgates had opened. By mid-2019, forty-four nations had passed laws or launched government initiatives to force further accountability on the social media giants. Australia went so far as to threaten Silicon Valley executives with three-year prison sentences if they did not promptly remove "abhorrent violent material" from their platforms.

Closer to home, U.S. legislators inaugurated the first major effort to regulate online political advertisements, especially the "dark ads" used by Russian propagandists to spread disinformation and by the Trump campaign to suppress minority voter turnout. It moved to subject them to the same Federal Election Commission disclosure rules that applied to broadcast television. Previously, political advertising on social media — a multibillion-dollar industry — had enjoyed all the same exemptions as skywriting.

For the titans of industry turned regulators of online war, it was an unexpected, unwanted, and often uncomfortable role to play. As Zuckerberg confessed in a 2018 interview, shortly before he was brought to testify before the U.S. Congress, "If you had asked me, when I got started with Facebook, if one of the central things I'd need to work on now is preventing governments from interfering in each other's elections, there's no way I thought that's what I'd be doing, if we talked in 2004 in my dorm room."

With each step the social media giants took as they waded deeper into political issues — tackling terrorism, extremism, and misinformation — they found themselves ever more bogged down by scandals that arose from the "gray areas" of politics and war. Sometimes, a new initiative to solve one problem might be exploited by a predatory government (Russia had a very different definition of "terrorism" than the United States) or well-meaning reporting systems gamed by trolls. Other times, it might lead to a clueless and costly mistake by a moderator expected to gauge the appropriateness of content from a country they'd never been to, amidst a political context they couldn't possibly understand.

One illustration of this problem was a Facebook rule, adopted to improve counterterrorism on the platform, that prohibited any positive mention of "violence to resist occupation of an internationally recognized state." From an engineering standpoint, it was an elegant solution — brief and broad. It was also one that any savvy political observer could have predicted would lead to massive problems. It led to mass deletions of user content from Palestine, Kashmir, and Western Sahara, each a political and cultural powder keg ruled by an occupying power.

These gray areas ran the gamut. A Chinese billionaire, taking refuge in the United States and vowing to reveal corruption among the highest ranks of the Communist Party, found his Facebook profile suspended for sharing someone else's "personally identifiable information" (which had kind of been the point). In Myanmar, when members of the Rohingya Muslim minority tried to use Facebook to document a government-led ethnic-cleansing campaign against them, some found their posts deleted — for the crime of detailing the very military atrocities they were suffering.

Throughout this messy and inexorable politicization, however, there was one rule that all of Silicon Valley made sure to enforce: the bottom line. The companies that controlled so much of modern life were themselves controlled by shareholders, their decision-making guided by quarterly earnings reports. When a Twitter engineer discovered evidence of massive Russian botnets as far back as 2015, he was told to ignore it. After all, every bot made Twitter look bigger and more popular. "They were more concerned with growth numbers than fake and compromised accounts," the engineer explained.

When Facebook employees confronted Mark Zuckerberg about then-candidate Trump's vow to bar all Muslims from entering the United States, he acknowledged that it was indeed hate speech, in violation of Facebook's policies. Nonetheless, he explained, his hands were tied. To remove the post would cost Facebook conservative users — and valuable business.

It was exactly as observed by writer Upton Sinclair a century earlier: "It is difficult to get a man to understand something when his salary depends on his not understanding it."

Today, the role of social media firms in public life is one that evades easy description. They are profit-motivated, mostly U.S.-based businesses that manage themselves like global governments. They are charmingly earnest, preaching inclusivity even as they play host to the world's most divisive forces. They are powerful entities that pretend to be powerless, inescapable political forces that insist they have no interest at all in politics. In essence, they are the mighty playthings of a small number of young adults, who have been given the unenviable task of shaping the nature of society, the economy, and now war and politics. And although the companies and those who lead them have matured an extraordinary amount in just a few years, the challenges they face only grow more complex.

But the most important part of their work is finding the answer to an obvious question — the kind of question that engineers like to hear. Assume that they've accepted the scope and complexity of their responsibilities, that they have decided to outlaw an unacceptable behavior and even defined exactly what that behavior looks like. How do they build the systems to stop it? What do those systems look like? Their answers have been to turn to the very same tools that created many of the problems in the first place: the online crowd and pitiless machines.

COMMUNITY WATCHES AND DIGITAL SERFS

America Online called them "community leaders," but this vague corporatese hardly described who they were or what they did. Nevertheless, a time traveler from thirteenth-century Europe would have recognized their role immediately. They were serfs — peasants who worked their feudal lord's land in return for a cut of the harvest. AOL's serfs just happened to toil through a dial-up modem. And their lord just happened to be the first true internet giant.

By the mid-1990s, AOL had evolved from a small internet service provider into a sprawling digital empire. For millions of users, AOL *was* the internet: an online chat service, a constellation of hosted websites (AOL partnered with everyone from CNN to the Library of Congress) and forum boards, and a general internet browser. AOL was both a piece of software and a massive media service, one that eventually reached 26 million subscribers. It marketed itself by carpet-bombing millions of homes with blue CDs emblazoned with the AOL logo, promising "500 Hours Free!" At one time, half of all the CDs produced on earth were used for AOL free trials.

Early in its corporate existence, AOL recognized two truths that every web company would eventually confront. The first was that the internet was a teeming hive of scum and villainy. The second was that there was no way AOL could afford to hire enough employees to police it. Instead, AOL executives stumbled upon a novel solution. Instead of trying to police their sprawling digital commonwealth, why not enlist their most passionate users to do it for them?

And so the AOL Community Leader Program was born. In exchange for free or reduced-price internet access, volunteers agreed to labor for dozens of hours each week to maintain the web communities that made AOL rich, ensuring that they stayed on topic and that porn was kept to a minimum. Given special screen names, or "uniforms," that filled them with civic pride, they could mute or kick out disruptive users.

As AOL expanded, the program grew more organized and bureaucratic. The Community Leader Program eventually adopted a formal three-month training process. Volunteers had to work a minimum of four hours each week and submit detailed reports of how they'd spent their time. At its peak, the program boasted 14,000 volunteers, including a "youth corps" of 350 teenagers. AOL had effectively doubled its workforce while subsidizing roughly 0.0005 percent of its subscriber base, all while maintaining a degree of plausible deniability if anything went wrong. It seemed to be the best investment AOL ever made.

Predictably, such a criminally good deal was bound for a criminal end. In 1999, two former community leaders sued AOL in a class-action lawsuit, alleging that they'd been employees in a "cyber-sweatshop" and that some were owed as much as $50,000 in back pay. A legal odyssey ensued. In 2005, AOL terminated the Community Leader Program, bestowing a free twelve-month subscription on any remaining volunteers. In 2008, AOL was denied its motion to dismiss the lawsuit. And at last, in 2010 — long after AOL had been eclipsed by the likes of Google and Facebook — the company suffered its final indignity, forced to pay its volunteer police force $15 million in back pay.

The rise and fall of AOL's digital serfs foreshadowed how all big internet companies would come to handle content moderation. If the internet of the mid-1990s had been too vast for paid employees to patrol, it was a mission impossible for the internet of the 2010s and beyond. Especially when social media startups were taking off, it was entirely plausible that there might have been more languages spoken on a platform than total employees at the company.

But as companies begrudgingly accepted more and more content moderation responsibility, the job still needed to get done. Their solution was to split the chore into two parts. The first part was crowdsourced to users (not just volunteers but *everyone*), who were invited to flag content they didn't like and prompted to explain why. The second part was outsourced to full-time content moderators, usually contractors based overseas, who could wade through as many as a thousand graphic images and videos each day. Beyond establishing ever-evolving guidelines and reviewing the most difficult cases in-house, the companies were able to keep their direct involvement in content moderation to a minimum. It was a tidy system tacked onto a clever business model. In essence, social media companies relied on their users to produce content; they sold advertising on that content and relied on other users to see that content in order to turn a profit. And if the content was inappropriate, they relied on still *other* users to find it and start the process of deletion.

When you report something on Facebook, for instance, you're propelled down a branching set of questions ("Is it a false story?" "Is it pornography?" "Is it just annoying?") that determine who reviews it and how seriously the report is taken. In this fashion, Facebook users flag more than a million pieces of content each day. The idea of user-based reporting has become so ingrained in the operations of the social media giants that it now carries a certain expectation. When Facebook came under fire in 2017 for allowing the livestreamed murder of a 74-year-old grandfather to remain viewable for more than two hours, it had a ready excuse: nobody had reported it. In effect, it was Facebook's users — not Facebook — who were at fault.

And then there are the people who sit at the other end of the pipeline, tech laborers who must squint their way through each beheading video, graphic car crash, or scared toddler in a dark room whose suffering has not yet been chronicled and added to Microsoft's horrifying child abuse database. There are an estimated 150,000 workers in these jobs around the world, most of them subcontractors scattered across India and the Philippines.

Like most outsourcing, it is competitive and decently compensated work, given the pay scales in these locales. Most of it is done by bright young college graduates who would otherwise find themselves under-employed. It takes brains and good judgment to decipher context in just a few seconds, applying all the proper policies and procedures. Thus, the most apt parallel to these jobs isn't the click farms where laborers endlessly repeat the rote process of SIM card swapping and account cre-ation, but Russia's sockpuppets and troll factories, which also recruit from the ranks of underemployed, English-speaking college graduates. In a way, the occupations are mirrors of each other. Professional trolls try to make the internet worse. Professional content moderators try to make it a little better.

Unsurprisingly, this work is grueling. It's obviously unhealthy to sit for eight or more hours a day, consuming an endless stream of all the worst that humanity has to offer. There's depression and anger, vom-iting and crying, even relationship trust issues and reduced libido. In the United States, companies that conduct this work offer regular psy-chological counseling to counter what they call "compassion fatigue" —a literal exhaustion of the brain's ability to feel empathy for others in harm's way. It may not be enough. In 2017, two former Microsoft em-ployees assigned to the Online Safety Team sued their former employer, alleging that they'd developed post-traumatic stress disorder. It was the first lawsuit of its type. One of the plaintiffs described how he'd devel-oped an "internal video screen" of horror that he couldn't turn off.

Aside from the problems of worker PTSD, this bifurcated system of content moderation is far from perfect. The first reason is that it comes at the cost of resources that might otherwise be plowed into profit gen-erators like new features, marketing, or literally anything else. Accord-ingly, companies will always view it as a tax on their business model. Af-ter all, no startup ever secured a round of funding by trotting out a shiny new gold-plated content moderation system.

The second problem is scale. To paraphrase Winston Churchill, never before has so much, posted by so many, been moderated by so few. When WhatsApp was being used by ISIS to coordinate the first battle for Mosul, the company had just 55 employees for its 900 million users. But even that made it a behemoth. When the newly launched video-hosting startup Vid.Me found itself infested by thousands of ISIS propaganda clips around the same time, the company had a total of just 6 people on staff, none of whom spoke Arabic.

Even these numbers pale in comparison with the true social media giants. Recall from chapter 3 the wealth of data that these services generate. Every minute, Facebook users post 500,000 new comments, 293,000 new statuses, and 450,000 new photos; YouTube users, more than 500 hours of video; and Twitter users, more than 350,000 tweets. Each of these posts is a Sword of Damocles hanging over the company. It can suffer devastating PR disasters if it allows any objectionable piece of content to stand for more than a few minutes before being deleted. But if a company acts rashly, the cries of censorship are liable to come just as fast.

Finally, if social media firms are to police their networks (which, remember, they don't really want to do), they must contend not just with millions of pieces of content, but also with adversaries who actively seek to thwart and confuse their content moderation systems. Think of the Islamic State's resilient, regenerating Twitter network, the Russian government's believable sockpuppets, or the smirking alt-right memes that straddle the line between hateful jokes and raw hate. When Facebook announced in 2017 that it was hiring 250 more people to review advertising on the platform, New York University business professor Scott Galloway rightly described it as "pissing in the ocean."

Under extraordinary pressure and facing an ever-expanding content moderation queue, the engineers of Silicon Valley have cast far and wide for an answer. Unsurprisingly, they think they've found that answer in more technology.

ROBO-REALITY WARS

"YOU LOOK LIKE A THING AND I LOVE YOU."

As a Tinder pickup line, it needed work. But it wasn't bad for something that hadn't even been written by a human. AI specialist Janelle Shane had done was compile a list of existing pickup lines and taught the computer to read them. After that, an artificial brain — a neural network — studied the list and invented a new pickup line all on its own.

Neural networks are a new kind of computing system: a calculating machine that hardly resembles a "machine" at all. Although such networks were theorized as far back as the 1940s, they've only matured during this decade as cloud processing has begun to make them practical. Instead of rule-based programming that relies on formal logic ("If A = yes, run process B; if A = no, run process C"), neural networks resemble living brains. They're composed of millions of artificial neurons, each of which draws connections to thousands of other neurons via "synapses." Each neuron has its own level of intensity, determined either by the initial input or by synaptic connections received from neurons farther up the stream. In turn, this determines the strength of the signal these neurons send down the stream through their own dependent synapses.

These networks function by means of pattern recognition. They sift through massive amounts of data, spying commonalities and making inferences about what might belong where. With enough neurons, it becomes possible to split the network into multiple "layers," each discovering a new pattern by starting with the findings of the previous layer. If a neural network is studying pictures, it might start by discovering the concept of "edges," sorting out all the edges from the non-edges. In turn, the next layer might discover "circles"; the layer after that, "faces"; the layer after that, "noses." Each layer allows the network to approach a problem with more and more granularity. But each layer also demands exponentially more neurons and computing power.

Neural networks are trained via a process known as "deep learning." Originally, this process was supervised. A flesh-and-blood human engineer fed the network a mountain of data (10 million images or a library of English literature) and slowly guided the network to find what the engineer was looking for (a "car" or a "compliment"). As the network went to work on its pattern-sorting and the engineer judged its performance and tweaked the synapses, it got a little better each time. Writer Gideon Lewis-Kraus delightfully describes the process as tuning a kind of "giant machine democracy."

Today, advanced neural networks can function without that human supervision. In 2012, engineers with the Google Brain project published a groundbreaking study that documented how they had fed a nine-layer neural network 10 million different screenshots from random YouTube videos, leaving it to play with the data on its own. As it sifted through the screenshots, the neural network — just like many human YouTube users — developed a fascination with pictures of cats. By discovering and isolating a set of cat-related qualities, it taught itself to be an effective cat detector. "We never told it during the training, 'This is a cat,'" explained one of the Google engineers. "It basically invented the concept of a cat."

Of course, the neural network had no idea what a "cat" was, nor did it invent the cat. The machine simply distinguished the pattern of a cat from all "not-cat" patterns. Yet this was really no different from the thought process of a human brain. Nobody is programmed from birth with a set, metaphysical definition of a cat. Instead, we learn a set of catlike qualities that we measure against each thing we perceive. Every time we spot something in the world — say, a dog or a banana — we are running a quick probabilistic calculation to check if the object is a cat.

Feed the network enough voice audio recordings, and it will learn to recognize speech. Feed it the traffic density of a city, and it will tell you where to put the traffic lights. Feed it 100 million Facebook likes and purchase histories, and it will predict, quite accurately, what any one person might want to buy or even whom they might vote for.

In the context of social media, the potential uses for neural networks are as diverse as they are tantalizing. The endless churn of content produced on the internet each day provides a limitless pipeline of data with which to train these increasingly intelligent machines.

Facebook is a fertile testing ground for such neural networks — a fact appreciated by none more than Facebook itself. By 2017, the social media giant had plunged into the field, running more than a million AI experiments each month on a dataset of more than a billion user-uploaded photographs. The system had far surpassed Facebook's already-uncanny facial recognition algorithm, learning to "see" hundreds of distinct colors, shapes, objects, and even places. It could identify horses, scarves, and the Golden Gate Bridge. It could even find every picture of a particular person wearing a black shirt. If such a system were unleashed on the open internet, it would be like having 10,000 Bellingcats at one's fingertips.

For the social media giants, an immediate application of this technology is solving their political and business problem — augmenting their overworked human content moderation specialists with neural network–based image recognition and flagging. In late 2017, Google announced that 80 percent of the violent extremist videos uploaded to YouTube had been automatically spotted and removed before a single user had flagged them.

Some at these companies believe the next stage is to "hack harassment," teaching neural networks to understand the flow of online conversation in order to identify trolls and issue them stern warnings before a human moderator needs to get involved. A Google system intended to detect online abuse — not just profanity, but toxic phrases and veiled hostility — has learned to rate sentences on an "attack scale" of 1 to 100. Its conclusions align with those of human moderators about 90 percent of the time.

Such neural network–based sentiment analysis can be applied not just to individual conversations, but to the combined activity of *every* social media user on a platform. In 2017, Facebook began testing an algorithm intended to identify users who were depressed and at risk for suicide. It used pattern recognition to monitor user posts, tagging those suspected to include thoughts of suicide and forwarding them to its content moderation teams. A suicidal user could receive words of support and link to psychological resources without any other human having brought the post to Facebook's attention (or even having seen it). It was a powerful example of a potential good — but also an obvious challenge to online privacy.

Social media companies can also use neural networks to analyze the links that users share. This is now being applied to the thorny problem of misinformation and "fake news." Multiple engineering startups are training neural networks to fact-check headlines and articles, testing basic statistical claims ("There were x number of illegal immigrants last month") against an ever-expanding database of facts and figures. Facebook's chief AI scientist turned many heads when, in the aftermath of the 2016 U.S. election, he noted that it was technically possible to stop viral falsehoods. The only problem, he explained, was in managing the "trade-offs" — finding the right mix of "filtering and censorship and free expression and decency." In other words, the same thorny political questions that have dogged Silicon Valley from the beginning.

Yet the most important applications of neural networks may be in simulating and replacing the very thing social networks were designed for: us. As we saw earlier, bots pose as humans online, pushing out rote messages. Their more advanced version, chatbots, are algorithms designed to convey the *appearance* of human intelligence by parroting scripts from a vast database. If a user says something to one of these "dumb" chatbots ("How's the weather?"), the chatbot will scan all previous instances in which the question appears, choosing a response whose other data points best align with those of the current conversation (if, for instance, the user has previously disclosed that her name is Sally or that she's from the United States and likes guns). No matter how convincing it is, though, each chatbot is basically reciting lines from a very, very long script.

By contrast, neural network–trained chatbots — also known as machine-driven communications tools, or MADCOMs — have no script at all, just the speech patterns deciphered by studying millions or billions of conversations. Instead of contemplating how MADCOMs might be used, it's easier to ask what one might *not* accomplish with intelligent, adaptive algorithms that mirror human speech patterns.

But the development of next-generation MADCOMs also illustrates a flaw inherent in all neural networks: they are only as good as their inputs — and only as moral as their users. In 2016, Microsoft launched Tay, a neural network–powered chatbot that adopted the speech patterns of a teenage girl. Anyone could speak to Tay and contribute to her dataset; she was also given a Twitter account. Trolls swarmed Tay immediately, and she was as happy to learn from them as from anyone else. Tay's bubbly personality soon veered into racism, sexism, and Holocaust denial. "RACE WAR NOW," she tweeted, later adding, "Bush did 9/11." After less than a day, Tay was unceremoniously put to sleep, her fevered artificial brain left to dream of electric frogs.

While the magic of neural networks might stem from their similarity to human brains, this is also one of their drawbacks. Nobody, their creators included, can fully comprehend how they work. When the network gets something wrong, there's no error log, just the knowledge that, with enough synaptic fiddling, the problem might be fixed. When there's no way to know if the network is wrong — if it's making a prediction of the future based on past data — users can either ignore it or take its prognostication at face value. The only way to understand a neural network is to steal a page from neuroscience, monitoring different groups of artificial neurons and testing different patterns to see what stimulates them. Ironically, neuroscientists who conduct similar experiments on human brains (like monitoring the electrical activity produced by each of 10,000 different words) have begun to use neural networks to map and model their results.

The greatest danger of neural networks, therefore, lies in their sheer versatility. Smart though the technology may be, it cares not how it's used. These networks are no different from a knife or a gun or a bomb — indeed, they're as double-edged as the internet itself.

Governments of many less-than-free nations salivate at the power of neural networks that can learn millions of faces, flag "questionable" speech, and infer hidden patterns in the accumulated online activity of their citizens. The most obvious candidate is China, whose keyword-filtering and social credit system will benefit greatly from the implementation of such intelligent algorithms. In 2016, Facebook was reported to be developing such a "smart" censorship system in a bid to allow it to expand into the massive Chinese market. This was an ugly echo of how Sun Microsystems and Cisco once conspired to build China's Great Firewall.

But it doesn't take an authoritarian state to turn a neural network toward evil ends. Anyone can build and train one using free, open-source tools. An explosion of interest in these systems has led to thousands of new applications. Some might be described as "helpful," others "strange." And a few — though developed with the best of intentions — are rightly described as nothing less than "mind-bendingly terrifying."

We've already seen how easy it is for obvious falsehoods ("The world is flat"; "The pizza parlor is a secret underage sex dungeon") to take hold and spread across the internet. Neural networks are set to massively compound this problem with the creation of what are known as "deep fakes."

Just as they can study recorded speech to infer meaning, these networks can also study a database of words and sounds to infer the *components* of speech — pitch, cadence, intonation — and learn to mimic a speaker's voice almost perfectly. Moreover, the network can use its mastery of a voice to approximate words and phrases that it's never heard. With a minute's worth of audio, these systems might make a good approximation of someone's speech patterns. With a few hours, they are essentially perfect.

One such "speech synthesis" startup, called Lyrebird, shocked the world in 2017 when it released recordings of an eerily accurate, entirely fake conversation between Barack Obama, Hillary Clinton, and Donald Trump. Another company unveiled an editing tool that it described as "Photoshop for audio," showing how one can tweak or add new bits of speech to an audio file as easily as one might touch up an image.

Neural networks can synthesize not just what we read and hear but also what we see. In 2016, a team of computer and audiovisual scientists demonstrated how, starting with a two-dimensional photograph, they could build a photorealistic, three-dimensional model of someone's face. They demonstrated it with the late boxing legend Muhammad Ali, transforming a single picture into a hyperrealistic face mask ready to be animated and placed in a virtual world — and able to rewrite the history of what Muhammad Ali did and said when he was alive.

This technology might also be used to alter the present or future. Using an off-the-shelf webcam, another team of scientists captured the "facial identity" of a test subject: the proportions of their features and the movement patterns of their mouth, brows, and jawline. Then they captured the facial identity of a different person in a prerecorded video, such as Arnold Schwarzenegger sitting for an interview or George W. Bush giving a speech. After that, they merged the two facial identities via "deformation transfer," translating movements of the first face into proportionate movements by the second. Essentially, the test subject could use their own face to control the expressions of the person onscreen, all in real time. If the petite female in front of the webcam opened her mouth, so did the faux Arnold Schwarzenegger. If the middle-aged guy with spiky hair and a goatee mouthed words in rapid succession and raised an eyebrow, so did the photorealistic George W. Bush. As the researchers themselves noted, "These results are hard to distinguish from reality, and it often goes unnoticed that the content is not real."

Neural networks can also be used to create deep fakes that aren't copies at all. Rather than just study images to learn the names of different objects, these networks can learn how to produce new, never-before-seen versions of the objects in question. They are called "generative networks." In 2017, computer scientists unveiled a generative network that could create photorealistic synthetic images on demand, all with only a keyword. Ask for "volcano," and you got fiery eruptions as well as serene, dormant mountains — wholly familiar-seeming landscapes that had no earthly counterparts. Another system created synthetic celebrities — faces of people who didn't exist, but whom real humans would likely view as being Hollywood stars.

Using such technology, users will eventually be able to conjure a convincing likeness of any scene or person they or the AI can imagine. Because the image will be truly *original,* it will be impossible to identify the forgery via many of the old methods of detection. And generative networks can do the same thing with video. They have produced eerie, looping clips of a "beach," a "baby," or even "golf." They've also learned how to take a static image (a man on a field; a train in the station) and generate a short video of a predictive future (the man walking away; the train departing). In this way, the figures in old black-and-white photographs may one day be brought to life, and events that never took place may nonetheless be presented online as real occurrences, documented with compelling video evidence.

And finally, there are the MADCOMs. The inherent promise of such technology — an AI that is essentially indistinguishable from a human operator — also sets it up for terrible misuse. Today, it remains possible for a savvy internet user to distinguish "real" people from automated botnets and even many sockpuppets (the Russified English helped us spot a few). Soon enough, even this uncertain state of affairs may be recalled fondly as the "good old days" — the last time it was possible to have *some* confidence that another social media user was a flesh-and-blood human being instead of a manipulative machine. Give a Twitter botnet to a MADCOM and the network might be able to distort the algorithmic prominence of a topic without anyone noticing, simply by creating realistic conversations among its many fake component selves. MADCOMs won't just drive news cycles, but will also trick and manipulate the people reacting to them. They may even grant interviews to unwitting reporters.

Feed a MADCOM enough arguments and it will never repeat itself.

Feed it enough information about a target population — such as the hundreds of billions of data points that reside in a voter database like Project Alamo — and it can spin a personalized narrative for every resident in a country. The network never sleeps, and it's always learning. In the midst of a crisis, it will invariably be the first to respond, commanding disproportionate attention and guiding the social media narrative in whichever direction best suites its human owners' hidden ends. Matthew Chessen, a senior technology policy advisor at the U.S. State Department, doesn't mince words about the inevitable MADCOM ascendancy. It will "determine the fate of the internet, our society, and our democracy," he writes. No longer will humans be reliably in charge of the machines. Instead, as machines steer our ideas and culture in an automated, evolutionary process that we no longer understand, they will "start programming us."

Combine all these pernicious applications of neural networks — mimicked voices, stolen faces, real-time audiovisual editing, artificial image and video generation, and MADCOM manipulation — and it's tough to shake the conclusion that humanity is teetering at the edge of a cliff. The information conflicts that shape politics and war alike are fought today by clever humans using viral engineering. The LikeWars of tomorrow will be fought by highly intelligent, inscrutable algorithms that will speak convincingly of things that never happened, producing "proof" that doesn't really exist. They'll seed falsehoods across the social media landscape with an intensity and volume that will make the current state of affairs look quaint.

Aviv Ovadya, chief technologist at the Center for Social Media Responsibility at the University of Michigan, has described this looming threat in stark, simple terms. "We are so screwed it's beyond what most of us can imagine," he said. "And depending how far you look into the future, it just gets worse."

For generations, science fiction writers have been obsessed with the prospect of an AI Armageddon: a *Terminator*-style takeover in which the robots scour puny human cities, flamethrowers and beam cannons at the ready. Yet the more likely takeover will take place on social media. If machines come to manipulate all we see and how we think online, they'll *already* control the world. Having won their most important conquest — the human mind — the machines may never need to revolt at all.

And yet, just as in the *Terminator* movies, if humans are to be spared from this encroaching, invisible robot invasion, their likely savior will be found in other machines. Recent breakthroughs in neural network training hint at what will drive machine evolution to the next level, but also save us from algorithms that seek to manipulate us: an AI survival of the fittest.

Newer, more advanced forms of deep learning involve the use of "generative adversarial networks." In this type of system, two neural networks are paired off against each other in a potentially endless sparring match. The first network strains to create something that seems real — an image, a video, a human conversation — while the second network struggles to determine if it's fake. At the end of each match, the networks receive their results and adjust to get just a little bit better. Although this process teaches networks to produce increasingly accurate forgeries, it also leaves open the potential for networks to get better and better at detecting fakes.

This all boils down to one important, extremely sci-fi question. If both networks are gifted with ever-improving calibration and processing power, which one — the "good" AI or the "bad" AI — will more often beat the other?

In the answer quite possibly lies not just the fate of content moderation policy, but also of future wars and elections, as well as democracy, civilization, and objective reality. Within a decade, Facebook, Google, Twitter, and every other internet company of scale will use neural networks to police their platforms. Dirty pictures, state-sponsored botnets, terrorist propaganda, and sophisticated disinformation campaigns will be hunted by machine intelligences that dwarf any now in existence. But they will be battled by other machine intelligences that seek to obfuscate and evade, disorient and mislead. And caught in the middle will be us — *all* of us — part of a conflict that we definitely started but whose dynamics we will soon scarcely understand.

It is a bizarre, science-fiction-seeming future. But for something that began with an SF-lovers email thread, it also seems strangely appropriate.

9

Conclusion

What Do We Know, What Can We Do?

> We are as gods and might as well get good at it.
>
> — STEWART BRAND, "We Are as Gods"

LONG BEFORE THE MILITARY CONVOY ARRIVED in the muggy town of Dara Lam, news of the meeting between the U.S. Army colonel and the unpopular governor of Kirsham province had seeped into social media. Angry with the American presence and the governor's corruption, local citizens organized for a demonstration. Their trending hashtag — #justice4all — soon drew the attention of international media. It also drew the eyes of some less interested in justice: the notorious Fariq terror network. Using sockpuppet accounts and false reports, the terrorists fanned the flames, calling for the protesters to confront the American occupiers.

But this wasn't the full extent of Fariq's plan. Knowing where a massive crowd of civilians would gather, the terrorists also set an ambush. They'd fire on the U.S. soldiers as they exited the building, and if the soldiers fired back, the demonstrators would be caught in the crossfire. Pre-positioned cameramen stood ready to record the bloody outcome: either dead Americans or dead civilians. A network of online proxies was prepared to drive the event to virality and use it for future propaganda and recruiting. Whatever the physical outcome, the terrorists would win this battle.

Luckily, other eyes were tracking the flurry of activity online: those of a U.S. Army brigade's tactical operations center. The center's task was to monitor the environment in which its soldiers operated, whether dense cities, isolated mountain ranges, or clusters of local blogs and social media influencers. The fast-moving developments were detected and then immediately passed up the chain of command. The officers might once have discounted internet chatter but now understood its importance. Receiving word of the protest's growing strength and fury, the colonel cut his meeting short and left discreetly through a back entrance. Fariq's plan was thwarted.

Try as you might, you won't find any record of this event in the news — and it is not because the battle never took place. It is because Dara Lam is a fake settlement in a fake province of a fake country, one that endures a fake war on a fake internet that breaks out every few months in the very real state of Louisiana.

The Joint Readiness Training Center at Fort Polk holds a special place in military history. It was created as part of the Louisiana Maneuvers, a series of massive training exercises held just before the United States entered World War II. When Hitler and his blitzkrieg rolled over Europe, the U.S. Army realized warfare was operating by a new set of rules. It had to figure out how to transition from a world of horses and telegraphs to one of mechanized tanks and trucks, guided by wireless communications. It was at Fort Polk that American soldiers, including such legendary figures as Dwight D. Eisenhower and George S. Patton, learned how to fight in a way that would preserve the free world.

Since then, Fort Polk has served as a continuous field laboratory where the U.S. Army trains for tomorrow's battles. During the Cold War, it was used to prepare for feared clashes with the Soviet Red Army and then to acclimatize troops to the jungles of Vietnam. After 9/11, the 72,000-acre site was transformed into the fake province of Kirsham, replete with twelve plywood villages, an opposing force of simulated insurgents, and scores of full-time actors playing civilians caught in the middle: in short, everything the Army thought it needed to simulate how war was changing. Today, Fort Polk boasts a brand-new innovation for this task: the SMEIR.

Short for Social Media Environment and Internet Replication, SMEIR simulates the blogs, news outlets, and social media accounts that intertwine to form a virtual battlefield atop the physical one. A team of defense contractors and military officers simulate the internet activity of a small city — rambling posts, innocuous tweets, and the occasional bit of viral propaganda — challenging the troops fighting in the Kirsham war games to navigate the digital terrain. For the stressed, exhausted soldiers dodging enemy bombs and bullets, it's not enough to safeguard the local population and fight the evil insurgents; they must now be mindful of the ebb and flow of online conversation.

From a military perspective, SMEIR is a surreal development. A generation ago, the internet was a niche plaything, one that the U.S. military itself had just walked away from. Only the most far-sighted futurists were suggesting that it might one day become a crucial battlefield. None imagined that the military would have to pay millions of dollars to simulate a second, *fake* internet to train for war on the real one.

But in the unbridled chaos of the modern internet, even an innovation like SMEIR is still playing catch-up. Thwarted by an eagle-eyed tactical operations officer, actual terrorists wouldn't just fade back into the crowd. They'd shoot the civilians anyway and simply manufacture evidence of U.S. involvement. Or they'd manufacture *everything* about the video and use armies of botnets and distant fanboys to overwhelm the best efforts of fact-checkers, manipulating the algorithms of the web itself.

Nor can such a simulation capture the most crucial parts of the battlefield. The digital skirmishes that would have determined who actually *won* this fight wouldn't have been limited to Louisiana or SMEIR. Rather, they would have been decided by the clicks of millions of people who've never met a person from Dara Lam and by whatever policy that social media company executives had chosen for how to handle Fariq propaganda. The reality of what took place in the (fake) battle would have been secondary to whatever aspects of it went viral.

Just as soldiers in Louisiana are struggling to adjust to this new information conflict, so are engineers in Silicon Valley. All the social media powers were founded on the optimistic premise that a more close-knit and communal world would be a better one. "[Facebook] was built to accomplish a social mission, to make the world more open and connected," wrote Mark Zuckerberg in a 2012 letter to investors, just as his company went public. Yet as we've seen, these companies must now address the fact that this very same openness and connection has also made their creations *the* place for continual and global conflicts.

This duality of the social media revolution touches the rest of us, too. The evolutionary advantages that make us such dynamic, social creatures — our curiosity, affinity for others, and desire to belong — also render us susceptible to dangerous currents of disinformation. It doesn't even help to be born *into* the world of the internet, as is the case for millennials and Generation Z. Study after study finds that youth is no defense against the dangers we've explored in this book. Regardless of how old they are, humans as a species are uniquely ill-equipped to handle both the instantaneity and the immensity of information that defines the social media age.

However, humans *are* unique in their ability to learn and evolve, to change the fabric of their surroundings. Although the maturation of the internet has produced dramatic new forces acting upon war and politics — and, by extension, upon all of society — these changes are far from unknown or unknowable. Even LikeWar has rules.

First, for all the sense of flux, the modern information environment is becoming stable. The internet is now the preeminent communications medium in the world; it will *remain* so for the foreseeable future. Through social media, the web will grow bigger in size, scope, and membership, but its essential form and centrality to the information ecosystem will not change. It has also reached a point of maturity whereby most of its key players will remain the same. Like them or hate them, the majority of today's most prominent social media companies and voices will continue to play a crucial role in public life for years to come.

Second, the internet is a battlefield. Like every other technology before it, the internet is not a harbinger of peace and understanding. Instead, it's a platform for achieving the goals of whichever actor manipulates it most effectively. Its weaponization, and the conflicts that then erupt on it, define both what happens on the internet and what we take away from it. Battle on the internet is continuous, the battlefield is contiguous, and the information it produces is contagious. The best and worst aspects of human nature duel over what truly matters most online: our attention and engagement.

Third, this battlefield changes how we must think about information itself. If something happens, we must assume that there's likely a digital record of it — an image, video, or errant tweet — that will surface seconds or years from now. However, an event only carries power if people also *believe* that it happened. The nature of this process means that a manufactured event can have real power, while a demonstrably true event can be rendered irrelevant. What determines the outcome isn't mastery of the "facts," but rather a back-and-forth battle of psychological, political, and (increasingly) algorithmic manipulation. Everything is now transparent, yet the truth can be easily obscured.

Fourth, war and politics have never been so intertwined. In cyberspace, the means by which the political or military aspects of this competition are "won" are essentially identical. As a result, politics has taken on elements of information warfare, while violent conflict is increasingly influenced by the tug-of-war for online opinion. This also means that the engineers of Silicon Valley, quite unintentionally, have turned into global power brokers. Their most minute decisions shape the battlefield on which both war and politics are increasingly decided.

Fifth, we're all part of the battle. We are surrounded by countless information struggles — some apparent, some invisible — all of which seek to alter our perceptions of the world. Whatever we notice, whatever we "like," whatever we share, becomes the next salvo. In this new war of wars, taking place on the network of networks, there is no neutral ground.

LikeWar isn't likeable. This state of affairs is certainly not what we were promised. And no matter how hard today's technologists try, their best efforts will never yield the perfect, glittering future once envisioned by the internet's early inventors.

Yet recognizing the new truths of the modern information environment and the eternal aspects of politics and war doesn't mean admitting defeat. Rather, it allows us to hone our focus and channel our energies into measures that can accomplish the most tangible good. Some of these initiatives can be undertaken by governments, others by social media companies, and still others by each of us on our own.

For governments, the first and most important step is to take this new battleground seriously. Social media now forms the foundation of commercial, political, and civic life. It is also a conflict space of immense consequence to both national and individual citizens' security. Just as the threat of cyberwar was recognized and then organized and prepared for over the past two decades, so, too, must this new front be addressed.

This advice is most urgent for democratic governments. As this book has shown, authoritarian leaders have long since attuned themselves to the potential of social media, both as a threat to their rule and as a new vector for attacking their foes. Although many democracies have formed national efforts to confront the resulting dangers, the United States — the very birthplace of the internet — has remained supremely ill-equipped. Indeed, in the wake of the episodes you have read about in this book, other nations now look to the United States as a showcase for all the developments they wish to *avoid.* So far, America has emerged as one of the clearest "losers" in this new kind of warfare. *Accordingly, the United States government must work quickly to develop a national strategy to address the challenges of LikeWar.* The wheels of government don't turn unless they have documents that establish a plan of action, laying out everything from the ends, ways, and means, to breaking down who has what responsibility. In 2019, the White House issued its first official new "Cybersecurity Strategy" in over fifteen years. It had nothing on the threats outlined in this book, leaving America's approach ad hoc and severely lacking.

A potential path forward can be seen in those nations that have moved beyond the previously discussed military organizations to launch "whole-of-nation" efforts, which are intended to inoculate their entire societies against information threats. It is not coincidental that among the first states to do so were Finland, Estonia, Latvia, Lithuania, and Sweden, all of which face a steady barrage of Russian information attacks, backed by the close proximity of Russian soldiers and tanks. Their inoculation efforts include citizen education programs, public tracking and notices of foreign disinformation campaigns, election protections and forced transparency of political campaign activities, and legal action to limit the effect of poisonous super-spreaders.

In many ways, such holistic responses to information threats have an American pedigree. One of the most useful efforts to foil Soviet operations during the Cold War was a comprehensive U.S. government effort called the Active Measures Working Group. It brought together people working in various government agencies — from spies to diplomats to broadcasters to educators — to collaborate on identifying and pushing back against KGB-planted false stories designed to fracture societies and undermine support for democracy. There is no such equivalent today. Neither is there an agency comparable to what the Centers for Disease Control and Prevention does for health — an information clearinghouse for government to connect with business and researchers in order to work together to battle dangerous viral outbreaks.

It would be easy to say that such efforts should merely be resurrected and reconstituted for the internet age — and doing so would be a welcome development. But we must also acknowledge a larger problem: *Today, a significant part of the American political culture is willfully denying the new threats to its cohesion. In some cases, it's colluding with them.*

Too often, efforts to battle back against online dangers emanating from actors at home and abroad have been stymied by elements within the U.S. government. After everything that played out in the 2016 election, it took two years for the Trump administration to hold its first cabinet-level meeting on election security and, even then, its focus was on the potential risk of hacked voting machines, not the actual online influence operations that had occurred. In turn, its State Department failed to increase efforts to battle online terrorist propaganda and Russian disinformation, even when Congress allocated hundreds of millions of dollars for the purpose. And when intelligence agencies and the U.S military tried to move forward in bolstering a response, officials described it "like pulling teeth" even to get meetings with the White House on the topic. Indeed, with Trump continuing to dismiss the threat as "a goddamn hoax," his own Secretary of Homeland Security was even told by the White House chief of staff to avoid discussing the issue in the president's presence.

As a consequence, the American election system remains remarkably vulnerable, not merely to hacking of the voting booth, but also to the foreign manipulation of U.S. voters' political dialogue and beliefs. Ironically, although the United States has contributed millions of dollars to help nations like Ukraine safeguard their citizens against these new threats, political paralysis has prevented the U.S. government from taking meaningful steps to inoculate its own population. Until this is reframed as a nonpartisan issue — akin to something as basic as health education — the United States will remain at grave risk.

Accordingly, information literacy is no longer merely an education issue but a national security imperative. Indeed, given how early children's thought patterns develop and their use of online platforms begins, the process cannot start early enough. Just as in basic health education, there are parallel roles for both families and schools in teaching children how to protect themselves online, as well as gain the skills needed to be responsible citizens. At younger ages, these include programs that focus on critical thinking skills, expose kids to false headlines, and encourage them to play with (and hence learn from) image-warping software. Nor should the education stop as students get older. As of 2017, at least a dozen universities offered courses targeting more advanced critical thinking in media consumption, including an aptly named one at the University of Washington: "Calling Bullshit: Data Reasoning in a Digital World." This small number of pilot programs point the way, but they also illustrate how far we have to go in making them more widely available.

As in public health, such efforts will have to be supported outside the classroom, targeting the broader populace. Just as in the case of viral outbreaks of disease, there is a need for everything from public awareness campaigns to explain the risks of disinformation efforts to mass media notifications that announce their detected onset.

Given the dangers, anger, and lies that pervade social media, there's a temptation to tell people to step away from it altogether. Sean Parker created one of the first file-sharing social networks, Napster, and then became Facebook's first president. However, he has since become a social media "conscientious objector," leaving the world that he helped make. Parker laments not just what social media has already done to us, but what it bodes for the next generation. "God only knows what it's doing to our children's brains," he said in 2017.

The problem is that not all of us either want to, or even can, make that choice. Like it or not, social media now plays a foundational role in public and private life alike; it can't be un-invented or simply set aside. Nor is the technology itself bad. As we've repeatedly seen in this book, its new powers have been harnessed toward both good and evil ends, empowering both wonderful and terrible people, often simultaneously. Finally, it is damned addictive. Any program advising people to "just say no" to social media will be as infamously ineffective as the original 1980s antidrug campaign.

Instead, part of the governance solution to our social media problem may actually be *more* social media, just of a different kind. While the technology has been used to foment a wide array of problems around the world, a number of leaders and nations have simultaneously embraced its participatory nature to do the opposite: to identify and enact shared policy solutions. Such "technocracy" views the new mass engagement that social networks allow as a mechanism to improve our civic lives. For instance, a growing number of elected governments don't use social media just to frighten or anger their followers; they also use it to expand public awareness and access to programs, track citizen wants and needs, even gather proposals for public spending. Some also are seeking to inject it more directly into the political process. Switzerland, for instance, may be the world's oldest continuous democracy, but it has been quick to use social networks to allow the digitization of citizen petitions and the insertion of online initiatives into its policy deliberations. In Australia and Brazil, the Flux movement seeks to use technology to return to true political representation, whereby elected leaders commit to a system allowing parliamentary submission of and digital voting on key issues, moving the power from the politician back to the people.

What is common across these examples of governance via network is the use of social media to learn and involve. It is the opposite of governance via trolling — the all-too-frequent use of social media to attack, provoke, and preen.

This points to perhaps the biggest challenge of all: it will be hard to overcome any system that incentivizes an opposite outcome. Not just our networks but our politics and culture have been swarmed by the worst aspects of social media, from lies and conspiracy theories to homophily and trolling. This has happened for the very reason that it works: it is rewarded with attention, and, as we have seen, this attention becomes power.

Super-spreaders play a magnified role in our world; there is no changing that fact now. But it is how they are rewarded that determines whether their influence is a malign or benevolent one. *When someone engages in the spread of lies, hate, and other societal poisons, they should be stigmatized accordingly.* It is not just shameful but dangerous that the purveyors of the worst behaviors on social media have enjoyed increased fame and fortune, all the way up to invitations to the White House. Stopping these bad actors requires setting an example and ensuring that repeat offenders never escape the gravity of their past actions and are excluded from the institutions and platforms of power that now matter most in our society. In a democracy, you have a right to your opinion, but no right to be celebrated for an ugly, hateful opinion, especially if you've spread lie after lie.

Indeed, social media actions need to be taken all the more seriously when their poisonous side infects realms like national security, where large numbers of people's lives are at stake. *Those who deliberately facilitate enemy efforts, whether it be providing a megaphone for terrorist groups or consciously spreading disinformation, especially that from foreign government offensives, have to be seen for what they are.* They are no longer just fighting for their personal brand or their political party; they are aiding and abetting enemies that seek to harm all of society.

We must also come to grips with the new challenge of free speech in the age of social media — what is known as "dangerous speech." This term has emerged from studies of what prompts communal violence. It describes public statements intended to inspire hate and incite violent actions, usually against minorities. Dangerous speech isn't merely partisan language or a bigoted remark. These are, alas, all too common. Rather, dangerous speech falls into one or more of five categories: *dehumanizing language* (comparing people to animals or as "disgusting" or subhuman in some way); *coded language* (using coy historical references, loaded memes, or terms popular among hate groups); *suggestions of impurity* (that a target is unworthy of equal rights, somehow "poisoning" society as a whole); *opportunistic claims of attacks on women, but by people with no concern for women's rights* (which allows the group to claim a valorous reason for its hate); and *accusation in a mirror* (a reversal of reality, in which a group is falsely told it is under attack, as a means to justify preemptive violence against the target). This sort of speech poses a mortal threat to a peaceable society, especially if crossed with the power of a super-spreader to give it both validation and reach.

Cloaking itself in ambiguity and spreading via half-truths, dangerous speech is uniquely suited to social media. Its human toll can be seen in episodes like the web-empowered anti-Muslim riots of India and the genocide of the Rohingya people in Myanmar. But what the researchers who focus on the problem have grown most disturbed by is how "dangerous speech" is increasingly at work in the U.S. Instances of dangerous speech are at an all-time high, spreading via deliberate information offenses from afar, as well as via once-scorned domestic extremists whose voices have become amplified and even welcomed into the mainstream. The coming years will determine whether these dangerous voices will continue to thrive in our social networks, and thus our politics, or be defeated.

This challenge takes us beyond governments and their voters to the accountability we should demand from the companies that now shape social media and the world beyond. It is a strange fact that the entities best positioned to police the viral spread of hate and violence are not legislatures, but social media companies. They have access to the data and patterns that evidence it, and they can more rapidly respond than governments. As rulers of voluntary networks, they determine the terms of service that reflect their communities' and stockholders' best interests. Dangerous speech is not good for either.

This is just one dimension of the challenges these companies must confront. *Put simply, Silicon Valley must accept more of the political and social responsibility that the success of its technology has thrust upon it.* "The more we connect, the better it gets" is an old Facebook slogan that remains generally representative of how social media companies see the world. As we've seen, that slogan is neither true nor an acceptable way for these firms to approach the new role they play in society.

Although figures like Mark Zuckerberg have protested at various times that they should not be considered "arbiters of the truth," this is exactly what they are. The information that spreads via *their* services — governed by *their* legal and software codes — shapes our shared reality. If they aren't the arbiters of truth, who is?

Accordingly, these companies must abandon the pretense that they are merely "neutral" platform providers. It is a weak defense that they outgrew many years ago. Bigots, racists, violent extremists, and professional trolls do not have to be accorded the same respect as marginalized peoples and democratic states. In turn, the authoritarian governments that exploit their networks and target their users must be treated as adversaries — not potential new markets.

In the process, Silicon Valley must also break the code of silence that pervades its own culture. Our past research has brought us into contact with soldiers, spies, mercenaries, insurgents, and hackers. In every case, they proved oddly more willing to speak about their work — and how they wrestle with the thorny dilemmas within it — than those employed at big social media companies. As technology reporter Lorenzo Franceschi-Bicchierai has similarly written of his experience reporting on Facebook, "In many cases, answers to simple questions — are the nude images blurred or not, for instance — are filtered to the point where the information Facebook gives journalists is not true."

These companies should walk their talk, embracing proactive information disclosures instead of just using the word "transparency" ad infinitum in vague press releases. This applies not just to the policies that govern our shared online spaces but also to the information these companies collect from those spaces. It's unacceptable that social media firms took nearly a year after the 2016 U.S. election to release data showing definitive proof of a Russian disinformation campaign — and even then, only after repeated demands by Congress.

It is perhaps especially troubling that, despite all the political and public pressure, most are still dragging their feet in disclosing the full extent of what played out across their networks. Of the major social media companies, Reddit is the only one that preserved the known fake Russian accounts for public examination. By wiping clean this crucial evidence, the firms are doing the digital equivalent of bringing a vacuum cleaner to the scene of a crime. They are not just preventing investigators and researchers from exploring the full extent of what occurred and how to prevent a repeat. They are destroying what should be a memorial to a moment of mass manipulation and misinformation that very much altered world history.

Just as all companies have a role in public health, so does Silicon Valley have a responsibility to help build public information literacy. Owning the platforms by which misinformation spreads, social media companies are in a powerful position to help inoculate the public. The most effective of these initiatives don't simply warn people about general misinformation (e.g., "Don't believe everything you read on the internet") or pound counterarguments into their heads (e.g., "Here are ten reasons why climate change is real"). Rather, effective information literacy education works by presenting the people being targeted with specific, proven instances of misinformation, encouraging them to understand how and why it worked against them. Here again, the firms have mostly buried this information instead of sharing it with victims. Social media firms should swallow the fear that people will abandon their services en masse if they engage in these sorts of initiatives (we're too addicted to quit anyway). In their quest to avoid liability and maintain the fiction that they're blameless, social media firms have left their customers unarmed for battle.

This battle will only become more intense. Therefore, these companies should steal a lesson from the fictional battlefields of Dara Lam. It's not enough to experiment and train for today's battles of LikeWar; they must look ahead to tomorrow's.

The companies must proactively consider the political, social, and moral ramifications of their services. It is telling that, across all the episodes described in this book, not a single social media firm tried to remedy the ills that played out on their networks until they became larger problems, even though executives could see these abuses unfolding in real time. Even when these firms' own employees sounded alarms about issues ranging from hate groups to harassment, they were consistently ignored. Similarly, when outside researchers raised concerns about emerging problems like neo-Nazi trolling and Russian disinformation campaigns during the 2016 U.S. election, the firms essentially dismissed them.

Changing to a proactive strategy will require the firms to alter their approach to product development. In the same way that social media companies learned to vet new features for technical bugs, any algorithmic tweak or added capability will require them to take a long, hard look at how it might be co-opted by bad actors or spark unintended consequences — *before* the feature is released to the masses in a chaotic "beta test." Much like the U.S. Army playing war games at Fort Polk, these companies should aggressively "game out" the potential legal, social, and moral effects of their products, especially in regard to how the various types of bad actors discussed in this book might use them. The next time a malicious group or state weaponizes a social media platform, these companies won't be able to beg ignorance. Nor should we allow them to use that excuse any longer.

Amid all this talk of taking responsibility, it's important to recognize that this is the appropriate moment in both the internet's and these companies' own maturation to do so. As internet sociologist Zeynep Tufekci noted in 2018, "Facebook is only 13 years old, Twitter 11, and even Google is but 19. At this moment in the evolution of the auto industry, there were still no seat belts, airbags, emission controls, or mandatory crumple zones." The critics of social media should remember that the companies aren't implacable enemies set on ruining the social fabric. They're just growing into their roles and responsibilities. *By acting less like angry customers and more like concerned constituents, we stand the best chance of guiding these digital empires in the right direction.*

And this points to our own individual role in a realm of escalating online war — that is, recognizing our burgeoning responsibilities as citizens and combatants alike.

Like any viral infection, information offensives work by targeting the most vulnerable members of a population — in this case, the least informed. The cascading nature of "likes" and "shares" across social networks, however, means that the gullible and ignorant are only the entry point. Ignorance isn't bliss; it just makes you a mark. It also makes you more likely to spread lies, which your friends and family will be more inclined to believe and spread in turn.

Yet the way to avoid this isn't some rote recommendation that we all simply "get smart." That would be great, of course, but it is unlikely to happen and still wouldn't solve most problems. *Instead, if we want to stop being manipulated, we must change how we navigate the new media environment.* In our daily lives, all of us must recognize that the intent of most online content is to subtly influence and manipulate. In response, we should practice a technique called "lateral thinking." In a study of information consumption patterns, Stanford University researchers gauged three groups — college undergraduates, history PhDs, and professional fact-checkers — on how they evaluated the accuracy of online information. Surprisingly, both the undergraduates and the PhDs scored low. While certainly intelligent, they approached the information "vertically." They stayed within a single worldview, parsing the content of only one source. As a result, they were "easily manipulated."

By contrast, the fact-checkers didn't just recognize online manipulation more often, they also detected it far more rapidly. The reason was that they approached the task "laterally," leaping across multiple other websites as they made a determination of accuracy. As the Stanford team wrote, the fact-checkers "understood the Web as a maze filled with trap doors and blind alleys, where things are not always what they seem." So they constantly linked to other locales and sources, "seeking context and perspective." In short, they networked out to find the truth. The best way to navigate the internet is one that reflects the very structure of the internet itself.

There is nothing inherently technological about this approach. Indeed, it's taught by one of the oldest and most widely shared narratives in human history: the parable of the blind men and the elephant. This story dates back to the earliest Buddhist, Hindu, and Jain texts, almost 4,000 years ago. It describes how a group of blind men, grasping at different parts of an elephant, imagine it to be many different things: a snake, a tree, a wall. In some versions of the story, the men fall to mortal combat as their disagreement widens. As the Hindu Rigveda summarizes the story, "Reality is one, though wise men speak of it variously."

When in doubt, seek a second opinion — then a third, then a fourth. If you're *not* in doubt, then you're likely part of the problem!

What makes social media so different, and so powerful, is that it is a tool of mass communication whose connections run both ways. Every act on it is simultaneously personal and global. *So in protecting ourselves online, we all, too, have broader responsibilities to protect others.* Think of this obligation as akin to why you cover your mouth when you cough. You don't do it because it directly protects you, but because it protects all those you come in contact with, and everyone whom they meet in turn. This ethic of responsibility makes us all safer in the end. It works the same way in social media.

That leads us to a final point as to how to handle a world of "likes" and lies gone viral online. Here again, to succeed in the digital future is to draw upon the lessons of the past, including some of the oldest recorded. Plato's *Republic,* written around 520 BCE, is one of the foundational works of Western philosophy and politics. One of its most important insights is conveyed through "The Allegory of the Cave." It tells the story of prisoners in a cave, who watch shadows dance across the wall. Knowing only that world, they think the shadows are reality, when actually they are just the reflections of a light they cannot see. (Note this ancient parallel to Zuckerberg's fundamental notion that Facebook was "a mirror of what existed in real life.")

The true lesson comes, though, when one prisoner escapes the cave. He sees real light for the first time, finally understanding the nature of his reality. Yet the prisoners inside the cave refuse to believe him. They are thus prisoners not just of their chains but also of their beliefs. They hold fast to the manufactured reality instead of opening up to the truth.

Indeed, it is notable that the ancient lessons of Plato's cave are a core theme of one of the foundational movies of the internet age, *The Matrix*. In this modern reworking, it is computers that hide the true state of the world from humanity, with the internet allowing mass-scale manipulation and oppression. *The Matrix* came out in 1999, however, before social media had changed the web into its new form. Perhaps, then, the new matrix that binds and fools us today isn't some machine-generated simulation plugged into our brains. It is just the way we view the world, filtered through the cracked mirror of social media.

But there may be something more. One of the underlying themes of Plato's cave is that power turns on perception and choice. It shows that if people are unwilling to contemplate the world around them in its actuality, they can be easily manipulated. Yet they have only themselves to blame. They, rather than the "ruler," possess the real power — the power to decide what to believe and what to tell others. So, too, in *The Matrix*, every person has a choice. You can pick a red pill (itself now an internet meme) that offers the truth. Or you can pick a blue pill, which allows you to "believe whatever you want to believe."

Social media is extraordinarily powerful, but also easily accessible and pliable. Across it play out battles for not just every issue you care about, but for the future itself. Yet within this network, and in each of the conflicts on it, we all still have the power of choice. Not only do we determine what role we play, but we also influence what others know and do, shaping the outcomes of all these battles. In this new world, the same basic law applies to us all:

You are now what you share.

And through what you choose, you share who you truly are.

Acknowledgments

This book is the culmination of a five-year journey that began with an idle conversation about social media and the future of war, long before the Islamic State had reared its ugly head and the Russian government transformed U.S. politics forever. In the years since, we've faced shock, surprise, discovery, and — as you might imagine — lots and lots of rewrites. We owe debts of gratitude to more than these pages can fill, but we'll try our best.

We'd like to start by thanking the whole team at Houghton Mifflin Harcourt, especially Barbara Jatkola, Rosemary McGuinness, Michael Dudding, Larry Cooper, and Michelle Triant. Thanks most of all to Eamon Dolan, who invested extraordinary time and effort in shepherding the book through publication. Our agent, Dan Mandel of Sanford J. Greenburger Associates, went above and beyond during this whole endeavor, for which we are eternally thankful. He didn't just represent us and the book; he also served as a sounding board, counselor, and all-around mensch from start to finish. All writers should be so lucky as to have someone like Dan on their team.

Our work was generously supported by the Smith Richardson Foundation, which enables the kind of deep-dive, long-term research so crucial to important issues of public policy. It was also supported by a fantastic team of young researchers, kindly offered by the Arizona State University Center on the Future of War and ably led by Daniel Rothenberg and Peter Bergen. Tremendous thanks to Bill McDonald, Hannah

Hallikainen, Erin Schulte, Joaquin Villegas, and Jaylia Yan, whose deft research and editing made this project possible. We hope this peek into "how the sausage is made" will serve them well in their great careers ahead.

We also want to recognize the scholars upon whose work we've built, most notably John Arquilla and David Ronfeldt, who kicked this all off. Our mentions in the book provide only a tiny taste of the vibrant and growing field of people working in this topic area, and we encourage you to check out more through the references in the endnotes.

We would like to thank the scores of interviewees who helped us navigate these complex issues. You met many in this book, and there are many you missed out on for reasons of space, but all generously gave of their time, for which we are deeply grateful. We also are appreciative of Walter Parkes giving us permission to quote from the movie *Sneakers* (1992) and of the inspiration his, Phil Robinson's, and Lawrence Lasker's movie has given to a generation of hackers and information warriors since. Thank you to Doan Trang for working with us to design the "LikeWarrior," off to battle with his RPG and smartphone in hand.

We'd be remiss if we didn't also tip our hats to "the internet" and all the hundreds of colorful personalities we interacted with over this long process, even — and especially — the haters. Without realizing it, you helped reveal larger truths amid all the noise and confusion.

Finally, we'd like to express a heartfelt appreciation for each other. Any long-term partnership poses challenges, but ours has fruitfully stretched into half a decade. From joint brainstorming sessions to shared horror over the latest online development to passionate arguments about comma placement, we've been through quite a lot together and are happier for it.

On a personal level, Peter: I would like to thank my colleagues at New America, ably led by Anne-Marie Slaughter. The organization is a unique place. It serves as an incredible platform to engage on the most crucial issues of today and tomorrow, offering both academic freedom and innovation. As the book demonstrates, these qualities are needed now more than ever. Finally, I would like to thank my family. The long haul of writing a book takes a toll not just on writers but on those they love. It is all the more true for projects like this, which involved delving into difficult, and even sometimes hateful and ugly, issues. Susan, you are my best friend, who keeps everything centered in the chaos. Owen and Liam, when I was on the computer, I was doing work, which I know for you little guys was sometimes hard to understand. But please know that you truly and always come first.

On a personal level, Emerson: I would like to thank the three people who made this achievement possible. Michael Horowitz, a brilliant and supportive professor who first opened my eyes to all this "policy stuff"; Janine Davidson, the perfect mentor and friend, who's been there for each step of my journey through the nation's gilded capital; and James Lindsay, whose advice and trust enabled me to take the critical leap from research assistant to bona fide expert. Thanks as well to the Council on Foreign Relations, whose generous research fellowship helped transform this manuscript from mostly theoretical to hard-won reality.

I'd also like to thank four people who *really* made it possible. David Frankenfield offered wise and steady counsel. From their home in the North Georgia wilderness, Emerson and Virginia Brooking followed every twist and turn of the book-writing process; they bolstered me with love and penned enough encouraging emails to equal several times the length of the final product. And thanks, finally, to Anubhuti Mishra. Her wisdom and patient ear were the rock upon which I built everything I did.

Notes

1. THE WAR BEGINS

page

1   *"It was an extraordinary"*: George Orwell, *Homage to Catalonia* (Houghton Mifflin Harcourt, 2015), 33.

"Be sure to tune in": Donald Trump (@realDonaldTrump), "Be sure to tune in and watch Donald Trump on Late Night with David Letterman as he presents the Top Ten List tonight!," Twitter, May 4, 2009, 11:54 a.m., https://twitter.com/realDonaldTrump/status/1698308935.

18 million users: "US Twitter Usage Surpasses Earlier Estimates," eMarketer, September 14, 2009, https://www.emarketer.com/Article/US-Twitter-Usage-Surpasses-Earlier-Estimates/1007271.

2   *surged to a record:* Maggie Shiels, "Web Slows After Jackson's Death," BBC News, June 26, 2009, http://news.bbc.co.uk/1/hi/technology/8120324.stm.

$1.2 billion: Russ Buettner and Charles V. Bagli, "How Donald Trump Bankrupted His Atlantic City Casinos, but Still Earned Millions," *New York Times,* June 11, 2016, https://www.nytimes.com/2016/06/12/nyregion/donald-trump-atlantic-city.html.

banished him: Andrew Bary, "More Troubles in Trump Land," *Barron's,* April 30, 2011, https://www.barrons.com/articles/SB50001424052970203579804576285341283000706?mg=prod/accounts-barrons.

75th most watched: Steve Johnson, "Donald Trump a 'Reality Star Genius'? TV Ratings Tell Different Story," *Chicago Tribune,* February 2, 2016,

http://www.chicagotribune.com/entertainment/tv/ct-donald-trump-not-a-reality-star-genius-20160201-column.html.

ratings were plummeting: "Final 2009–10 Broadcast Primetime Show Average Viewership," TV by the Numbers, June 16, 2010, http://tvby thenumbers.zap2it.com/broadcast/final-2009-10-broadcast-primetime-show-average-viewership/54336/.

beneath his shock-blond: Ashley Feinberg, "Is Donald Trump's Hair a $60,000 Weave? A Gawker Investigation [Updated]," *Gawker,* May 24, 2016, http://gawker.com/is-donald-trump-s-hair-a-60-000-weave-a-gawker-invest-1777581357.

"Don't be afraid": Donald Trump (@realDonaldTrump), "'Don't be afraid of being unique — it's like being afraid of your best self.' — Donald J. Trump http://tinyurl.com/pqpfvm," Twitter, May 17, 2009, 8:00 a.m., https://twitter.com/realdonaldtrump/status/1826225450.

Trump's Twitter messages: In 2009, 59 messages; in 2010, 142; in 2011, 774; and in 2012, 3,531. Trump Twitter Archive, http://www.trumptwitter archive.com/archive.

soon reached into: David Robinson, "Text Analysis of Trump's Tweets Confirms He Writes Only the (Angrier) Android Half," *Variance Explained* (blog), August 9, 2016, http://varianceexplained.org/r/trump-tweets/.

3   *Trump issued screeds:* Donald Trump (@realDonaldTrump), "What a convenient mistake: @BarackObama issued a statement for Kwanza but failed to issue one for Christmas," Twitter, December 28, 2011, 8:02 a.m., https://twitter.com/realdonaldtrump/status/152056935712169984.

whom he'd praised: Friga Garza, "Remember When Donald Trump Said 'I Really Like' President Obama in 2009?," *Complex,* July 13, 2015, http://www.complex.com/pop-culture/2015/07/donald-trump-i-really-like-obama.

directing his Twitter followers: Donald Trump (@realDonaldTrump), "Busy doing phoners this week with Neil Cavuto, Wolf Blitzer, Fox & Friends, and Larry Kudlow... check out http://shouldtrumprun.com/," Twitter, January 21, 2011, 9:20 a.m., https://twitter.com/realdonaldtrump/status/28502098983260160.

"Let's take a closer look": Donald Trump (@realDonaldTrump), "Let's take a closer look at that birth certificate. @BarackObama was described in 2003 as being 'born in Kenya,'" Twitter, May 18, 2012, 12:31 p.m., https://twitter.com/realdonaldtrump/status/203568571148800001.

tiny bursts of dopamine: Amy B. Wang, "Former Facebook VP Says Social Media Is Destroying Society with 'Dopamine-Driven Feedback Loops,'" *Washington Post,* December 12, 2017, https://www.washington post.com/news/the-switch/wp/2017/12/12/former-facebook-vp-says-social-media-is-destroying-society-with-dopamine-driven-feedback-loops/?utm_term=.7fab7098c0aa.

some 15,000 tweets: Compiled from Trump's use of "Twitter for Android." See Trump Twitter Archive, http://www.trumptwitterarchive.com/archive.

nine-tenths of Americans: Bruce Mehlman, "Washington Update: The Race Gets Real," Mehlman Castagnetti Rosen & Thomas, mailer to clients and friends, April 15, 2016.

4   300 million active users: Lara O'Reilly, "Twitter Beats on Revenue and Earnings but Confirms Layoffs," Business Insider, October 27, 2016, http://www.businessinsider.com/twitter-q3-earnings-2016-10.

"I am honored": Daniel Politi, "Trump Deletes One of First Tweets as President After Writing He Is 'Honored' to Serve," *The Slatest* (blog), *Slate,* January 21, 2017, http://www.slate.com/blogs/the_slatest/2017/01/21/trump_deletes_tweet_after_writing_he_is_honored_to_serve.html.

dusty pickup trucks: Matthew Mosk, Brian Ross, and Alex Hosenball, "US Officials Ask How ISIS Got So Many Toyota Trucks," ABC News, October 6, 2015, http://abcnews.go.com/International/us-officials-isis-toyota-trucks/story?id=34266539.

5   *boosting the invaders' messages:* J. M. Berger, "How ISIS Games Twitter," *The Atlantic,* June 16, 2014, http://www.theatlantic.com/international/archive/2014/06/isis-iraq-twitter-social-media-strategy/372856/.

top-trending hashtag: Ibid.

three-quarters of Iraqis: Ghayth Ali Jarad, "The Potential of Developing Iraq Smartphone Market as an Emerging and Lucrative Market," *British Journal of Marketing Studies* 2, no. 2 (2014): 37–42, http://www.eajournals.org/wp-content/uploads/The-potential-of-developing-Iraq-smartphone-market-as-an-emerging-and-lucrative-market.pdf.

nearly 4 million: "Iraq Internet Users," Internet Live Stats, accessed March 15, 2018, http://www.internetlivestats.com/internet-users/iraq/.

metropolis of 1.8 million: Mustafa Habib, "Did They or Didn't They? Iraqi Army Did Not Desert Mosul, They Were Ordered to Leave," Niqash, http://www.niqash.org/en/articles/security/3461/.

6   *25,000-strong garrison:* Ned Parker, Isabel Coles, and Raheem Salman, "Special Report: How Mosul Fell — An Iraqi General Disputes Baghdad's Story," Reuters, October 14, 2014, http://www.reuters.com/article/us-mideast-crisis-gharawi-special-report-idUSKCN0I30Z820141014.

the roughly 10,000: Ibid.

Thousands of soldiers: Habib, "Did They or Didn't They?"

Humvees: Nick Robins-Early, "Iraqi Prime Minister Says ISIS Seized 2,300 Humvees When It Took Mosul," Huffington Post, June 1, 2015, http://

www.huffingtonpost.com/2015/06/01/iraq-isis-humvees_n_7487254. html.

M1A1 battle tanks: Richard Sisk, "ISIS Captures Hundreds of U.S. Vehicles and Tanks in Ramadi from Iraqis," Military.com, http://www.mili tary.com/daily-news/2015/05/20/isis-captures-hundreds-of-us-vehicles-and-tanks-in-ramadi-from-i.html.

half dozen Black Hawk helicopters: "Terror's New Headquarters," *The Economist,* June 14, 2014, http://www.economist.com/news/ leaders/21604160-iraqs-second-city-has-fallen-group-wants-create-state-which-wage-jihad.

7   *5 million soldiers:* Hugh Schofield, "The WW2 Soldiers France Has Forgotten," BBC News, June 4, 2015, http://www.bbc.com/news/maga zine-32956736.

60 massive fortresses: "Maginot Line," History.com, 2009, http://www. history.com/topics/world-war-ii/maginot-line.

precise new battle plans: Robert O. Paxton, "It Wasn't Just Morale," *New York Review of Books,* February 15, 2007, http://www.nybooks.com/arti cles/2007/02/15/it-wasnt-just-morale/.

8   *"Many instructions":* Marc Bloch, *Strange Defeat* (Stellar Books, 2013), 132.

by one of us: See P. W. Singer and Allan Friedman, *Cybersecurity and Cyberwar: What Everyone Needs to Know* (Oxford University Press, 2014).

9   *30,000 foreigners:* Martin Chulov, Jamie Grierson, and Jon Swaine, "Isis Faces Exodus of Foreign Fighters as Its 'Caliphate' Crumbles," *The Guardian,* April 26, 2017, https://www.theguardian.com/world/2017/apr/26/isis-exodus-foreign-fighters-caliphate-crumbles.

more than a dozen: Tim Lister et al., "ISIS Goes Global: 143 Attacks in 29 Countries Have Killed 2,043," CNN, February 12, 2018, http://www. cnn.com/2015/12/17/world/mapping-isis-attacks-around-the-world/ index.html.

more frightened of terrorism: Andrew McGill, "Americans Are More Worried About Terrorism Than They Were After 9/11," *The Atlantic,* September 8, 2016, https://www.theatlantic.com/politics/archive/2016/09/ american-terrorism-fears-september-11/499004/.

"Twitter wars": "Israel and Hamas Wage Twitter War over Gaza Conflict," BBC News, November 15, 2012, http://www.bbc.com/news/technol ogy-20339546.

influenced the targets: Thomas Zeitzoff, "Does Social Media Influence Conflict? Evidence from the 2012 Gaza Conflict," *Journal of Conflict Resolution* 62, no. 1 (2016): 35.

10   *portable cellphone towers:* Ben Kesling and Ali A. Nabhan, "Iraqi Forces
Seek Help by Restoring Mosul Cellphone Service," *Wall Street Journal,* November 1, 2016, https://www.wsj.com/articles/iraqis-seek-help-in-mosul-by-restoring-cell-phone-service-1478030760?tesla=y.

grinning selfies: Iraqi PMU English (@pmu_english), "#Iraqi soldiers
taking #selfies as an #ISIS suicide truck is detonated during their mission
to liberate #Mosul from #ISIS. #selfiewars," Twitter, October 20, 2016, 3:38
p.m. (tweet deleted).

#FreeMosul: Kurdistan Regional Government, "Mosul Liberation
Operation Starts — Statement by Kurdistan Region General Command
of Peshmerga Forces," news release, October 17, 2016, http://cabinet.gov.
krd/a/d.aspx?s=040000&l=12&a=55018.

pretend ISIS propagandists: Patrick Tucker, "How Special Operators
Trained for Psychological Warfare Before the Mosul Fight," Defense One,
November 14, 2016, http://www.defenseone.com/technology/2016/11/
how-special-operators-trained-Psychological-warfare-mosul-fight
/133166/.

shot down a drone: Iraqi News, "#Iraqi forces bring down #ISIS drone
near #Mosul," Facebook, November 2, 2016, November 2, 2016, https://
www.facebook.com/IraqNews/posts/1497246236957690?comment_
id=1497262060289441&comment_tracking=%7B%22tn%22%3A%22R0%
22%7D.

livestreamed the whole thing: Hemin Lihony, "Rudaw: A Pioneer in War
Coverage Through Livestream," Rudaw, October 19, 2016, http://www.rudaw.net/english/opinion/19102016.

tens of thousands: Ibid.

11   *"I helped rescue them":* Hannah Lynch, "Twitter-Sourced Rescue Ops
Saving Civilian Lives in Mosul," Rudaw, March 16, 2017, http://www.rudaw.net/english/middleeast/iraq/160320172.

12   *He'd made disparaging:* Jeff Mayes, "Germel Dossie Now Charged with
Murder for Allegedly Shooting Clifton Frye over Facebook Posts," *Chicago Sun-Times,* June 30, 2015, http://homicides.suntimes.com/2015/06/30/
germel-dossie-now-charged-with-murder-for-allegedly-shooting-clifton-frye-over-facebook-posts/.

killed by gang violence: Madison Park, "Chicago Police Count Fewer
Murders in 2017, but Still 650 People Were Killed," CNN, January 1, 2018,
https://www.cnn.com/2018/01/01/us/chicago-murders-2017-statistics/
index.html.

"settling personal scores": Larry Yellen, "'Cyberbanging': Personal Attacks Online Contributing to Chicago Gang Violence," Fox 32, October

20, 2015, http://www.fox32chicago.com/news/chicago-at-the-tipping-point/36651505-story.

"cybertag" and "cyberbang": "Gangs Rely on Social Media to Communicate," UPI, January 27, 2012, http://www.upi.com/Science_News/Technology/2012/01/27/Gangs-rely-on-social-media-to-communicate/61461327673197/.

13   *The "cyber" version:* Annie Sweeney, "Gangs Increasingly Challenge Rivals Online with Postings, Videos," *Chicago Tribune,* August 17, 2015, http://www.chicagotribune.com/news/local/breaking/ct-gangs-violence-internet-banging-met-20150814-story.html.

Digital sociologists describe: Eric Gordon and Adriana de Souza e Silva, *Net Locality: Why Location Matters in a Networked World* (Wiley-Blackwell, 2011), 3.

The difference in being online: Sweeney, "Gangs Increasingly Challenge Rivals Online."

80 percent of the fights: Ben Austen, "Public Enemies: Social Media Is Fueling Gang Wars in Chicago," *Wired,* September 17, 2013, https://www.wired.com/2013/09/gangs-of-social-media/.

regional variants: Sweeney, "Gangs Increasingly Challenge Rivals Online."

"I'm going to catch you": Ibid.

"haters ten miles north": Austen, "Public Enemies."

allows any individual: Sweeney, "Gangs Increasingly Challenge Rivals Online."

14   *"your dead boy's candles":* Ibid.

"causing people to die": Kim Brunhuber, "L.A. Gangs Using Social Media to Terrorize Communities," CBC News, August 8, 2015, http://www.cbc.ca/news/world/l-a-gangs-using-social-media-to-terrorize-communities-1.3181678.

shareable music videos: Manuel Roig-Franzia, "Mexican Drug Cartels Leave a Bloody Trail on YouTube," *Washington Post,* April 9, 2007, http://www.washingtonpost.com/wp-dyn/content/article/2007/04/08/AR2007040801005.html.

dueling Instagram posts: Alasdair Baverstock, "Narcos at (Instagram) War! From Guns and Girls to Big Cats and Big Piles of Cash, El Chapo's Sons Spark Social Media Battles Between Mexican Cartel Members Showing Off Their Sickening Wealth," *Daily Mail,* September 15, 2015, http://www.dailymail.co.uk/news/article-3226232/Narcos-Instagram-war-guns-girls-big-cats-big-piles-cash-El-Chapo-s-sons-spark-social-media-battles-Mexican-cartel-members-showing-sickening-wealth.html.

El Salvadoran drug gangs: Robert Muggah and Steven Dudley, "Digital Tough Guys," *Foreign Affairs,* November 2, 2015, https://www.foreignaffairs.com/articles/el-salvador/2015-11-02/digital-tough-guys.

fragile 2016 peace: Charlotte Mitchell, "Colombia: Fragile Peace a Year After FARC Referendum," Al Jazeera, October 2, 2017, http://www.aljazeera.com/indepth/features/2017/10/colombia-fragile-peace-year-farc-referendum-171002065629390.html.

rifles for smartphones: Ramón Campos Iriarte, "The Revolution Will Be Whatsapped," Mobilisation Lab, August 15, 2017, https://mobilisationlab.org/colombia-path-revolution-whatsapp/.

15 *"with data plans":* Ibid.

first "social media president": "Rappler, Twitter Launch #PresidentDuterte Emoji," Rappler, July 1, 2016, https://www.rappler.com/technology/social-media/138238-president-duterte-twitter-emoji-hashflag.

a custom emoji: Ibid.

a brutal crackdown: Adrian Chen, "When a Populist Demagogue Takes Power," *The New Yorker,* November 21, 2016, https://www.newyorker.com/magazine/2016/11/21/when-a-populist-demagogue-takes-power.

discrediting journalists: Lauren Etter, "What Happens When the Government Uses Facebook as a Weapon?," *Bloomberg Businessweek,* December 7, 2017, https://www.bloomberg.com/news/features/2017-12-07/how-rodrigo-duterte-turned-facebook-into-a-weapon-with-a-little-help-from-facebook.

sowed false stories: Maria A. Ressa, "Propaganda War: Weaponizing the Internet," Rappler, October 3, 2016, https://www.rappler.com/nation/148007-propaganda-war-weaponizing-internet.

more than 12,000 people: "Philippines: Duterte's 'Drug War' Claims 12,000+ Lives," Human Rights Watch, January 18, 2018, https://www.hrw.org/news/2018/01/18/philippines-dutertes-drug-war-claims-12000-lives.

Israelis and the Palestinian Authority: Hayes Brown, "So It Looks Like the Israelis and Palestinians Have Taken Their Beef to Twitter," BuzzFeed, April 5, 2016, https://www.buzzfeed.com/hayesbrown/so-it-looks-like-the-israelis-and-palestinians-have-taken-th?utm_term=.rbRkqQAqZ2#.bpyjaROaGN.

16 *dueling "Facebook militias":* Francisco Perez, "Graphic Campaign Puts Violence in Kashmir in Social Media Fore," *Deutsche Welle,* July 26, 2016, http://www.dw.com/en/graphic-campaign-puts-violence-in-kashmir-in-social-media-fore/a-19428356.

launching online "expeditions": Anne Henochowicz, "Minitrue: Trolling Tsai Ing-Wen Beyond the Great Firewall," *China Digital Times,* January 22, 2016, https://chinadigitaltimes.net/2016/01/minitrue-trolling-tsai-ing-wen-beyond-great-firewall/.

constantly pushing: Bethany Allen-Ebrahimian and Fergus Ryan, "'Stop Boasting and Fight,'" *Tea Leaf Nation* (blog), *Foreign Policy,* October 27, 2015, https://foreignpolicy.com/2015/10/27/china-south-china-sea-nationalism-united-states-navy-lassen/.

Attending a U.S. military: Chinese crisis simulation, Washington, DC, 2016.

limiting leaders' options: Roundtable (not for attribution), Washington, DC, 2016. See also Center for Strategic and International Studies, "Global Security Forum 2013: A Simulated Crisis in East Asia" (Washington, DC, November 6, 2013).

17   *"continuation of political intercourse":* Carl von Clausewitz, *On War,* ed. and trans. Michael Howard and Peter Paret (Princeton University Press, 2008), 605.

"In essentials": Ibid.

"into the subsequent peace": Ibid.

18   *act of force:* Joseph L. Strange and Richard Iron, "Center of Gravity: What Clausewitz Really Meant" (report, National Defense University, Washington, DC, 2004), http://www.dtic.mil/dtic/tr/fulltext/u2/a520980.pdf.

"center of gravity": Clausewitz, *On War,* 184.

"The moral elements": Ibid.

6.5 million tons: "General Statistics Vietnam War," 103 Field Battery RAA, http://www.103fieldbatteryraa.net/documents/74.html.

killed tens of thousands: Tom Valentine, "How Many People Died in the Vietnam War?," Vietnam War, April 11, 2014, http://thevietnamwar.info/how-many-people-died-in-the-vietnam-war/.

loved to laugh: Harold Graves, *War on the Short Wave* (Foreign Policy Association, 1941), 26.

19   *Twitter cofounder:* David Streitfeld, "'The Internet Is Broken': @ev Is Trying to Salvage It," *New York Times,* May 20, 2017, https://www.nytimes.com/2017/05/20/technology/evan-williams-medium-twitter-internet.html.

20   *some light treason:* Michael D. Shear and Adam Goldman, "Michael Flynn Pleads Guilty to Lying to the F.B.I. and Will Cooperate with Russia Inquiry," *New York Times,* December 1, 2017, https://www.nytimes.com/2017/12/01/us/politics/michael-flynn-guilty-russia-investigation.html.

2. EVERY WIRE A NERVE

24   *"You ask what I am?":* Ray Bradbury, "I Sing the Body Electric!," in *I Sing the Body Electric! Stories by Ray Radbury* (Random House, 1969).
"What is Internet": "1994: 'Today Show': 'What Is the Internet, Anyway?,'" YouTube video, 01:26, uploaded by Jason Miklacic, January 28, 2015, https://www.youtube.com/watch?v=UlJku_CSyNg.
Roughly half the world's: Assuming 7.6 billion. "Current World Population," Worldometers, accessed March 16, 2018, http://www.worldometers.info/world-population/.
never stop being online: Andrew Perrin and Jingjing Jiang, "About a Quarter of U.S. Adults Say They Are 'Almost Constantly' Online," *FactTank* (blog), Pew Research Center, March 14, 2018, http://www.pewresearch.org/fact-tank/2015/12/08/one-fifth-of-americans-report-going-online-almost-constantly/.

25   *"In a few years":* J.C.R. Licklider and Robert W. Taylor, "The Computer as a Communication Device," *Science and Technology,* April 1968, http://urd.let.rug.nl/~welling/cc/licklider-taylor%5B1%5D.pdf.

26   *Intergalactic Computer Network:* Ibid.
"surely the boon": Ibid.
"Every particle": John Archibald Wheeler, "It from Bit," quoted in James Gleick, *The Information: A History, a Theory, a Flood* (Pantheon, 2011), 356.
"competitor to ourselves": Johnny Ryan, *A History of the Internet and the Digital Future* (Reaktion Books, 2010), loc. 207, Kindle.

27   *350 miles:* Ibid., loc. 489.

28   *two Bibles a year:* Max Roser, "Books," published online at OurWorldInData.org, 2015, retrieved from https://ourworldindata.org/data/media-communication/books/, accessed March 16, 2018, https://web.archive.org/web/20160412190622/https://ourworldindata.org/data/media-communication/books/.
first invented in China: Johanna Neuman, *Lights, Camera, War: Is Media Technology Driving International Politics?* (St. Martin's, 1996), 55.
his mass-produced Bibles: Ibid.
resorted to printing: Clay Shirky, *Here Comes Everybody: The Power of Organizing Without Organizations* (New York: Penguin, 2008), 68..
some 200 million: Eltjo Buringh and Jan Luiten van Zanden, "Charting the 'Rise of the West': Manuscripts and Printed Books in Europe, a Long-Term Perspective from the Sixth Through Eighteenth Centuries" (unpublished paper, n.d.), https://socialhistory.org/sites/default/files/docs/projects/books500-1800.pdf.

sold 300,000 copies: Johanna Neuman, "The Media's Impact on International Affairs, Then and Now," *SAIS Review* 16, no. 1 (1996): 109–23.

collection of "news advice": Johannes Weber, "Straßburg, 1605: The Origins of the Newspaper in Europe," *German History* 24 (2006): 387–412.

29   *"Mrs. Silence Dogood":* "The Birth of Silence Dogood," Massachusetts Historical Society, https://www.masshist.org/online/silence_dogood/essay.php?entry_id=203.

26.2-mile distance: "The History of the Marathon," Exercise the Right to Read, http://www.exercisetherighttoread.org/historyofmarathon.pdf.

"Nikomen!": Plutarch, "De Gloria Atheniensium," in *Moralia,* vol. 4, trans. F. C. Babbitt (Loeb Classical Library Edition, 1936), http://penelope.uchicago.edu/Thayer/E/Roman/Texts/Plutarch/Moralia/De_gloria_Atheniensium*.html.

"bursting his heart": Robert Browning, "Pheidippides," PoemHunter.com, https://www.poemhunter.com/poem/pheidippides-2/.

fifty miles per day: A. M. Ramsay, "The Speed of the Roman Imperial Post," *Journal of Roman Studies* 15, no. 1 (1925): 60–74.

meaning "far writer": Tom Standage, *The Victorian Internet: The Remarkable Story of the Telegraph and the Nineteenth Century's On-Line Pioneers* (Walker, 2014), 9, e-book.

Morse spent years: Tom Wheeler, "The First Wired President," *Opinionator* (blog), *New York Times,* May 24, 2012, https://opinionator.blogs.nytimes.com/2012/05/24/the-first-wired-president/?_r=0.

30   *650,000 miles of wire:* Standage, *The Victorian Internet,* 102.

"ten million hands": Neuman, *Lights, Camera, War,* 30.

"the peacemaker of the age": Standage, *The Victorian Internet,* 186.

"diffuse religion": Nigel Linge, "The Trans-Atlantic Telegraph Cable 150th Anniversary Celebration, 1858–2008," University of Salford, http://www.cntr.salford.ac.uk/comms/transatlanticstory.php.

31   *"It gives you":* Standage, *The Victorian Internet,* 153.

yellow-colored comics: J. Hoberman, "When the Yellow Press Got Color," *NYR Daily* (blog), *New York Review of Books,* December 31, 2013, http://www.nybooks.com/daily/2013/12/31/early-comics-society-is-nix/.

"You furnish the pictures": Neuman, *Lights, Camera, War,* 43.

"No Fake War News": Adrienne LaFrance, "How the Fake News Crisis of 1896 Explains Trump," *The Atlantic,* January 19, 2017, https://www.theatlantic.com/technology/archive/2017/01/the-fake-news-crisis-120-years-ago/513710/?utm_source=feed.

number to call: "This Day in History: 1877, Hayes Has First Phone Installed in White House," History.com, http://www.history.com/this-day-in-history/hayes-has-first-phone-installed-in-white-house.

32   *first to build:* Anthony Brown, *Great Ideas in Communication* (D. White, 1969), 141.

aggressively peddled it: Marc Raboy, *Marconi: The Man Who Networked the World* (Oxford University Press, 2016), 423.

played "O Holy Night": Neuman, *Lights, Camera, War,* 137.

20 million radio listeners: Ibid., 139.

just ten minutes: Ibid., 140.

four-fifths of American households: Ibid., 149.

divert the news: Ibid., 151.

33   *"It would not":* Joseph Goebbels, "The Radio as the Eighth Great Power," in *Signals of the New Era: 25 Selected Speeches by Dr. Joseph Goebbels* (Munich: Zentralverlag der NSDAP, 1938).

nearly a thousand: Neuman, *Lights, Camera, War,* 148.

"propagandistic casus belli": Roy Godson and James J. Wirtz, *Strategic Denial and Deception: The Twenty-First Century Challenge* (Transaction, 2011), 100.

"Around the world": Donald M. Bishop, "Classic Quotable: Robert D. Leigh on the Aims of Broadcasting During World War II (1944)," Public Diplomacy Council, http://www.publicdiplomacycouncil.org/commentaries/10-13-15/classic-quotable-robert-d-leigh-aims-broadcasting-during-world-war-ii-1944 (page deleted).

ventriloquist's dummy: Andrew Liszewski, "TVs in the 1920s Had Bottle Cap–Sized Screens, with Just 30 Lines of Resolution," Gizmodo, January 16, 2017, http://gizmodo.com/in-1929-tvs-had-bottle-cap-sized-screens-with-just-30-l-1791250849?utm_campaign=socialflow_gizmodo_twitter&utm_source=gizmodo_twitter&utm_medium=socialflow.

nine of ten American homes: Jordan Winthrop, *The Americans* (McDougal Littell, 1996), 798.

34   *"the most trusted man":* "Final Words: Cronkite's Vietnam Commentary," *All Things Considered,* NPR, July 18, 2009, http://www.npr.org/templates/story/story.php?storyId=106775685.

"I've lost Middle America": Louis Menand, "Seeing It Now," *The New Yorker,* July 9, 2012, https://www.newyorker.com/magazine/2012/07/09/seeing-it-now.

35   *fifteen university computer labs:* Ryan, *A History of the Internet,* loc. 490.

its first international connection: Ibid., loc. 613.

36   *The "@" symbol:* Ian Peter, "The History of Email," NetHistory, http://www.nethistory.info/History%20of%20the%20Internet/email.html.

The email subject: Judy Malloy, "The Origins of Social Media," in *Social Media Archeology and Poetics,* edited by Judy Malloy (Cambridge, MA: MIT Press, 2016), 10.

"we had a social medium": Vint Cerf, phone interview with authors, May 23, 2016.

Yumyum: Ibid.

37   *a good stress test*: Ryan, *A History of the Internet*, loc. 1446.

the humble emoticon: "Original Bboard Thread in Which :-) Was Proposed," Carnegie Mellon University School of Computer Science, http://www.cs.cmu.edu/~sef/Orig-Smiley.htm.

old and familiar things: This is not an uncommon phenomenon. See Andrew Chadwick, *The Hybrid Media System* (New York: Oxford University Press, 2017).

70 institutions: Ryan, *A History of the Internet*, loc. 1673.

38   *AT&T said "no thanks"*: Ibid., loc. 1645.

grew to nearly 160,000: Ibid., loc. 1778.

"Lay down thy packet": Alex Scroxton, "CW@50: The Story of the Internet, and How It Changed the World," *Computer Weekly*, July 2016, http://www.computerweekly.com/feature/CW50-The-story-of-the-internet-and-how-it-changed-the-world.

39   *"information superhighway"*: High Performance Computing Act of 1991, Pub. L. No. 102-194, 105 Stat. 1594 (1991).

NSFNET formally closed: Ryan, *A History of the Internet*, loc. 2367.

reached 360 million: "Internet Users," Internet Live Stats, accessed March 16, 2018, http://www.internetlivestats.com/internet-users/.

worth $3 billion: Adam Lashinsky, "Netscape IPO 20-Year Anniversary: Read Fortune's 2005 Oral History of the Birth of the Web," *Fortune*, http://fortune.com/2015/08/09/remembering-netscape/.

40   *"Google" symbolized*: "Our Story: From the Garage to the Googleplex," Google, https://www.google.com/about/our-story/.

Twelve thousand soldiers: David Ronfeldt et al., *The Zapatista Social Netwar in Mexico* (monograph, RAND, 1998), 2, https://www.rand.org/content/dam/rand/pubs/monograph_reports/1998/MR994.pdf.

41   *more than 130 countries*: Ibid., 117.

"a war on the internet": Ibid., 4.

"rumor first reported": "Scandalous Scoop Breaks Online," BBC News, January 25, 1998, http://news.bbc.co.uk/2/hi/special_report/1998/clinton_scandal/50031.stm.

"comparable to": Manuel Castells, *The Information Age: Economy, Society and Culture*, vol. 1, *The Rise of the Networked Society*, quoted in Paul DiMaggio et al., "Social Implications of the Internet," *Annual Review of Sociology* 27 (2001): 309.

Bowie waxed philosophical: Matt Novak, "Watching David Bowie Argue with an Interviewer About the Future of the Internet Is Beautiful," *Paleofu-*

ture, January 10, 2017, https://paleofuture.gizmodo.com/watching-david-bowie-argue-with-an-interviewer-about-th-1791017656.

42   *"The goal wasn't":* Mark Zuckerberg, "Facebook Interview," YouTube video, 04:49, uploaded by Derek Franzese, March 26, 2013, https://www.youtube.com/watch?v=—APdD6vejI.

had built ZuckNet: Phillip Tracy, "Before Facebook, There Was 'Zuck-Net,' the Chat Service 12-Year-Old Mark Zuckerberg Built for His Family," The Daily Dot, https://www.dailydot.com/debug/mark-zuckerberg-messaging-service-zucknet/.

a bold, crude proclamation: Bari Schwartz, "Hot or Not? Website Briefly Judges Looks," *Harvard Crimson,* November 4, 2003, http://www.thecrimson.com/article/2003/11/4/hot-or-not-website-briefly-judges/?page=single.

43   *some 22,000 votes:* Ibid.

20,000 students: Rachel Feintzeig, "Students Flock to Join College On-line Facebook," *Daily Pennsylvanian,* March 18, 2004, https://web.archive.org/web/20110825022008/http://thedp.com/node/41990.

"I've been paralyzed": Ibid.

44   *3.5 million registered members:* Julia Angwin, *Stealing MySpace: Battle to Control the Most Popular Website in America* (Random House, 2009), 52.

shortened to "blogs": Tim Dowling, "Should We Ban the Word 'Blog'?" *Books* (blog), *The Guardian,* March 22, 2007, https://www.theguardian.com/books/booksblog/2007/mar/22/shouldwebanthewordblog.

100 million active blogs: Jenna Wortham, "After 10 Years of Blogs, the Future's Brighter Than Ever," *Wired,* December 17, 2007, https://www.wired.com/2007/12/after-10-years-of-blogs-the-futures-brighter-than-ever/.

$2.5 trillion: Adjusted for inflation. David Kleinbard, "The $1.7 Trillion Dot.Com Lesson," CNNMoney, November 9, 2000, http://cnnfn.cnn.com/2000/11/09/technology/overview/.

820 million: "Internet Users."

50 percent each year: Jakob Nielsen, "Nielsen's Law of Internet Bandwidth," Nielsen Norman Group, April 5, 1998, https://www.nngroup.com/articles/law-of-bandwidth/.

45   *voice of Salem: Sabrina, the Teenage Witch,* IMDb, http://www.imdb.com/title/tt0115341/.

"Web 2.0": Tim O'Reilly, "What Is Web 2.0: Design Patterns and Business Models for the Next Generation of Software," O'Reilly (website), September 30, 2005, http://www.oreilly.com/pub/a/web2/archive/what-is-web-20.html.

more than 2 million articles: "Wikipedia Publishes 2-Millionth Article," Reuters, September 12, 2007, https://www.reuters.com/article/

us-wikipedia-growth/wikipedia-publishes-2-millionth-articleid
USN1234286820070912.

3 million users: Gary Rivlin, "Wallflower at the Web Party," *New York Times,* October 15, 2006, http://www.nytimes.com/2006/10/15/business/
yourmoney/15friend.html?_r=1&mtrref=en.wikipedia.org.

46   *more than a million active accounts:* Ami Sedghi, "Facebook: 10 Years of
Social Networking, in Numbers," *The Guardian,* February 4, 2014, https://
www.theguardian.com/news/datablog/2014/feb/04/facebook-in-num
bers-statistics.

58 million users: Ibid.

"300 million stories a day": Tom Loftus, "Mark Zuckerberg's Best
Quotes," *Digits* (blog), *Wall Street Journal,* February 1, 2012, http://blogs.
wsj.com/digits/2012/02/01/mark-zuckerbergs-best-quotes/.

2 billion users: Kaya Yurieff, "Facebook Hits 2 Billion Monthly Users,"
CNNMoney, June 27, 2017, http://money.cnn.com/2017/06/27/technol
ogy/facebook-2-billion-users/index.html.

He would show off: Sarah Perez, "Mark Zuckerberg Meets Pope Francis,
Gives Him a Drone," TechCrunch, August 29, 2016, https://techcrunch.
com/2016/08/29/mark-zuckerberg-meets-pope-francis-gives-him-a-
drone/.

arbitrate the pleas: Vitaly Shevchenko, "Ukrainians Petition Facebook
Against 'Russian Trolls,'" BBC News, May 13, 2015, http://www.bbc.com/
news/world-europe-32720965.

47   *"Apple is reinventing":* Mic Wright, "The Original iPhone Announcement Annotated: Steve Jobs' Genius Meets Genius," The Next Web, September 9, 2015, https://thenextweb.com/apple/2015/09/09/genius-anno
tated-with-genius/.

$10,000: John F. Clark, "History of Mobile Applications," http://www.
uky.edu/~jclark/mas490apps/History%20of%20Mobile%20Apps.pdf.

as far back as 1997: Abdulrauf M. Ahmad, "The World's First 'Smartphone' Was Launched on 1997," LinkedIn, April 9, 2016, https://www.
linkedin.com/pulse/worlds-first-smartphone-launched-1997-abdulrauf-
m-ahmad.

Only 200 were produced: Ibid.

the list of features: Wright, "The Original iPhone Announcement Annotated."

making the entire internet: Taylor Martin, "The Evolution of the Smartphone," Pocketnow, July 28, 2014, http://pocketnow.com/2014/07/28/the-
evolution-of-the-smartphone.

smartphone could be used: Mehul Rajput, "Tracing the History and

Evolution of Mobile Apps," Tech.Co, November 27, 2015, https://tech.co/mobile-app-history-evolution-2015-11.

48 *2.5 million such apps:* "Number of Apps Available in Leading App Stores as of March 2017," Statista, accessed March 17, 2018, https://www.statista.com/statistics/276623/number-of-apps-available-in-leading-app-stores/.
Google's Android operating system: Kent German, "A Brief History of Android Phones," CNET, August 2, 2011, https://www.cnet.com/news/a-brief-history-of-android-phones/.
there were some 2 billion: "Ericsson Mobility Report" (Ericsson, November 2017), https://www.ericsson.com/assets/local/mobility-report/documents/2017/ericsson-mobility-report-november-2017.pdf.
reach 8 billion: Ibid.
three-quarters of Americans: "Smartphones Are More Common in Europe, U.S., Less So in Developing Countries," Pew Research Center, February 23, 2016, http://www.pewglobal.org/2016/02/22/smartphone-ownership-and-internet-usage-continues-to-climb-in-emerging-economies/2-23-2016-10-31-58-am-2/.
long since replaced televisions: "Daily Dose: Smartphones Have Become a Staple of the U.S. Media Diet," Nielsen, April 21, 2016, http://www.nielsen.com/us/en/insights/news/2016/daily-dose-smartphones-have-become-a-staple-of-the-us-media-diet.html.
"The definition was": David Sarno, "Twitter Creator Jack Dorsey Illuminates the Site's Founding Document. Part 1," *Technology* (blog), *Los Angeles Times,* February 18, 2009, http://latimesblogs.latimes.com/technology/2009/02/twitter-creator.html.
5,000 tweets per day: Claudine Beaumont, "Twitter Users Send 50 Million Tweets Per Day," *The Telegraph,* February 23, 2010, http://www.telegraph.co.uk/technology/twitter/7297541/Twitter-users-send-50-million-tweets-per-day.html.
50 million: Ibid.

49 *500 million:* "Twitter Usage Statistics," Internet Live Stats, accessed March 17, 2018, http://www.internetlivestats.com/twitter-statistics/.
Journalists took to using: Benjamin Mullin, "Report: Journalists Are Largest, Most Active Verified Group on Twitter," Poynter, May 26, 2015, https://www.poynter.org/2015/report-journalists-are-largest-most-active-group-on-twitter/346957/.
"build and gut-check": Farhad Manjoo, "How Twitter Is Being Gamed to Feed Misinformation," *New York Times,* May 31, 2017, https://www.nytimes.com/2017/05/31/technology/how-twitter-is-being-gamed-to-feed-misinformation.html.

"owning your own newspaper": Michael Barbaro, "Pithy, Mean and Powerful: How Donald Trump Mastered Twitter for 2016," *New York Times*, October 5, 2015, https://www.nytimes.com/2015/10/06/us/politics/donald-trump-twitter-use-campaign-2016.html.

330 million users: "Q3 2017 Letter to Shareholders," Twitter, October 26, 2017, http://files.shareholder.com/downloads/AMDA-2F526X/545891 8398x0x961121/3D6E4631-9478-453F-A813-8DAB496307A1/Q3_17_Share holder_Letter.pdf.

60 million photographs: "Simply Measured Q3 2014 Instagram Study," Simply Measured, http://get.simplymeasured.com/rs/simplymeasured2/images/InstagramStudy2014Q3.pdf?mkt_tok=3RkMMJWWfF9wsRol ua%252FAZKXonjHpfsX56%252BgtXaColMI%252F0ER3fOvrPUfGjI4C TsViI%252BSLDwEYGJlv6SgFQrDEMal41bgNWRM%253D.

50   *$19 billion*: Nicholas Carlson, "Facebook Is Buying Huge Messaging App WhatsApp for $19 Billion!," Business Insider, February 19, 2014, http://www.businessinsider.com/facebook-is-buying-whatsapp-2014-2.

Thailand and the Philippines: Adam Minter, "Emerging Markets Can't Quit Facebook," Bloomberg, April 19, 2018, https://www.bloomberg.com/view/articles/2018-04-19/emerging-markets-can-t-quit-facebook.

"The web that many": Tim Berners-Lee, "The Web Is Under Threat. Join Us and Fight for It," Web Foundation, March 12, 2018, https://web foundation.org/2018/03/web-birthday-29/.

51   *nearly a billion users*: Emma Lee, "WeChat Nears 1 Billion Users," Tech-Node, August 17, 2017, https://technode.com/2017/08/17/wechat-nears-1-billion-users/.

On WeChat: Jonah M. Kessel and Paul Mozur, "How China Is Changing Your Internet," *New York Times*, August 9, 2016, https://www.nytimes.com/video/technology/100000004574648/china-internet-wechat.html.

not allowed to delete: "Why Am I Unable to Delete My WeChat Account?," WeChat Help Center, accessed March 17, 2018, https://help.we-chat.com/cgi-bin/micromsg-bin/oshelpcenter?opcode=2&plat=android &lang=en&id=161028miE7fI161028Qjiii2&Channel=helpcenter.

More than half: "Internet Usage Statistics," Internet World Stats, accessed March 17, 2018, https://www.internetworldstats.com/stats.htm.

"Statistically, we're likely": Jakob Nielsen, "One Billion Internet Users," Nielsen Norman Group, December 19, 2005, https://www.nngroup.com/articles/one-billion-internet-users/.

two-thirds of the online population: *Measuring the Information Society Report 2014* (International Telecommunications Union, 2014), http://www.itu.int/en/ITU-D/Statistics/Documents/publications/mis2014/MIS2014_without_Annex_4.pdf.

double in the next five years: GSMA, "The Mobile Economy 2018," https://www.gsma.com/mobileeconomy/wp-content/uploads/2018/02/ The-Mobile-Economy-Global-2018.pdf.

more people in sub-Saharan Africa: Global Trends 2030: Alternative Worlds (report, National Intelligence Council, 2012), 52, https://www.dni.gov/ files/documents/GlobalTrends_2030.pdf.

52 *Mount Everest base camp:* Daniel Oberhaus, "How to Use the Internet on the Summit of Everest," *Motherboard* (blog), *Vice,* July 31, 2016, https:// motherboard.vice.com/en_us/article/4xa4zp/when-the-internet-came- to-everest.

hundreds of feet beneath: Author's visit to ICBM Vision Lab, Ogden, UT, July 13, 2017.

passed 1 billion: "There Are Now over 1 Billion Websites on the Internet," Business Insider, September 17, 2014, http://www.businessinsider. com/afp-number-of-websites-explodes-past-a-billion-and-counting- 2014-9.

3. THE TRUTH IS OUT THERE

53 *"For nothing is secret":* Luke 8:17 (King James Version).

@ReallyVirtual wasn't: Jethro Mullen, "Whatever Happened to Guy Who Tweeted About Raid That Killed Osama bin Laden?," CNN, January 20, 2016, http://www.cnn.com/2016/01/20/asia/osama-bin-laden- raid-tweeter-sohaib-athar-rewind/.

the more pleasant town: Paul McNamara, "Catching Up with the Guy Who 'Live-Blogged' bin Laden Raid," *BuzzBlog* (blog), Network World, May 1, 2014, https://www.networkworld.com/article/2226829/software/ catching-up-with-the-guy-who-live-blogged-bin-laden-raid.html.

54 *"Helicopter hovering above":* Sohaib Athar (@ReallyVirtual), "Helicopter hovering above Abbottabad at 1 AM (is a rare event)," Twitter, May 1, 2011, 12:58 p.m., https://twitter.com/ReallyVirtual/status/64780730286358528.

"Go away helicopter": Sohaib Athar (@ReallyVirtual), "Go away helicopter — before I take out my giant swatter :-/," Twitter, May 1, 2011, 1:05 p.m., https://twitter.com/reallyvirtual/status/64782523485528065?lang=e n.

"I hope its not": Sohaib Athar (@ReallyVirtual), "A huge window shaking bang here in Abbottabad Cantt. I hope its not the start of something nasty :-S," Twitter, May 1, 2011, 1:09 p.m., https://twitter.com/ReallyVir tual/status/64783440226168832.

Just as Donald Trump: Matthew Dessem, "Fact Check: What *Was* Donald Trump Doing During the bin Laden Raid?" *Browbeat* (blog), *Slate,*

October 20, 2016, http://www.slate.com/blogs/browbeat/2016/10/20/
what_was_donald_trump_doing_during_the_bin_laden_raid.html.

"Justice has been done": Macon Phillips, "Osama bin Laden Dead,"
Home (blog), The White House, May 2, 2011, https://obamawhitehouse.
archives.gov/blog/2011/05/02/osama-bin-laden-dead.

"now I'm the guy": Sohaib Athar (@ReallyVirtual), "Uh oh, now I'm the
guy who liveblogged the Osama raid without knowing it," Twitter, May 1,
2011, 9:41 p.m., https://twitter.com/reallyvirtual/status/649124403532349
44?lang=en.

Twitter follower count jumped: Steve Myers, "How 4 People and Their
Social Network Turned an Unwitting Witness to bin Laden's Death Into
a Citizen Journalist," Poynter, May 3, 2011, http://www.poynter.org/2011/
how-4-people-their-social-network-turned-an-unwitting-witness-to-bin-
ladens-death-into-a-citizen-journalist/130724/.

Local journalists sped: Mullen, "Whatever Happened to Guy?"

only way that Athar: Ibid.

"disintermediation": Robert Gellman, "Disintermediation and the Internet," *Government Information Quarterly* 13, no. 1 (1996): 1–8.

55   *just 6 percent:* "Internet Users (Per 100 People): Pakistan," UNdata,
http://data.un.org/Data.aspx?d=WDI&f=Indicator_Code%3AIT.NET.
USER.P2.

"Secrets now come": Authors' interview with CIA official, northern Virginia, September 10, 2016.

"Welcome to America": "George Allen Introduces Macaca," YouTube
video, 01:02, uploaded by zkman, August 15, 2006, https://www.youtube.
com/watch?v=r9oz0PMnKwI.

made exploratory trips: Tim Craig, "The 'What If' of Allen Haunts the
GOP Race," *Washington Post,* February 6, 2008, http://www.washington
post.com/wp-dyn/content/article/2008/02/05/AR2008020503237.html.

56   *the rally's 100 attendees:* Tim Craig and Michael D. Shear, "Allen Quip
Provokes Outrage, Apology," *Washington Post,* August 15, 2006, http://
www.washingtonpost.com/wp-dyn/content/article/2006/08/14/
AR2006081400589.html.

referring to Sidarth's hair: Ibid.

57   *Salon's person of the year:* Michael Scherer, "Salon Person of the Year: S.
R. Sidarth," Salon, December 16, 2006, http://www.salon.com/2006/12/16/
sidarth/.

9 billion digital devices: Peter Newman, "The Internet of Things 2018
Report: How the IoT Is Evolving to Reach the Mainstream with Businesses and Consumers," Business Insider, February 26, 2018, http://www.
businessinsider.com/the-internet-of-things-2017-report-2018-2-26-1.

soar to 50 billion: "The Sensor-Based Economy," *Wired,* January 2017, https://www.wired.com/brandlab/2017/01/sensor-based-economy/.

almost a trillion sensors: Ibid.

sixty-five different elements: Rebecca Hill, "Shocker: Cambridge Analytica Scandal Touch-Paper Aleksandr Kogan Tapped Twitter Data Too," The Register, April 30, 2018, https://www.theregister.co.uk/2018/04/30/aleksandr_kogan_also_slurped_twitter_data/.

Argus Panoptes: "Argus," *Encyclopedia Britannica,* accessed March 18, 2018, https://www.britannica.com/topic/Argus-Greek-mythology.

the Panopticon: Thomas McMullan, "What Does the Panopticon Mean in the Age of Digital Surveillance?," *The Guardian,* July 23, 2015, https://www.theguardian.com/technology/2015/jul/23/panopticon-digital-surveillance-jeremy-bentham.

filled with "telescreens": George Orwell, *Nineteen Eighty-Four* (Harcourt, Brace & World, 1977).

58 *more than a dozen:* Michael Bauman, "How Beto O'Rourke Explains America," The Ringer, February 28, 2018, https://www.theringer.com/2018/2/28/16898726/beto-orourke-ted-cruz-texas-senate-race-2018-midterm-elections.

500,000 new comments: "The Top 20 Valuable Facebook Statistics—Updated March 2018," Zephoria Digital Marketing, accessed March 18, 2018, https://zephoria.com/top-15-valuable-facebook-statistics/.

500 hours of video: Alexandria Walden, "Hate Crimes and the Rise of White Nationalism," testimony before the House Judiciary Committee, April 9, 2019, https://judiciary.house.gov/sites/democrats.judiciary.house.gov/files/documents/Testimony%20of%20Alexandria%20Walden.pdf.

350,000 tweets: "Twitter Usage Statistics," Internet Live Stats, accessed April 23, 2019, https://www.internetlivestats.com/twitter-statistics/.

"digital universe" doubles: John Gantz and David Reinsel, "The Digital Universe in 2020: Big Data, Bigger Digital Shadows, and Biggest Growth in the Far East—United States," IDC Country Brief (IDC, February 2013), https://www.emc.com/collateral/analyst-reports/idc-digital-universe-united-states.pdf.

agents' daily jogs: "Exercise App Shows Why Anonymous Data Can Still Be Dangerous," CBC Radio, February 2, 2018, http://www.cbc.ca/radio/spark/383-dangerous-data-libraries-and-more-1.4516637/exercise-app-shows-why-anonymous-data-can-still-be-dangerous-1.4516651.

59 *"For the first time":* Mark Milley, speech (Future of War Conference 2017, New America Foundation, Washington, DC, March 21, 2017).

the Allies amassed: William Mahoney, "Before the Beaches: The Logistics of Operation Overlord and D-Day" (undergraduate thesis, University of Indiana, 2014), https://spea.indiana.edu/doc/undergraduate/ugrd_thesis2014_mgmt_mahoney.pdf.

Its algorithms mine: Erica Fink et al., "Ashley Madison: Life After the Hack," CNN (2017), http://money.cnn.com/mostly-human/click-swipe-cheat/?playvid=3.

Russians used it to send: Associated Press, "Ukraine Soldiers Bombarded by 'Pinpoint Propaganda' Texts," ABC News, May 11, 2017, http://abcnews.go.com/amp/Technology/wireStory/sinister-text-messages-reveal-high-tech-front-ukraine-47341695.

started back in 2006: Kate Knibbs, "How Facebook's Design Has Changed over the Last 10 Years," The Daily Dot, February 4, 2014, https://www.dailydot.com/debug/old-facebook-profiles-news-feeds/.

60   *26,000 selfies:* Lauren Buchanan, "Staggering Stats on Selfies," *Best Beauty* (blog), Luster Premium White, November 6, 2015, http://blog.lusterpremiumwhite.com/staggering-stats-on-selfies.

Refugees take selfies: "Paddling to Europe," Reuters, September 18, 2015, http://www.reuters.com/news/picture/paddling-to-europe-idUSRTS1S1K.

victim of an airplane hijacking: Danielle Wiener-Bronner, "The British Bloke Who Took a Photo with the EgyptAir Hijacker Is History's Greatest Hero," Splinter, March 29, 2016, https://fusion.kinja.com/the-british-bloke-who-took-a-photo-with-the-egyptair-hi-1793855879.

leaders of 178 countries: "Twinplomacy Study 2017," Twinplomacy, May 31, 2017, http://twiplomacy.com/blog/twiplomacy-study-2017/.

"Let's all love": Erin Cunningham, "Former Iranian President Ahmadinejad Banned Twitter. Then He Joined It," *Washington Post,* March 6, 2017, https://www.washingtonpost.com/news/worldviews/wp/2017/03/06/former-iranian-president-ahmadinejad-banned-twitter-then-he-joined-it/?utm_term=.12a3f4eb8193.

#TalkOIR: OIR Spokesman (@OIRSpox), "Send YOUR questions about #CJTFOIR #Iraq and #Syria on Thurs. May 26 at 9pm in #Baghdad, 2pm EDT using #TalkOIR," Twitter, March 23, 2016, 9:43 a.m., https://twitter.com/OIRSpox/status/734786795859283968.

A U.S. military officer: Steve Warren, "Hey Reddit," Reddit, https://www.reddit.com/r/IAmA/comments/4i5r4h/hey_reddit_im_col_steve_warren_spokesman_for/.

"the end of forgetting": Jeffrey Rosen, "The Web Means the End of Forgetting," *New York Times,* July 21, 2010, http://www.nytimes.com/2010/07/25/magazine/25privacy-t2.html?pagewanted=all.

61   *a thousand hours:* "Trump Archive," Internet Archive, accessed March 28, 2018, http://archive.org/details/trumparchive&tab=about.

some 40,000 messages: "Trump Twitter Archive," accessed March 28, 2018, http://trumptwitterarchive.com/.

whose very essence: Nahal Toosi, "Is Trump's Twitter Account a National Security Threat?," Politico, December 13, 2016, http://www.politico.com/story/2016/12/trump-twitter-national-security-232518.

"Solid gold info": Noor Al-Sibai, "Naval War College Prof Explains How Trump's 'Stress' Tweets Are a Roadmap for America's Enemies," Raw Story, May 8, 2017, http://www.rawstory.com/2017/05/naval-war-college-prof-explains-how-trumps-stress-tweets-are-a-roadmap-for-americas-enemies/.

Russian intelligence services: Bill Neely, "Russia Compiles Psychological Dossier on Trump for Putin," NBC News, February 20, 2017, http://www.nbcnews.com/news/world/russia-compiles-psychological-dossier-trump-putin-n723196.

"If you had pictures": "Obama Avoids Partisanship in First Post–White House Appearance," CBS News, April 24, 2017, http://www.cbsnews.com/news/obama-speaks-univeristy-of-chicago-community-organizing-live-updates/.

"something much more akin": Olivia Solon, "'This Oversteps a Boundary': Teenagers Perturbed by Facebook Surveillance," *The Guardian,* May 2, 2017, https://www.theguardian.com/technology/2017/may/02/facebook-surveillance-tech-ethics.

62   *6 million Twitter users:* Stephanie Busari, "Tweeting the Terror: How Social Media Reacted to Mumbai," CNN, November 28, 2008, http://www.cnn.com/2008/WORLD/asiapcf/11/27/mumbai.twitter/.

"I have just heard": Kapil (@kapilb), "I have just heard 2 more loud blasts around my house in colaba," Twitter, November 26, 2008, 9:09 a.m., https://twitter.com/kapilb/status/1024849394.

"Grenades thrown": Romi (@romik), "grenades thrown at colaba," Twitter, November 26, 2008, 9:31 a.m., https://twitter.com/romik/status/1024888964.

"people have been evacuated": Sunil Verma (@skverma), "I just spoke with my friends at the Taj and Oberoi — people have been evacuated or are barracaded in their Rooms," Twitter, November 26, 2008, 10:55 a.m., https://twitter.com/skverma/status/1025031065.

Mumbai's online community: Robert Mackey, "Tracking the Mumbai Attacks," *The Lede* (blog), *New York Times,* November 26, 2008, https://thelede.blogs.nytimes.com/2008/11/26/tracking-the-mumbai-attacks/?pagemode=print&_r=0.

He posted them: Charles Arthur, "How Twitter and Flickr Recorded the Mumbai Terror Attacks," *The Guardian,* November 27, 2008, http://www.theguardian.com/technology/2008/nov/27/mumbai-terror-attacks-twitter-flickr.

63   *more than 1,800 times:* "2008 Mumbai Attacks: Revision History," Wikipedia, accessed March 18, 2018, https://en.wikipedia.org/w/index.php?title=2008_Mumbai_attacks&dir=prev&offset=20081129144458&limit=250&action=history.

Google Maps would: "Map of Mumbai Attacks," Google Maps, accessed March 18, 2018, https://www.google.com/maps/d/viewer?ll=18.91740000000004%2C72.82687799999997&spn=0.007054%2C0.007864&hl=en&msa=0&z=15&ie=UTF8&mid=1I6SuyXRZLDapOIK8ViEQ3j608Tw.

in cellphone contact: Thomas Elkjer Nissen, *#TheWeaponizationOf SocialMedia: @Characteristics_of_Contemporary_Conflicts* (Royal Danish Defence College, 2015), 93, https://www.stratcomcoe.org/thomas-nissen-weaponization-social-media; Manish Agrawal, Onook Oh, and H. Raghav Rao, "Information Control and Terrorism: Tracking the Mumbai Terrorist Attack Through Twitter," *Information Systems Frontiers* 13, no. 1 (2011): 33–43.

64   *Indian government had said:* Busari, "Tweeting the Terror."

"Die, die, die": Ibid.

begging for blood donations: Noah Schatman, "Mumbai Attack Aftermath Detailed, Tweet by Tweet," *Wired,* November 26, 2008, https://www.wired.com/2008/11/first-hand-acco/.

spread word of tip lines: Tamar Weinberg, *The New Community Rules: Marketing on the Social Web* (O'Reilly, 2009), 127.

to work collectively: Jeff Howe, "The Rise of Crowdsourcing," *Wired,* June 1, 2006, https://www.wired.com/2006/06/crowds/.

65   *$218 million:* Clare Foran, "Bernie Sanders's Big Money," *The Atlantic,* March 1, 2016, https://www.theatlantic.com/politics/archive/2016/03/bernie-sanders-fundraising/471648/.

"to crowdfund their war": "Why an Ordinary Man Went to Fight Islamic State," *The Economist,* December 24, 2016, https://www.economist.com/news/christmas-specials/21712055-when-islamic-state-looked-unbeatable-ordinary-men-and-women-went-fight-them-why.

fundamentalist donors: Elizabeth Dickinson, "Private Gulf Donors and Extremist Rebels in Syria" (panel presentation, Brookings Institution, Washington, DC, December 19, 2013).

"financial jihad": Lisa Daftari, "Hezbollah's New Crowdfunding Campaign: 'Equip a Mujahid,'" The Foreign Desk, February 9, 2017, http://www.foreigndesknews.com/world/middle-east/hezbollahs-new-crowd

funding-campaign-equip-mujahid/?utm_content=buffer08b58&utm_
medium=social&utm_source=twitter.com&utm_campaign=buffer.

rocket-propelled grenade: Dickinson, "Private Gulf Donors."

their religious obligations: Daftari, "Hezbollah's New Crowdfunding
Campaign."

66 *"Thanks for vote":* Adam Linehan, "This Controversial Instagram Account Lets You Decide Whether 'ISIS Fighters' Live or
Die," Task & Purpose, March 28, 2016, http://taskandpurpose.com/
instagram-account-lets-decide-whether-isis-fighters-live-die/?utm_
source=twitter&utm_medium=social&utm_campaign=share&utm_
content=tp-share.

"A guy on the toilet": Ibid.

*took about thirty seconds: After Action Report for the Response to the 2013
Boston Marathon Bombings* (Massachusetts Emergency Management
Agency et al., December 2014), http://www.mass.gov/eopss/docs/mema/
after-action-report-for-the-response-to-the-2013-boston-marathon-bomb-
ings.pdf.

"Holy shit!": Kristen Surman (@KristenSurman), "Holy shit! Explosion!," Twitter, April 15, 2013, 2:50 p.m., https://twitter.com/KristenSur
man/status/323871059499683840.

first photo of the attack: Dan Lampariello (@Boston_to_a_T), "Explosion at coply," Twitter, April 15, 2013, 2:50 p.m., https://twitter.com/Dan
LampNews/status/323871088532668416.

Fox Sports Radio: Fox Sports 1380/95.3 (@KRKO1380), "BREAKING:
Per our man on the ground at the Boston Marathon, @tooblackdogs, there
was an explosion. More to follow," Twitter, April 15, 2013, 2:52 p.m., https://
twitter.com/KRKO1380/status/323871355860840450.

nearly an hour: Hong Qu, "Social Media and the Boston Bombings:
When Citizens and Journalists Cover the Same Story," Nieman Lab, April
17, 2013, http://www.niemanlab.org/2013/04/social-media-and-the-bos
ton-bombings-when-citizens-and-journalists-cover-the-same-story/.

more than doubled: "Number of Smartphone Users Worldwide from
2014 to 2020 (Billions)," Statista, accessed March 18, 2018, https://www.
statista.com/statistics/330695/number-of-smartphone-users-worldwide/.

an incomprehensibly vast now: "Time," *Stanford Encyclopedia of Philosophy,* updated January 24, 2014, accessed March 18, 2018, https://plato.stan
ford.edu/entries/time/#PreEteGroUniThe.

67 *"present shock":* Douglas Rushkoff, *Present Shock: When Everything Happens Now* (Current, 2014).

ripe old age: Ryan Broderick, "What It's Like to Live-Tweet the Day
Your Neighborhood Becomes a War Zone," BuzzFeed, August 30, 2016,

https://www.buzzfeed.com/ryanhatesthis/from-complexo-da-alemao-
to-torchbearer?utm_term=.ev4KvD1v5E#.kdy0XexXgb.

68   *hyperlocal reporting:* Ibid.

a truly local paper: Selinsgrove, like the other small towns of Pennsylvania's central Susquehanna Valley, is served by the *Daily Item,* circulation
roughly 14,000.

"I may be nine": Hilde Kate Lysiak, "Yes, I'm a Nine-Year-Old Girl. But
I'm Still a Serious Reporter," *The Guardian,* April 6, 2016, https://www.
theguardian.com/commentisfree/2016/apr/06/nine-year-old-reporter-
orange-street-news-truth.

nearly 800 documented attacks: Paul Imison, "Journalists in Mexico
Killed in Record Numbers — Along with Freedom of Speech," Fox News,
April 4, 2017, http://www.foxnews.com/world/2017/04/04/journalists-
in-mexico-killed-in-record-numbers-along-with-freedom-speech.html.

The Norte *newspaper:* Associated Press, "Mexican Newspaper Closes
Citing Insecurity for Journalists," Fox News, April 2, 2017, http://www
.foxnews.com/world/2017/04/02/mexican-newspaper-closes-citing-inse
curity-for-journalists.html.

69   *"You do it":* Dana Priest, "Censor or Die: The Death of Mexican News
in the Age of Drug Cartels," *Washington Post,* December 11, 2015, https://
www.washingtonpost.com/investigations/censor-or-die-the-death-of-
mexican-news-in-the-age-of-drug-cartels/2015/12/09/23acf3ae-8a26-11e5
-9a07-453018f9a0ec_story.html?utm_term=.e783173fb136.

But even that: "'Adios!': Mexican Newspaper Norte Closes After Murder of Journalist," *The Guardian,* April 3, 2017, https://www.theguardian.
com/world/2017/apr/03/adios-mexican-newspaper-norte-closes-after-of-
journalist.

Catwoman: Jason McGahan, "She Tweeted Against the Mexican Cartels, They Tweeted Her Murder," The Daily Beast, October 21, 2014, http://
www.thedailybeast.com/she-tweeted-against-the-mexican-cartels-they-
tweeted-her-murder?via=desktop&source=twitter.

killed over 15,000 people: Alasdair Baverstock, "Revealed, America's
Most Fearful City Where Texans Live Next to a 'War Zone,'" *Daily Mail,*
October 8, 2015, http://www.dailymail.co.uk/news/article-3263226/Reve
aled-America-s-fearful-city-Texans-live-war-zone-McAllen-two-mur
ders-year-mile-away-Mexican-border-Reynosa-15-000-cut-five-years-vor
tex-cartel-murders-extortion-torture.html.

70   *"tweeted her murder":* McGahan, "She Tweeted Against the Mexican
Cartels."

a group of seventeen: Alice Speri, "Raqqa Is Being Slaughtered Silently, and These Guys Are Risking Their Lives to Document It," *Vice,* September 25, 2014, https://news.vice.com/article/raqqa-is-being-slaughtered-silently-and-these-guys-are-risking-their-lives-to-document-it.

more powerful than: David Remnick, "Telling the Truth About ISIS and Raqqa," *The New Yorker,* November 22, 2015, https://www.newyorker.com/news/news-desk/telling-the-truth-about-isis-and-raqqa.

paraded in front of: Mansour Al-Hadj, "Anti-ISIS Activists in Al-Raqqa Vow to Remain Resolute Despite Constant Death Threats, Assassinations," Middle East Media Research Institute, February 3, 2016, https://www.memri.org/jttm/anti-isis-activists-al-raqqa-vow-remain-resolute-despite-constant-death-threats-assassinations.

ten members of the network: David Remnick, "The Tragic Legacy of Raqqa Is Being Slaughtered Silently," *The New Yorker,* October 21, 2017, https://www.newyorker.com/news/as-told-to/the-tragic-legacy-of-raqqa-is-being-slaughtered-silently.

"It's okay": Elahe Izadi and Liz Sly, "Female Activist Killed by the Islamic State Posted This Final Defiant Message," *Washington Post,* January 7, 2016, https://www.washingtonpost.com/news/worldviews/wp/2016/01/07/female-activist-killed-by-the-islamic-state-posted-this-final-defiant-message/.

71 *"here's what it looks like":* Cor Pan, "Mocht hij verdwijnen, zo ziet hij d'r uit," Facebook, July 17, 2014, accessed March 18, 2018, https://www.facebook.com/photo.php?fbid=465419050262262&set=a.121009184703252.21333.100003825135026&type=3&theater.

They were a mix: Christopher Miller, "Field of Death: How MH17 and Its Passengers Became Victims of a Distant War," Mashable, July 16, 2015, http://mashable.com/2015/07/16/mh17-crash-field-of-death/#krJ3QrQ8Xiqo.

thirty milliseconds: Crash of Malaysia Airlines Flight MH17 (report, Dutch Safety Board, October 2015), 115.

over 7,600 pieces: "A Detailed Description of the BUK SA-11 Which Could Have Shot Down MH17," WhatHappenedToFlightMH17.com, March 21, 2015, http://www.whathappenedtoflightmh17.com/a-detailed-description-of-the-buk-sa-11-which-could-have-shot-down-mh17/.

separated into three pieces: Crash of Malaysia Airlines Flight MH17, 162.

For ninety seconds: Ibid., 165.

under five minutes: Amber Dawson et al., "As It Happened: Malaysian Plane Crash in Ukraine," BBC News, July 17, 2014, http://www.bbc.com/news/world-28354787.

"just fell": Miller, "Field of Death."

72   *World of Warcraft addict:* Patrick Radden Keefe, "Rocket Man," *The New Yorker,* November 25, 2013, https://www.newyorker.com/maga zine/2013/11/25/rocket-man-2.

"Brown Moses": Ibid.

his knowledge: Matthew Weaver, "How Brown Moses Exposed Syrian Arms Trafficking from His Front Room," *The Guardian,* March 21, 2013, https://www.theguardian.com/world/2013/mar/21/frontroom-blogger-analyses-weapons-syria-frontline.

had used nerve gas: Brown Moses, "Who Was Responsible for the August 21st Attack?," *Brown Moses Blog,* September 16, 2013, http://brown-moses.blogspot.com/2013/09/who-was-responsible-for-august-21st.html.

73   *published its first report:* Eliot Higgins, "Buk Transporter Filmed 'Heading to Russia' Sighted in an Earlier Photograph," Bellingcat, July 18, 2014, https://www.bellingcat.com/news/uk-and-europe/2014/07/18/buk-transporter-filmed-heading-to-russia-sighted-in-an-earlier-photograph/.

pattern of shrapnel damage: Eliot Higgins, "The Latest Open Source Theories, Speculation and Debunks on Flight MH17," Bellingcat, July 22, 2014, https://www.bellingcat.com/news/uk-and-europe/2014/07/22/the-latest-open-source-theories-speculation-and-debunks-on-flight-mh17/.

"a lot of obsessive people": Keefe, "Rocket Man."

His in-laws thought: Aric Toler, interview with authors, Washington, DC, March 10, 2016.

soon after the crash: Higgins, "Buk Transporter."

74   *the equivalent of:* "Origin of the Separatists' Buk: A Bellingcat Investigation," Bellingcat, November 8, 2014, https://www.bellingcat.com/news/uk-and-europe/2014/11/08/origin-of-the-separatists-buk-a-belling cat-investigation/.

mapping out the odyssey: Bellingcat interactive map, Mapbox, accessed March 18, 2018, https://www.mapbox.com/labs/bellingcat/index .html.

even snapped a picture: Toler interview.

Worried about their loved ones: "MH17 — Potential Suspects and Witnesses from the 53rd Anti-Aircraft Missile Brigade," Bellingcat, 2016, https://www.bellingcat.com/wp-content/uploads/2016/02/53rd-report-public.pdf.

75   *It included the names:* Janene Pieters, "Twenty Russians Wanted for Questioning in MH17 Downing," NL Times, January 4, 2016, http://nl times.nl/2016/01/04/twenty-russians-wanted-questioning-mh17-down ing.

One OSINT analyst: Email to authors, February 8, 2016.

76 *GVA Dictator Alert:* Amar Toor, "This Twitter Bot Is Tracking Dictators' Flights in and out of Geneva," The Verge, October 13, 2016, https://www.theverge.com/2016/10/13/13243072/twitter-bot-tracks-dictator-planes-geneva-gva-tracker.

one of every five: Eric Gomez, "How Collectible Medals and Facebook Likes Encouraged Cheaters in the Mexico City Marathon," ESPN.com, April 16, 2018, http://www.espn.com/blog/onenacion/post/_/id/8439/how-collectible-medals-and-likes-encouraged-cheaters-in-the-mexico-city-marathon.

chemical weapons in Syria: "Open Source Survey of Alleged Chemical Attacks in Douma on 7th April 2018," Bellingcat, April 11, 2018, https://www.bellingcat.com/news/mena/2018/04/11/open-source-survey-alleged-chemical-attacks-douma-7th-april-2018/.

indicted Mahmoud Al-Werfalli: "Situation in Libya in the Case of *The Prosecutor v. Mahmoud Mustafa Busayf Al-Werfalli,*" International Criminal Court, August 15, 2017, https://www.icc-cpi.int/CourtRecords/CR2017_05031.PDF.

"virtual kidnappings": Daniel Borunda, "'Virtual Kidnapping' Cases Spread from Mexico to US, FBI Says," *El Paso Times,* October 19, 2017, https://www.elpasotimes.com/story/news/crime/2017/10/19/virtual-kidnapping-cases-spread-mexico-us-fbi-says/780847001/.

Scouring Libyan Facebook groups: C. J. Chivers, "Facebook Groups Act as Weapons Bazaars for Militias," *New York Times,* April 6, 2016, https://www.nytimes.com/2016/04/07/world/middleeast/facebook-weapons-syria-libya-iraq.html.

77 *traced to stocks:* Ibid.

"By carefully gathering": Sangwon Yoon, "This Startup Is Predicting the Future by Decoding the Past," *Bloomberg, Markets,* April 6, 2016, https://www.bloomberg.com/news/articles/2016-04-06/this-startup-is-predicting-the-future-by-decoding-the-past.

even ballistic missile tests: "Beyond Parallel," Center for Strategic and International Studies, https://beyondparallel.csis.org/signals/.

"The exponential explosion": Michael Flynn, phone interview with authors, May 26, 2016.

78 *reading the obituary:* Adam Rawnsley, "The Open-Source Spies of World War II: U.S. Intelligence Analysts Helped Shape Modern Spycraft," *War Is Boring* (blog), Medium, March 2, 2015, https://medium.com/war-is-boring/the-open-source-spies-of-world-war-ii-7943bd5b663c.

roughly 45,000 pages: Anthony Olcott, *Open Source Intelligence in a Networked World* (Continuum, 2012), 16.

Foreign Broadcast Monitoring Service: Kalev Leetaru, "The Scope of

FBIS and BBC Open-Source Media Coverage, 1979–2008," *Studies in Intelligence* 54, no. 1 (2010): 17–37.

over a thousand Soviet journals: Olcott, *Open Source Intelligence in a Networked World.*

a thousand hours of television: Leetaru, "The Scope of FBIS and BBC Open-Source Media Coverage."

they suspected trickery: Olcott, *Open Source Intelligence in a Networked World.*

had risen to 50,000: Ibid., 90.

79   *"Whether you're a CEO":* Flynn interview.

Flynn joined: Nicholas Schmidle, "Michael Flynn, General Chaos," *The New Yorker,* February 27, 2017, http://www.newyorker.com/maga zine/2017/02/27/michael-flynn-general-chaos.

They would eschew: Ibid.

80   *He envisioned:* Ibid.

Before the rise: Flynn interview.

"unwanted pregnancy": Ibid.

"redheaded stepchild": Ibid.

alarmed the DIA's bureaucracy: Patrick Tucker, "The Other Michael Flynn," Defense One, November 21, 2016, https://cdn.defenseone. com/b/defenseone/interstitial.html?v=8.8.0&rf=http%3A%2F%2F www.defenseone.com%2Fpolitics%2F2016%2F11%2Fother-michael-flynn%2F133337%2F.

after thirty-three years: Schmidle, "Michael Flynn."

$530,000 deal: Fredreka Schouten, "Turkish Client Paid $530,000 to Michael Flynn's Consulting Firm," *USA Today,* March 8, 2017, https:// www.usatoday.com/story/news/politics/2017/03/08/michael-flynn-received-530000-from-turkish-client-during-trump-campaign/98917184/.

81   *"he was going to be":* Schmidle, "Michael Flynn."

"Fear of Muslims": Michael Flynn (@GenFlynn), "Fear of Muslims is RATIONAL: please forward this to others: the truth fears no questions . . . ," Twitter, February 26, 2016, 5:14 p.m., accessed March 18, 2018, https://twitter.com/genflynn/status/703387702998278144?lang=en.

"Not anymore, Jews": Kristen East, "Flynn Retweets Anti-Semitic Remark," Politico, July 24, 2016, http://www.politico.com/story/2016/07/michael-flynn-twitter-226091.

a "jihadi" who "laundered": Bryan Bender and Andrew Hanna, "Flynn Under Fire for Fake News," Politico, December 5, 2016, https://www.polit ico.com/story/2016/12/michael-flynn-conspiracy-pizzeria-trump-232227.

"Sex Crimes w Children": Lauren Carroll, "Michael Flynn's Troubling Penchant for Conspiracy Theories," Politifact, February 14, 2017, http://

www.politifact.com/truth-o-meter/article/2017/feb/14/michael-flynns-troubling-penchant-conspiracy-thoer/.

outlaw Christianity: Ibid.

human blood and semen: Bender and Hanna, "Flynn Under Fire for Fake News."

"making sure that": Flynn interview.

82 *"We are going":* Michael Flynn (@GenFlynn), "We are going to win and win and win at everything we do. It is going to be tough, but Team Trump-Pence will #MAGA," Twitter, December 1, 2016, 7:33 p.m., https://twitter.com/GenFlynn/status/804528907978412033?ref_src=twsrc%5Etfw&ref_url=https%3A%2F%2F; http://www.washingtonpost.com%2Fblogs%2Fpost-partisan%2Fwp%2F2016%2F12%2F02%2Fsorry-lt-gen-flynn-its-unrealistic-to-win-and-win-and-win-at-everything%2F.

"false, fictitious": United States of America v. Michael T. Flynn, United States District Court for the District of Columbia, filed November 30, 2017, https://www.politico.com/f/?id=00000160-128a-dd6b-afeb-37afd8000000.

piercing through the "fog": Flynn interview.

4. THE EMPIRES STRIKE BACK

83 *"'Truth' is a lost cause":* Peter Pomerantsev and Michael Weiss, "The Menace of Unreality: How the Kremlin Weaponizes Information, Culture and Money" (report, Institute of Modern Russia, 2014), http://www.interpretermag.com/wp-content/uploads/2014/11/The_Menace_of_Unreality_Final.pdf.

"Information wants": Steven Levy, "'Hackers' and 'Information Wants to Be Free,'" *Backchannel* (blog), Medium, November 21, 2014, https://medium.com/backchannel/the-definitive-story-of-information-wants-to-be-free-a8d95427641c.

"The Net interprets": Philip Elmer-Dewitt, "First Nation in Cyberspace," *Time,* December 6, 1993, http://kirste.userpage.fu-berlin.de/outerspace/internet-article.html.

"the Japanese guy": Bruce Sterling, "Triumph of the Plastic People," *Wired,* January 1, 1995, https://www.wired.com/1995/01/prague/.

84 *first so-called internet revolution:* Olesya Tkacheva et al., *Internet Freedom and Political Space* (RAND, 2013), 121.

government censors: Lev Grossman, "Iran Protests: Twitter, the Medium of the Movement," *Time,* June 17, 2009, http://content.time.com/time/world/article/0,8599,1905125,00.html.

98 percent of the links: "Iran and the 'Twitter Revolution,'" Pew Research Center, June 25, 2009, http://www.journalism.org/2009/06/25/iran-and-twitter-revolution/.

"The Revolution": Andrew Sullivan, "The Revolution Will Be Twittered," *The Daily Dish* (blog), *The Atlantic,* June 13, 2009, http://www.theatlantic.com/daily-dish/archive/2009/06/the-revolution-will-be-twittered/200478/.

Nobel Peace Prize: Lewis Wallace, "*Wired* Backs Internet for Nobel Peace Prize," *Wired,* November 20, 2009, https://www.wired.com/2009/11/internet-for-peace-nobel/.

85   *Mohamed Bouazizi:* Yasmine Ryan, "The Tragic Life of a Street Vendor," Al Jazeera, January 20, 2011, http://www.aljazeera.com/indepth/features/2011/01/201111684242518839.html.

"Is Egypt about to have": Abigail Hauslohner, "Is Egypt About to Have a Facebook Revolution?," *Time,* January 24, 2011, http://content.time.com/time/world/article/0,8599,2044142,00.html.

Mubarak's resignation: Leila Fadel, "With Peace, Egyptians Overthrow a Dictator," *Washington Post,* February 11, 2011, http://www.washingtonpost.com/wp-dyn/content/article/2011/02/11/AR2011021105709.html.

"just give them": Jeffrey Ghannam, "In the Middle East, This Is Not a Facebook Revolution," *Washington Post,* February 20, 2011, http://www.washingtonpost.com/wp-dyn/content/article/2011/02/18/AR2011021806964.html.

"I want to meet": Sajid Farooq, "Organizer of 'Revolution 2.0' Wants to Meet Mark Zuckerberg," NBC Bay Area, March 5, 2011, https://www.nbcbayarea.com/blogs/press-here/Egypts-Revolution-20-Organizer-Wants-to-Thank-Mark-Zuckerberg-115924344.html.

naming his firstborn: Alexia Tsotsis, "To Celebrate the #Jan25 Revolution, Egyptian Names His Firstborn 'Facebook,'" TechCrunch, February 20, 2011, https://techcrunch.com/2011/02/19/facebook-egypt-newborn/.

the Arab Spring: Kentaro Toyama, "Malcolm Gladwell Is Right: Facebook, Social Media and the Real Story of Political Change," Salon, June 6, 2015, http://www.salon.com/2015/06/06/malcolm_gladwell_is_right_facebook_social_media_and_the_real_story_of_political_change/.

86   *"organize without organizations":* Clay Shirky, *Here Comes Everybody: The Power of Organizing Without Organizations* (Penguin, 2008).

"the liberating power": Roger Cohen, "Revolutionary Arab Geeks," *New York Times,* January 27, 2011, http://www.nytimes.com/2011/01/28/opinion/28iht-edcohen28.html.

"We had an arsenal": Evgeny Morozov, *The Net Delusion: The Dark Side of Internet Freedom* (PublicAffairs, 2011), loc. 250, Kindle.

"an enthusiastic belief": Ibid., loc. 223–31.

87   *Liu was a new arrival:* Charles Liu, "Chinese Guy, Angry at Embarrassing Photos Circulating Online, Tries to Destroy Internet, " Nanfang, August 26, 2016, https://thenanfang.com/man-tries-prevent-online-humiliation-destroying-public-internet-routers/.

Liu was sent to prison: "The Man Was Sneered by the Jump Square Dance Maliciously Disrupting the Communications Cable," trans. Google Translate, website in Chinese, August 24, 2016, http://news.163.com/16/0824/11/BV7TGLJS00014SEH.html.

88   *some 2,000 satellites:* Virgil Labrador, "Satellite Communication," *Encyclopedia Britannica,* accessed March 19, 2018, https://www.britannica.com/technology/satellite-communication.

Just a few ISPs: "Internet Service Providers (ISPs) — The World Factbook — CIA," *Encyclopedia of the Nations,* accessed March 19, 2018, http://www.nationsencyclopedia.com/WorldStats/CIA-Internet-Service-Providers-ISPs.html.

two-thirds of all ISPs: Ibid.

sixty-one countries: Jim Cowie, "Could It Happen in Your Country?," *VantagePoint* (blog), Dyn, November 30, 2012, http://dyn.com/blog/could-it-happen-in-your-countr/.

cut off the internet: Jim Cowie, "Syrian Internet Shutdown," *VantagePoint* (blog), Dyn, June 3, 2011, https://dyn.com/blog/syrian-internet-shutdown/.

Many Algerians suspected: Elvis Boh, "Algeria's Decision to Block Social Media Highly Criticized," Africanews, June 21, 2016, http://www.africanews.com/2016/06/21/algeria-s-decision-to-block-social-media-highly-criticised//.

89   *Algeria's economy lost:* Darrell M. West, "Internet Shutdowns Cost Countries $2.4 Billion Last Year" (report, Center for Technology Information at Brookings, Brookings Institution, October 2016), https://www.brookings.edu/wp-content/uploads/2016/10/intenet-shutdowns-v-3.pdf.

$190 million: Ibid.

"internet curfew": Bill Marczak, "Time for Some Internet Problems in Duraz: Bahraini ISPs Impose Internet Curfew in Protest Village," Bahrain Watch, August 3, 2016, https://bahrainwatch.org/blog/2016/08/03/bahrain-internet-curfew/.

authorities narrowed their focus: Ibid.

every time a protest: "First Evidence of Iranian Internet Throttling as a Form of Censorship," *MIT Technology Review,* June 24, 2013, https://www.technologyreview.com/s/516361/first-evidence-of-iranian-internet-throttling-as-a-form-of-censorship/.

a "clean" internet: "Tightening the Net: Internet Security and Censorship in Iran. Part 1: The National Internet Project" (Article 19, Free Word Center, London, 2012), https://www.article19.org/data/files/me dialibrary/38315/The-National-Internet-AR-KA-final.pdf; Corin Faife, "Iran's 'National Internet' Offers Connectivity at the Cost of Censorship," *Motherboard* (blog), *Vice*, March 29, 2016, https://motherboard.vice.com/ en_us/article/yp3pxg/irans-national-internet-offers-connectivity-at-the-cost-of-censorship.

90   *solar-powered phone chargers:* Donia Al-Watan, "Seeking Internet Access, Syrians Turn to Turkey's Wireless Network," *Al Monitor*, April 19, 2015, http://www.al-monitor.com/pulse/originals/2015/04/aleppo-rebel-control-internet-networks-syria-turkey.html.

about thirty websites: E. Tammy Kim, "Two Koreas, Two Cults, Two Internets," *The New Yorker*, November 3, 2016, https://www.newyorker.com/ tech/elements/two-koreas-two-cults-two-internets.

Nissenbaum is a journalist: Dion Nissenbaum, email to author, April 15, 2017.

91   *attempted military coup:* David Cenciotti, "Exclusive: All the Details About the Air Ops and Aerial Battle over Turkey During the Military Coup to Depose Erdogan," *The Aviationist* (blog), July 18, 2016, https:// theaviationist.com/2016/07/18/exclusive-all-the-details-about-the-aerial-battle-over-turkey-during-the-military-coup/.

"RT HERKES SOKAGA": Dion Nissenbaum, "Turkish President Foiled Coup with Luck, Tech Savvy," *Wall Street Journal*, July 17, 2016, https:// www.wsj.com/articles/coup-plotters-targeted-turkish-president-with-daring-helicopter-raid-1468786991.

digital content coordinator: Emre Kizilkaya, "FaceTime Beats WhatsApp in Turkey's Failed Coup," *U.S. News and World Report*, July 25, 2016, http://www.al-monitor.com/pulse/originals/2016/07/turkey-coup-attempt-whatsapp-facetime.html#ixzz4iDFRBe2B.

game of hide-and-seek: Ibid.

"This insurgency": Natasha Bertrand, "The Coup Attempt in Turkey Has Presented the US and Europe with a Huge Dilemma," Business Insider, July 15, 2016, http://www.businessinsider.com/erdogan-statement-after-coup-attempt-2016-7.

over 45,000 people: Gareth Jones and Ercan Gurses, "Turkey's Erdogan Shuts Schools, Charities in First State of Emergency Decree," Reuters, July 23, 2016, http://www.reuters.com/article/us-turkey-security-emergency-idUSKCN1030BC.

few of these subsequent arrests: Loveday Morris, Thomas Gibbons-Neff, and Souad Mekhennet, "Turkey Is Expected to Curb Military Power as Purge Expands," *Washington Post,* July 19, 2016, https://www.washingtonpost.com/world/turkey-jails-generals-as-post-coup-purge-widens/2016/07/19/db076c84-4d1f-11e6-bf27-405106836f96_story.html?utm_term=.dae46a54ad4f.

92   *over 135,000 civil servants:* Jones and Gurses, "Turkey's Erdogan Shuts Schools."

were increasingly restricted: "Facebook, Twitter, YouTube, and WhatsApp Shutdown in Turkey," Turkey Blocks, November 4, 2016, https://turkeyblocks.org/2016/11/04/social-media-shutdown-turkey/.

Journalists saw their accounts: Mahir Zeynalov (@MahirZeynalov), "Twitter is now withholding accounts of Turkish journalists even without a court order. The number of blockaded accounts is mind-boggling," Twitter, August 11, 2016, 5:32 a.m., https://twitter.com/MahirZeynalov/status/763714666040291328.

satirical Instagram caption: Rod Nordland, "Turkey's Free Press Withers as Erdogan Jails 120 Journalists," *New York Times,* November 17, 2016, https://www.nytimes.com/2016/11/18/world/europe/turkey-press-erdogan-coup.html?_r=0.

clicked the "retweet" button: Dion Nissenbaum, "Detained in Turkey: A Journal Reporter's Story," *Wall Street Journal,* January 6, 2017, https://www.wsj.com/articles/detained-in-turkey-a-journal-reporters-story-1483721224.

As Nissenbaum explained: Ibid.

93   *he had just become:* Nissenbaum email.

"cost of the retweet": Ibid.

polices its "clean" internet: Faife, "Iran's 'National Internet.'"

In Saudi Arabia: Ben Elgin and Peter Robison, "How Despots Use Twitter to Hunt Dissidents," Bloomberg, October 27, 2016, https://www.bloomberg.com/news/articles/2016-10-27/twitter-s-firehose-of-tweets-is-incredibly-valuable-and-just-as-dangerous.

Pakistan became the first: Yasmeen Sherhan, "A Death Penalty for Alleged Blasphemy on Social Media," *The Atlantic,* June 12, 2017, https://www.theatlantic.com/news/archive/2017/06/pakistan-facebook-death-penalty/529968/.

94   *a crop top:* Jay Akbar, "Thailand to Prosecute Anyone That Even LOOKS at Material Considered Insulting to the Monarchy in Extension to Strict Internet Censorship After King Was Pictured in Crop Top," *Daily Mail,* May 22, 2017, http://www.dailymail.co.uk/news/article-4529788/Thailand-prosecute-internet-insult-monarchy-king-crop-top.html.

"We'll send you": Adam Seft et al., "Information Controls During Thailand's 2014 Coup," The Citizen Lab, July 9, 2014, https://citizenlab.ca/2014/07/information-controls-thailand-2014-coup/.

"Cyber Scouts" program: David Gilbert, "Thailand's Government Is Using Child 'Cyber Scouts' to Monitor Dissent," *Vice*, September 19, 2016, https://news.vice.com/article/thailands-royal-family-is-using-child-cyber-scouts-to-monitor-dissent.

A Kazakh visiting: Catherine Putz, "Kazakh Man Given 3 Years for Insulting Putin," *The Diplomat*, December 28, 2016, https://thediplomat.com/2016/12/kazakh-man-given-3-years-for-insulting-putin/.

"discrediting the political order": Tetyana Lokot, "Hard Labor for Woman Who Reposted Online Criticism of Russia's Actions in Ukraine," Global Voices, February 22, 2016, https://globalvoices.org/2016/02/22/hard-labor-for-woman-who-reposted-online-criticism-of-russias-actions-in-ukraine/.

Durov sold his shares: Amar Toor, "How Putin's Cronies Seized Control of Russia's Facebook," The Verge, January 31, 2014, https://www.theverge.com/2014/1/31/5363990/how-putins-cronies-seized-control-over-russias-facebook-pavel-durov-vk.

95   *"spiral of silence"*: Elizabeth Stoycheff, "Under Surveillance: Examining Facebook's Spiral of Silence Effects in the Wake of NSA Internet Monitoring," *Journalism and Mass Communication Quarterly* 93, no. 2 (2016): 296–311.

actual *majority opinion*: Ibid.

first email ever sent: Jeremy Goldhorn, "The Internet," in *The China Story,* Australian Centre on China in the World, August 2, 2012, https://www.thechinastory.org/keyword/the-internet/.

China passed the United States: David Barboza, "China Surpasses U.S. in Number of Internet Users," *New York Times,* July 26, 2008, http://www.nytimes.com/2008/07/26/business/worldbusiness/26internet.html.

nearly 800 million: Steven Millward, "China Now Has 731 Million Internet Users, 95% Access from Their Phones," Tech in Asia, January 22, 2017, https://www.techinasia.com/china-731-million-internet-users-end-2016.

96   *"harmonious society"*: Maureen Fan, "China's Party Leadership Declares New Priority: 'Harmonious Society,'" *Washington Post,* October 12, 2006, http://www.washingtonpost.com/wp-dyn/content/article/2006/10/11/AR2006101101610.html.

"correct guidance": David Bandurski, "Chinese Leaders Meditate Loudly on the Philosophy of Censorship as 17th Congress Nears," China Media Project, August 30, 2007, http://cmp.hku.hk/2007/08/30/as-

the-17th-national-congress-nears-party-meditations-on-the-philosophy-of-censorship/.

hand in hand with: Jack Linchuan Qiu, "Virtual Censorship in China: Keeping the Gate Between the Cyberspaces," *International Journal of Communications Law and Policy,* no. 4 (Winter 1999/2000): 11.

97 *Golden Shield Project:* Zixue Tai, "Casting the Ubiquitous Net of Information Control: Internet Surveillance in China from Golden Shield to Green Dam," *International Journal of Advanced Pervasive and Ubiquitous Computing* 2, no. 1 (2010): 239.

Sun Microsystems and Cisco: Qiu, "Virtual Censorship in China."

fail to reach: Lotus Ruan, Jeffrey Knockel, and Masashi Crete-Nishihata, "We (Can't) Chat: '709 Crackdown' Discussions Blocked on Weibo and WeChat," The Citizen Lab, April 13, 2017, https://citizenlab.ca/2017/04/we-cant-chat-709-crackdown-discussions-blocked-on-weibo-and-wechat/.

so-called Panama Papers: Tom Phillips, "All Mentions of Panama Papers Banned from Chinese Websites," *The Guardian,* April 5, 2016, https://www.theguardian.com/news/2016/apr/05/all-mention-of-panama-papers-banned-from-chinese-websites.

"Delete Report" was dispatched: Samuel Wade, "Minitrue: Panama Papers and Foreign Media Attacks," *China Digital Times,* April 4, 2016, https://chinadigitaltimes.net/2016/04/minitrue-panama-papers-foreign-media-attacks/.

98 *nation of Panama:* Phillips, "All Mentions of Panama Papers Banned."

"river crab'd": "Harmonization," Know Your Meme, accessed March 19, 2018, http://knowyourmeme.com/memes/harmonization-%E6%B2%B3%E8%9F%B9.

Winnie-the-Pooh was disappeared: Yuan Yang, "Winnie the Pooh Blacklisted by China's Online Censors," *Financial Times,* July 16, 2017, https://amp.ft.com/content/cf7fd22e-69d5-11e7-bfeb-33fe0c5b7eaa.

"cleanse the web": David Wertime, "Chinese Websites Deleted One Billion Posts in 2014, State Media Says," *Tea Leaf Nation* (blog), *Foreign Policy,* January 17, 2015, http://foreignpolicy.com/2015/01/17/chinese-websites-deleted-one-billion-posts-in-2014-state-media-says/.

elimination of nearly 300: Nikhil Sonnad, "261 Ways to Refer to the Tiananmen Square Massacre in China," Quartz, June 3, 2016, https://qz.com/698990/261-ways-to-refer-to-the-tiananmen-square-massacre-in-china/.

Baidu Baike: Malcolm Moore, "Tiananmen Massacre 25th Anniversary: The Silencing Campaign," *The Telegraph,* May 18, 2014, http://www

.telegraph.co.uk/news/worldnews/asia/china/10837992/Tiananmen-Massacre-25th-anniversary-the-silencing-campaign.html.

"disturbing public order": Oiwan Lam, "Chinese Police Arrested a Man for Complaining About Hospital Food. Netizens Say It's Police Abuse," *Advox* (blog), Global Voices, August 25, 2017, https://advox.globalvoices.org/2017/08/25/chinese-police-arrested-a-man-for-complaining-about-hospital-food-netizens-say-its-police-abuse/?utm_content=buffer7e970&utm_medium=social&utm_source=twitter.com&utm_campaign=buffer.

too much grassroots support: Gary King, Jennifer Pan, and Margaret E. Roberts, "How Censorship in China Allows Government Criticism but Silences Collective Expression," *American Political Science Review* 107, no. 2 (2013): 1–18.

99   *movement to ban plastic bags:* Ibid.

"it's not true that": Geremie R. Barmé, "Burn the Books, Bury the Scholars!," *The Interpreter,* Lowy Institute, August 23, 2017, https://www.lowyinstitute.org/the-interpreter/burn-books-bury-scholars.

authorities have ruled: Qiu, "Virtual Censorship in China."

spate of corruption scandals: "China Bans Internet News Reporting as Media Crackdown Widens," Bloomberg, July 25, 2016, https://www.bloomberg.com/news/articles/2016-07-25/china-slaps-ban-on-internet-news-reporting-as-crackdown-tightens.

5,000 internet users: "China Threatens Tough Punishment for Online Rumor Spreading," Reuters, September 9, 2013, http://www.reuters.com/article/us-china-internet-idUSBRE9880CQ20130909.

online personalities were "invited": Angus Grigg, "How China Stopped Its Bloggers," *Financial Review,* July 4, 2015, http://www.afr.com/technology/social-media/how-china-stopped-its-bloggers-20150701-gi34za.

like hotel reviews: Ibid.

"I forgot who I am": William Wan, "China Broadcasts Confession of Chinese-American Blogger," *Washington Post,* September 15, 2013, https://www.washingtonpost.com/world/china-broadcasts-confession-of-chinese-american-blogger/2013/09/15/3f2d82da-1e1a-11e3-8459-657e0c72fec8_story.html?utm_term=.e9e6afb7a72e.

100   *anything digital:* Sui-Lee Wee, "Chinese Police Arrest 15,000 for Internet Crimes," Reuters, August 18, 2015, http://www.reuters.com/article/us-china-internet-idUSKCN0QN1A520150818.

speech of each group member: Lulu Yilun Chen and Keith Zhai, "China's Latest Crackdown on Message Groups Chills WeChat Users," Bloomberg, September 12, 2017, https://www.bloomberg.com/news/

articles/2017-09-12/china-s-latest-crackdown-on-message-groups-chills-wechat-users.

armies of bureaucrats: Zhang Lei, "Invisible Footprints of Online Commentators," *Global Times,* February 5, 2010, http://www.globaltimes.cn/special/2010-02/503820.html.

leaked government memo: Gary King, Jennifer Pan, and Margaret E. Roberts, "How the Chinese Government Fabricates Social Media Posts for Strategic Distraction, Not Engaged Argument," *American Political Science Review* 111, no. 3 (August): 501.

its own pay scales: Lei, "Invisible Footprints of Online Commentators."

official job certifications: David Bandurski, "China's Guerrilla War for the Web," *Home Is Where the Heart Dwells* (blog), Harvard University, September 24, 2008, https://blogs.harvard.edu/guorui/2008/09/24/chinas-guerrilla-war-for-the-web/.

"50-Cent Army": Lei, "Invisible Footprints of Online Commentators."

ban the term "50 cents": Christina Sterbenz, "China Banned the Term '50 Cents' to Stop Discussion of an Orwellian Propaganda Program," Business Insider, October 17, 2014, http://www.businessinsider.com/chinas-50-cent-party-2014-10?IR=T.

"awards in municipal publicity": Lei, "Invisible Footprints of Online Commentators."

roughly 280,000 members: Bandurski, "China's Guerrilla War for the Web."

2 million members: King, Pan, and Roberts, "How the Chinese Government Fabricates Social Media Posts."

also been mimicked: Lei, "Invisible Footprints of Online Commentators."

101 *When Mao broke:* Mao Tsetung, *A Critique of Soviet Economics,* trans. Moss Roberts (Monthly Review, 1977).

Mao envisioned: "Short Definitions of the 'Mass Line' and a 'Mass Perspective,'" taken from Scott Harrison, *The Mass Line and the American Revolutionary Movement,* The Mass Line, http://massline.info/sum1p.htm.

hammered into a single vision: Ibid.

has made a comeback: David Cohen, "A Mass Line for the Digital Age," *China Brief* (Jamestown Foundation) 16, no. 8 (2016), https://jamestown.org/program/a-mass-line-for-the-digital-age/.

Jingwang (web-cleansing) app: Oiwan Lam, "China's Xinjiang Residents Are Being Forced to Install Surveillance Apps on Mobile Phones," Global

Voices, July 19, 2017, https://globalvoices.org/2017/07/19/chinas-xinji
ang-residents-are-being-forced-to-install-surveillance-apps-on-mobile-
phones/.

"electronic handcuffs": Ibid.

"mutually helpful social atmosphere": CCP Central Committee General Office, "Opinions Concerning Accelerating the Construction of
Credit Supervision, Warning and Punishment Mechanisms for Persons
Subject to Enforcement for Trust-Breaking," China Copyright and Media, September 25, 2016, https://chinacopyrightandmedia.wordpress.
com/2016/09/25/opinions-concerning-accelerating-the-construction-of-
credit-supervision-warning-and-punishment-mechanisms-for-persons-
subject-to-enforcement-for-trust-breaking/.

unwavering loyalty: Jacob Silverman, "China's Troubling New Social
Credit System — and Ours," *New Republic,* October 29, 2015, https://new
republic.com/article/123285/chinas-troubling-new-social-credit-system-
and-ours; "Planning Outline for the Construction of a Social Credit System (2014–2020)," China Copyright and Media, April 25, 2015, https://
chinacopyrightandmedia.wordpress.com/2014/06/14/planning-outline-
for-the-construction-of-a-social-credit-system-2014-2020/.

reflecting their "trustworthiness": Silverman, "China's Troubling New Social Credit System."

"trustworthiness" score: Ibid.

mobile services like WeChat: Jonah M. Kessel and Paul Mozur, "How
China Is Changing Your Internet," *New York Times,* August 9, 2016, https://
www.nytimes.com/video/technology/100000004574648/china-internet-
wechat.html.

102   *a staggering amount:* Ibid.

Buying too many: Celia Hatton, "China's 'Social Credit': Beijing Sets Up
Huge System," BBC News, October 26, 2015, http://www.bbc.com/news/
world-asia-china-34592186.

regularly buying diapers: Ibid.

"breaks social trust": Clinton Nguyen, "China Might Use Data to Create
a Score for Each Citizen Based on How Trustworthy They Are," Business
Insider, October 26, 2016, http://www.businessinsider.com/china-social-
credit-score-like-black-mirror-2016-10?r=UK&IR=T.

"report acts": "Planning Outline for the Construction of a Social Credit
System."

you can lose access: Nguyen, "China Might Use Data."

online matchmaking service: Celia Hatton, "China 'Social Credit': Beijing Sets Up Huge System," BBC News, October 26, 2015, http://www.bbc. com/news/world-asia-china-34592186.

Thailand: Michael de Waal-Montgomery, "Thailand Reportedly Close to Introducing Its Own China-Style Internet Firewall," VentureBeat, September 23, 2015, https://venturebeat.com/2015/09/23/thailand-reportedly-close-to-introducing-its-own-china-style-internet-firewall/.

Vietnam: Ian Timberlake, "Vietnam Steps Up China-Style Internet Censorship," *Sydney Morning Herald,* July 1, 2010, http://www.smh. com.au/technology/vietnam-steps-up-chinastyle-internet-censorship-20100701-zpg0.html.

Zimbabwe: Elin Box, "Zimbabwe to Implement China-Style Internet Censorship Regime," *Global Marketing News* (blog), Webcertain, April 11, 2016, http://blog.webcertain.com/zimbabwe-internet-censorship-like-china/11/04/2016/.

Cuba: Mauricio Claver-Carone, "When Helping 'the Cuban People' Means Bankrolling the Castros," *Wall Street Journal,* June 23, 2015, https:// www.wsj.com/articles/SB11021741326745413664304581020103630034440.

Putin has even gone: Andrei Soldatov and Irina Borogan, "Putin Brings China's Great Firewall to Russia in Cybersecurity Pact," *The Guardian,* November 29, 2016, https://www.theguardian.com/world/2016/nov/29/ putin-china-internet-great-firewall-russia-cybersecurity-pact.

at least 36 countries: Adrian Shabaz, "Freedom on the Net 2018: The Rise of Digital Authoritarianism," Freedom House, https://freedomhouse.org/ report/freedom-net/freedom-net-2018/rise-digital-authoritarianism.

103 *"It was difficult":* Katie Davies, "Revealed: Confessions of a Kremlin Troll," *Moscow Times,* April 18, 2017, https://themoscowtimes.com/ articles/revealed-confessions-of-a-kremlin-troll-57754.

more than 200 blog posts: Ibid.

One story (possibly apocryphal): Ion Mihai Pacepa and Ronald J. Rychlak, *Disinformation: Former Spy Chief Reveals Secret Strategies for Undermining Freedom, Attacking Religion, and Promoting Terrorism* (WND Books, 2013), loc. 284, Kindle.

more than 10,000: Thomas Rid, "Disinformation: A Primer in Russian Active Measures and Influence Campaigns," testimony before the Senate Committee on Intelligence, March 30, 2017, https://www.intelligence.sen ate.gov/sites/default/files/documents/os-trid-033017.pdf.

104 *Operation INFEKTION:* Thomas Boghardt, "Operation INFEKTION: Soviet Bloc Intelligence and Its AIDS Disinformation Campaign," *Studies in Intelligence* 53, no. 4 (2009).

Indian newspaper Patriot: Ibid.

"well-known American scientist": David Robert Grimes, "Russian Fake News Is Not New: Soviet AIDS Propaganda Cost Countless Lives," *The Guardian,* June 14, 2017, https://www.theguardian.com/science/blog/2017/jun/14/russian-fake-news-is-not-new-soviet-aids-propaganda-cost-countless-lives.

Lyndon LaRouche movement: Boghardt, "Operation INFEKTION."

"Everyone shall have": Constitution of the Russian Federation, art. 29.4.

105 *"spoke in grave":* "1984 in 2014," *The Economist,* March 29, 2014, https://www.economist.com/news/europe/21599829-new-propaganda-war-underpins-kremlins-clash-west-1984-2014.

A pop star garbed: Christine Friar, "Russia's Using Pop Music on YouTube to Ridicule Millennial Protesters," The Daily Dot, May 19, 2017, https://www.dailydot.com/upstream/russia-youtube-propoganda-pop-music/?tw=dd.

constant drumbeat of anxiety: Gary Shteyngart, "'Out of My Mouth Comes Unimpeachable Manly Truth,'" *New York Times Magazine,* February 18, 2015, https://www.nytimes.com/2015/02/22/magazine/out-of-my-mouth-comes-unimpeachable-manly-truth.html?_r=0.

"Imagine you have": Evan Osnos, David Remnick, and Joshua Yaffa, "Trump, Putin, and the New Cold War," *The New Yorker,* March 6, 2017, http://www.newyorker.com/magazine/2017/03/06/trump-putin-and-the-new-cold-war.

He once proposed: Diana Bruk, "The Best of Vladimir Zhirinovsky, the Clown Prince of Russian Politics," *Vice,* August 10, 2013, https://www.vice.com/en_us/article/the-best-of-vladimir-zhirinovsky-russias-craziest-politician.

Boris Nemtsov was not: Joshua Yaffa, "The Unaccountable Death of Boris Nemtsov," *The New Yorker,* February 26, 2016, http://www.newyorker.com/news/news-desk/the-unaccountable-death-of-boris-nemtsov.

at least thirty-eight: Oren Dorell, "Mysterious Rash of Russian Deaths Casts Suspicion on Vladimir Putin," *USA Today,* May 2, 2017, https://www.usatoday.com/story/news/world/2017/05/02/dozens-russian-deaths-cast-suspicion-vladimir-putin/100480734/.

106 *arranging "scandals":* Jill Dougherty, "How the Media Became One of Putin's Most Powerful Weapons," *The Atlantic,* April 21, 2015, https://www.theatlantic.com/international/archive/2015/04/how-the-media-became-putins-most-powerful-weapon/391062/.

dozens of independent journalists: "Journalists Killed in Russia Between 1992 and 2018/Motive Confirmed," Committee to Protect Journal-

ists, https://cpj.org/data/killed/europe/russia/?status=Killed&motive
Confirmed%5B%5D=Confirmed&type%5B%5D=Journalist&cc_
fips%5B%5D=RS&start_year=1992&end_year=2018&group_by=year.

illusion of free speech: Dougherty, "How the Media Became."

"all forms of political discourse": Peter Pomerantsev, *Nothing Is True and
Everything Is Possible: The Surreal Heart of the New Russia* (PublicAffairs,
2014), 64.

the most serious protests: Ellen Barry, "Rally Defying Putin's Party
Draws Tens of Thousands," *New York Times,* December 10, 2011, http://
www.nytimes.com/2011/12/11/world/europe/thousands-protest-in-mos
cow-russia-in-defiance-of-putin.html.

"the role of": Mark Galeotti, "The 'Gerasimov Doctrine' and Russian
Non-Linear War," *In Moscow's Shadows* (blog), July 6, 2014, https://inmos
cowsshadows.wordpress.com/2014/07/06/the-gerasimov-doctrine-and-
russian-non-linear-war/.

the Gerasimov Doctrine: As Mark Galeotti notes, the Gerasimov Doctrine was neither Gerasimov's nor was it presented at the time as a doctrine. Nonetheless, this is the name that stuck. Galeotti, "The 'Gerasimov
Doctrine.'"

enshrined in Russian military theory: See Embassy of the Russian Federation to the United Kingdom of Great Britain and Northern Island, "The
Military Doctrine of the Russian Federation," news release, June 29, 2015
(policy adopted December 25, 2014), https://rusemb.org.uk/press/2029;
Ministry of Foreign Affairs of the Russian Federation, "Doctrine of Information Security of the Russian Federation," December 5, 2016, http://
www.mid.ru/en/foreign_policy/official_documents/-/asset_publisher/
CptICkB6BZ29/content/id/2563163.

"war on information warfare": Jolanta Darczewska, *The Anatomy of Russian Information Warfare: The Crimean Operation, a Case Study,* Point of
View, no. 42 (Centre for Eastern Studies, May 2014), 10https://www.osw
.waw.pl/sites/default/files/the_anatomy_of_russian_information_war
fare.pdf, 13.

107 *conglomerate of nearly seventy-five:* Ibid., 10.

the "4 Ds": Ben Nimmo, "Anatomy of an Info-War: How Russia's Propaganda Machine Works, and How to Counter It," StopFake, May 19, 2015,
https://www.stopfake.org/en/anatomy-of-an-info-war-how-russia-s-pro
paganda-machine-works-and-how-to-counter-it/.

identity and mission shifted: Dougherty, "How the Media Became."

$30 million: Simon Shuster, "Russia Today: Inside Putin's On-Air Ma-
chine," *Time,* March 5, 2015, http://time.com/rt-putin/.

approximately $400 million: Gabrielle Tetrault-Farber, "Looking West, Russia Beefs Up Spending on Global Media Giants," *Moscow Times,* September 23, 2014, https://themoscowtimes.com/articles/looking-west-russia-beefs-up-spending-on-global-media-giants-39708.

"weapons system": Shuster, "Russia Today."

"The phone exists": Ibid.

more YouTube subscribers: "Assessing Russian Activities and Intentions in Recent US Elections" (Intelligence Community Assessment, Office of the Director of National Intelligence, January 6, 2017), 10, https://www.dni.gov/files/documents/ICA_2017_01.pdf.

108 *RT has promoted:* Matthew Bodner, Matthew Kupfer, and Bradley Jardine, "Welcome to the Machine: Inside the Secretive World of RT," *Moscow Times,* June 1, 2017, https://themoscowtimes.com/articles/welcome-to-the-machine-inside-the-secretive-world-of-rt-58132.

"'Question More' is not about": Matthew Armstrong, "RT as a Foreign Agent: Political Propaganda in a Globalized World," War on the Rocks, May 4, 2015, https://warontherocks.com/2015/05/rt-as-a-foreign-agent-political-propaganda-in-a-globalized-world/.

Sputnik International: "Major News Media Brand 'Sputnik' Goes Live November 10," Sputnik, October 11, 2014, https://sputniknews.com/russia/201411101014569630/.

Baltica targets audiences: Inga Springe et al., "Sputnik's Unknown Brother," Re:Baltica, April 6, 2017, https://en.rebaltica.lv/2017/04/sputniks-unknown-brother/.

first source of this false report: Ben Nimmo, "Three Thousand Fake Tanks," @DFRLLab (blog), Medium, January 12, 2017, https://medium.com/@DFRLab/three-thousand-fake-tanks-575410c4f64d.

109 *all-out assault:* Matthew Sparkes, "Russian Government Edits Wikipedia on Flight MH17," *The Telegraph,* July 18, 2014, http://www.telegraph.co.uk/technology/news/10977082/Russian-government-edits-Wikipedia-on-flight-MH17.html.

"Questions over Why": Paul Szoldra, "Here's the Ridiculous Way Russia's Propaganda Channel Is Covering the Downed Malaysia Airliner," Business Insider Australia, July 19, 2014, https://www.businessinsider.com.au/rt-malaysia-airlines-ukraine-2014-7#JhJsCOWZzphQooIG.99.

Russian Union of Engineers: Eliot Higgins, "SU-25, MH17 and the Problems with Keeping a Story Straight," Bellingcat, January 10, 2015, https://www.bellingcat.com/news/uk-and-europe/2015/01/10/su-25-mh17-and-the-problems-with-keeping-a-story-straight/.

110 *bad photoshop job:* Veli-Pekka Vivimäki, "Russian State Television Shares Fake Images of MH17 Being Attacked," Bellingcat, November 14,

2014, https://www.bellingcat.com/news/2014/11/14/russian-state-televi-sion-shares-fake-images-of-mh17-being-attacked/.

"It came from": Max Seddon, "Russian TV Airs Clearly Fake Image to Claim Ukraine Shot Down MH17," BuzzFeed, November 15, 2014, https://www.buzzfeed.com/maxseddon/russian-tv-airs-clearly-fake-image-to-claim-ukraine-shot-dow?utm_term=.vhnM2Yn2y4#.yvpq59Z5Q6.

at least a half dozen: Aric Toler, "The Kremlin's Shifting, Self-Contra-dicting Narratives on MH17," Bellingcat, January 5, 2018, https://www.bellingcat.com/news/uk-and-europe/2018/01/05/kremlins-shifting-self-contradicting-narratives-mh17/.

doctored satellite images: Eliot Higgins, "Russia's Colin Powell Mo-ment—How the Russian Government's MH17 Lies Were Exposed," Bellingcat, July 16, 2015, https://www.bellingcat.com/news/uk-and-europe/2015/07/16/russias-colin-powell-moment-how-the-russian-gov ernments-mh17-lies-were-exposed/.

"news division": Charles Maynes, "The trolls are winning, says Russian troll hunter," PRI, March 13, 2019, https://www.pri.org/stories/2019-03-13/trolls-are-winning-says-russian-troll-hunter.

111   *"This is information war"*: "An Ex St. Petersburg 'Troll' Speaks Out," Meduza, October 15, 2017, https://meduza.io/en/feature/2017/10/15/an-ex-st-petersburg-troll-speaks-out?utm_source=Sailthru&utm_medium=email&utm_campaign=Newpercent20Campaign&utm_term=percent2ASituationpercent20Report.

solicited Russian advertisers: Ilya Klishin, "How Putin Secretly Con-quered Russia's Social Media over the Past 3 Years," Global Voices, Janu-ary 30, 2015, https://globalvoices.org/2015/01/30/how-putin-secretly-con quered-russias-social-media-over-the-past-3-years/.

Nearly a dozen: Ibid.

Special Counsel Robert Mueller: Priscilla Alvarez and Taylor Hosking, "The Full Text of Mueller's Indictment of 13 Russians," *The Atlantic,* Feb-ruary 16, 2018, https://www.theatlantic.com/politics/archive/2018/02/rosenstein-mueller-indictment-russia/553601/.

Internet Research Agency: Adrian Chen, "The Agency," *New York Times Magazine,* June 7, 2015, https://www.nytimes.com/2015/06/07/magazine/the-agency.html.

the "Facebook desk": Associated Press, "Ex-Workers at Russian 'Troll Factory' Trust U.S. Indictment," *Los Angeles Times,* February 19, 2018, http://www.latimes.com/politics/la-na-pol-russian-troll-factory-20180219-story.html.

"I really only stayed": Davies, "Revealed:" Confessions of a Kremlin Troll.

112 *"expected to manage":* Max Seddon, "Documents Show How Russia's Troll Army Hit America," BuzzFeed, June 2, 2014, https://www.buzzfeed.com/maxseddon/documents-show-how-russias-troll-army-hit-america?utm_term=.kaWolQvoO#.tsawEvpw1.

retweeted 1,213,506 times: Ben Popken, "Twitter Deleted 200,000 Russian Troll Tweets. Read Them Here," NBC News, February 14, 2018, https://www.nbcnews.com/tech/social-media/now-available-more-200-000-deleted-russian-troll-tweets-n844731.

seventh most retweeted account: Kevin Poulsen, "Exclusive: Russia Activated Twitter Sleeper Cells for 2016 Election Day Blitz," The Daily Beast, November 7, 2017, https://www.thedailybeast.com/exclusive-russia-activated-twitter-sleeper-cells-for-election-day-blitz?via=twitter_page.

Flynn followed at least: Kevin Poulsen and Ben Collins, "Michael Flynn Followed Russian Troll Accounts, Pushed Their Messages in Days Before Election," The Daily Beast, November 1, 2017, https://www.thedailybeast.com/michael-flynn-followed-russian-troll-accounts-pushed-their-messages-in-days-before-election.

@tpartynews: Drew Griffin and Donnie O'Sullivan, "The Fake Tea Party Twitter Account Linked to Russia and Followed by Sebastian Gorka," CNN, September 22, 2017, https://www.cnn.com/2017/09/21/politics/tpartynews-twitter-russia-link/index.html.

seemingly trustworthy individuals: John D. Gallacher et al., "Junk News on Military Affairs and National Security: Social Media Disinformation Campaigns Against US Military Personnel and Veterans" (data memo 2017.9, Computational Propaganda Project, University of Oxford, October 9, 2017), http://comprop.oii.ox.ac.uk/wp-content/uploads/sites/93/2017/10/Junk-News-on-Military-Affairs-and-National-Security-1.pdf.

"First you had to be": "Former Russian Troll Describes Night Shift as 'Bacchanalia,'" *Moscow Times,* October 27, 2017, https://themoscowtimes.com/news/former-russian-troll-describes-night-shift-as-bacchanalia-59398.

113 *"vote for Jill Stein":* Donie O'Sullivan and Dylan Byers, "Exclusive: Fake Black Activist Accounts Linked to Russian Government," CNNMoney, September 28, 2017, http://money.cnn.com/2017/09/28/media/blacktivist-russia-facebook-twitter/index.html.

103.8 million times: Craig Timberg, "Russian Propaganda May Have Been Shared Hundreds of Millions of Times, New Research Says," *The Switch* (blog), *Washington Post,* October 5, 2017, https://www.washingtonpost.com/news/the-switch/wp/2017/10/05/russian-propaganda-may-have-been-shared-hundreds-of-millions-of-times-new-research-says/?utm_term=.b14ae0521f56.

as high as 24 percent: "HPSCI Minority Exhibit A," U.S. House of Representatives Permanent Select Committee on Intelligence Democrats, accessed March 19, 2018, https://democrats-intelligence.house.gov/hpsci-11-1/.

over 800 million users: Jonathan Albright, "Instagram, Meme Seeding, and the Truth About Facebook Manipulation, Pt. 1," Berkman Klein Center for Internet & Society at Harvard University, November 8, 2017, https://medium.com/berkman-klein-center/instagram-meme-seeding-and-the-truth-about-facebook-manipulation-pt-1-dae4d0b61db5.

an astounding 145 million: Ibid.

@Jenn_Abrams: Ben Collins and Joseph Cox, "Jenna Abrams, Russia's Clown Troll Princess, Duped the Mainstream Media and the World," The Daily Beast, November 2, 2017, https://www.thedailybeast.com/jenna-abrams-russias-clown-troll-princess-duped-the-mainstream-media-and-the-world.

114 *Facebook automatically steered:* Nicholas Confessore and Daisuke Wakabayashi, "How Russia Harvested American Rage to Reshape U.S. Politics," *New York Times,* October 9, 2017, https://www.nytimes.com/2017/10/09/technology/russia-election-facebook-ads-rage.html.

Secured Borders: Scott Shane, "Purged Facebook Page Tied to the Kremlin Spread Anti-Immigrant Bile," *New York Times,* September 12, 2017, https://www.nytimes.com/2017/09/12/us/politics/russia-facebook-election.html.

journalist Jessikka Aro: Jessikka Aro, "My Year as a Pro-Russia Troll Magnet: International Shaming Campaign and an SMS from Dead Father," *Yle Kioski,* September 11, 2015, http://kioski.yle.fi/omat/my-year-as-a-pro-russia-troll-magnet.

labeled a "pornographer": Jeff Stein, "How Russia Is Using LinkedIn as a Tool of War Against Its U.S. Enemies," *Newsweek,* August 3, 2017, http://www.newsweek.com/russia-putin-bots-linkedin-facebook-trump-clinton-kremlin-critics-poison-war-645696.

115 *pushing the theme of #UniteTheRight:* Conspirado Norteño (@conspiratoro), "David Jones' Locker: Where Truth Goes to Die #TrumpRussia," Twitter, August 22, 2017, 5:51 p.m., https://twitter.com/conspiratoro/status/900158639884955648.

disputing the report: For a lively conversation, see the Twitter account Logic Reason (@gsobjc), https://twitter.com/gsobjc?lang=en.

In Venezuela: Peter Pomerantsev, introduction to *The New Authoritar-*

ians: Ruling Through Disinformation, Beyond Propaganda (Transitions Forum, Legatum Institute, June 2015), 6, https://lif.blob.core.windows.net/lif/docs/default-source/publications/the-new-authoritarians — ruling-through-disinformation-june-2015-pdf.pdf?sfvrsn=4.

In Azerbaijan: Arzu Geybulla, "In the Crosshairs of Azerbaijan's Patriotic Trolls," Open Democracy, November 22, 2016, https://www.open democracy.net/od-russia/arzu-geybulla/azerbaijan-patriotic-trolls; Claudio Guarnieri, Joshua Franco, and Collin Anderson, "False Friends: How Fake Accounts and Crude Malware Targeted Dissidents in Azerbaijan," *Amnesty Global Insights* (blog), Medium, March 9, 2017, https://medium.com/amnesty-insights/false-friends-how-fake-accounts-and-crude-malware-targeted-dissidents-in-azerbaijan-9b6594cafe60.

116   *Even in democratic India:* Sohini Mitter, "India's Ruling Party Has a Troll Army to Silence Opponents Online, Book Claims," Mashable, December 27, 2016, http://mashable.com/2016/12/27/bjp-planned-online-trolling/#go_GF.fD8Gqo.

at least forty-eight regimes: Samantha Bradshaw and Philip Howard, "Challenging Truth and Trust: A Global Inventory of Organized Social Media Manipulation," Oxford Internet Institute, http://comprop.oii.ox.ac.uk/wp-content/uploads/sites/93/2018/07/ct2018.pdf.

at least thirty national-level elections: Adrian Shabaz, "Freedom on the Net 2018: The Rise of Digital Authoritarianism," Freedom House, https://freedomhouse.org/report/freedom-net/freedom-net-2018/rise-digital-authoritarianism.

denying the truth: George Simon, "Gaslighting as a Manipulation Tactic: What It Is, Who Does It, and Why," Counselling Resource, November 8, 2011, http://counsellingresource.com/features/2011/11/08/gaslighting/.

"Facts . . . become interchangeable": Lauren Ducca, "Donald Trump Is Gaslighting America," *Teen Vogue,* December 10, 2016, https://www.teen vogue.com/story/donald-trump-is-gaslighting-america.

5. THE UNREALITY MACHINE

118   *"When all think alike":* Walter Lippmann, *The Stakes of Diplomacy* (Henry Holt, 1915), 51.

Moët champagne: Samanth Subramanian, "Inside the Macedonian Fake-News Complex," *Wired,* February 15, 2017, https://www.wired.com/2017/02/veles-macedonia-fake-news/.

slick wardrobes: Alexander Smith and Vladimir Banic, "Fake News: How a Partying Macedonian Teen Earns Thousands Publishing Lies," NBC News, December 9, 2016, http://www.nbcnews.com/news/world/

fake-news-how-partying-macedonian-teen-earns-thousands-publishing-lies-n692451.

old industrial town: "About Veles," Macedonia Information, accessed March 19, 2018, http://makedonija.name/cities/veles.

four times the advertising: Josh Constine, "Facebook Swells to 1.65B Users and Beats Q1 Estimates with $5.38B Revenue," TechCrunch, April 27, 2016, https://techcrunch.com/2016/04/27/facebook-q1-2016-earnings/.

With each click: Subramanian, "Inside the Macedonian Fake-News Complex."

tens of thousands of dollars: Ibid.

119 *swelled into the hundreds:* Craig Silverman and Lawrence Alexander, "How Teens in the Balkans Are Duping Trump Supporters with Fake News," BuzzFeed, November 3, 2016, https://www.buzzfeed.com/craigsilverman/how-macedonia-became-a-global-hub-for-pro-trump-misinfo?utm_term=.mqxmBEGNRa#.panz3vD86O.

would hold special events: Smith and Banic, "Fake News."

"Dmitri": Ibid.

a "clickbait coach": Isa Soares et al., "The Fake News Machine: Inside a Town Gearing Up for 2020," CNNMoney, http://money.cnn.com/interactive/media/the-macedonia-story/.

"Since fake news started": Smith and Banic, "Fake News."

"some white people": Silverman and Alexander, "How Teens in the Balkans."

false reports received: Craig Silverman, "This Analysis Shows How Viral Fake Election News Outperformed Real News on Facebook," BuzzFeed, November 16, 2016, https://www.buzzfeed.com/craigsilverman/viral-fake-election-news-outperformed-real-news-on-facebook?utm_term=.apjBaw3rL#.tezr61jzN.

120 *"You see they like":* Smith and Banic, "Fake News."

Of the top twenty: Silverman, "This Analysis Shows."

"Pope Francis Shocks World": Ibid.

Three times as many: Ibid.

Pope Francis didn't: Philip Pullella, "Pope Warns Media over 'Sin' of Spreading Fake News, Smearing Politicians," Reuters, December 7, 2016, https://www.reuters.com/article/us-pope-media/pope-warns-media-over-sin-of-spreading-fake-news-smearing-politicians-idUSKBN13W1TU.

"I didn't force anyone": Smith and Banic, "Fake News."

"They're not allowed": Ibid.

President Obama himself: David Remnick, "Obama Reckons with a Trump Presidency," *The New Yorker,* November 28, 2016, http://www.

newyorker.com/magazine/2016/11/28/obama-reckons-with-a-trump-presidency.

121   *"access to their own"*: James Breiner, "What Freedom of the Press Means for Those Who Own One," MediaShift, December 10, 2014, http://mediashift. org/2014/12/what-freedom-of-the-press-means-for-those-who-own-one/.
"axiom of political science": Alexis de Tocqueville, *The Republic of the United States of America, and Its Political Institutions, Reviewed and Examined,* trans. Henry Reeves (A. S. Barnes, 1851), 199.
"Imagine a future": Nicholas Negroponte, *Being Digital* (Knopf, 1995), 153.
the "Daily Me": Ibid.

122   *the "Daily We"*: Cass Sunstein, "The Daily We," *Boston Review,* June 1, 2001, http://bostonreview.net/cass-sunstein-internet-democracy-daily-we.
"You're the only person": Eli Pariser, *The Filter Bubble: How the New Personalized Web Is Changing What We Read and How We Think* (Penguin, 2011), 9.

123   *"YouTube cannot contain"*: Aric Toler, "'No Safe Spaces on the Flat Earth' — Emerging Alt-Right Inspires Flat Earth Online Communities," Bellingcat, June 7, 2017, https://www.bellingcat.com/resources/articles/2017/06/07/flat-earth-online-communities/.
The best predictor: We consulted several excellent sources for a primer on homophily. See Aris Anagnostopoulos et al., "Viral Misinformation: The Role of Homophily and Polarization," arXiv:1411.2893 [cs.SI], November 2014; Walter Quattrociocchi, Antonia Scala, and Cass R. Sunstein, "Echo Chambers on Facebook" (discussion paper no. 877, Harvard Law School, Cambridge, MA, September 2016), http://www.law.harvard.edu/programs/olin_center/papers/pdf/Sunstein_877.pdf; Michela Del Vicario et al., "The Spreading of Misinformation Online," *PNAS* 113, no. 3 (2016): 554–59, https://www.researchgate.net/publication/289263634_The_spreading_of_misinformation_online; Delia Mocanu et al., "Collective Attention in the Age of (Mis)information," arXiv:1403.3344 [cs.SI], March 2014.
"love of the same": Aaron Recitca, "Homophily," *New York Times Magazine,* December 10, 2006, https://www.nytimes.com/2006/12/10/magazine/10Section2a.t-4.html.

124   *Yale University researchers:* Gordon Pennycook, Tyrone Cannon, and David G. Rand, "Implausibility and Illusory Truth: Prior Exposure Increases Perceived Accuracy of Fake News but Has No Effect on Entirely Implausible Statements" (working paper, March 16, 2018), https://papers. ssrn.com/sol3/papers.cfm?abstract_id=2958246.

a second "Holocaust": Orac, "The Violent Rhetoric of the Antivaccine Movement: 'Vaccine Holocaust' and Potential Impending Attacks on Journalists," *Respectful Insolence* (blog), May 17, 2017, https://respectfulinso lence.com/2017/05/17/the-violent-rhetoric-of-the-antivaccine-movement-vaccine-holocaust-and-potential-impending-attacks-on-journalists/.

125   *lower-tier celebrities:* For a blunt take on the subject, see "Anti-Vaccine Body Count," http://www.jennymccarthybodycount.com/.

"Healthy young child": Donald Trump (@realDonaldTrump), "Healthy young child goes to doctor, gets pumped with massive shot of many vaccines, doesn't feel good and changes—AUTISM. Many such cases!," Twitter, March 28, 2014, 5:35 a.m., https://twitter.com/realdonaldtrump/status/449525268529815552?lang=en.

"personal belief exception": Anna Merlan, "Meet the New, Dangerous Fringe of the Anti-Vaccination Movement," Jezebel, June 29, 2015, https://jezebel.com/meet-the-new-dangerous-fringe-of-the-anti-vaccina tion-1713438567.

sixty-year high: Rong-Gong Lin, "Latest Measles Outbreak Highlights a Growing Problem in California," *Los Angeles Times,* January 7, 2015, http://www.latimes.com/local/california/la-me-aa2-snapshot-measles-whooping-cough-20150108-story.html.

sickened 147 children: "Year in Review: Measles Linked to Disneyland," *Public Health Matters Blog,* Centers for Disease Control and Prevention, December 2, 2015, https://blogs.cdc.gov/publichealthmatters/2015/12/year-in-review-measles-linked-to-disneyland/.

law requiring *kindergarten vaccinations:* Erin Hare, "Facts Alone Won't Convince People to Vaccinate Their Kids," FiveThirtyEight, June 12, 2017, https://fivethirtyeight.com/features/facts-alone-wont-convince-people-to-vaccinate-their-kids/?utm_content=buffer21fe5&utm_medium=social&utm_source=twitter.com&utm_campaign=buffer.

dig in their heels: Charles G. Lord, Lee Ross, and Mark R. Lepper, "Biased Assimilation and Attitude Polarization: The Effects of Prior Theories on Subsequently Considered Evidence," *Journal of Personality and Social Psychology* 37, no. 11 (1979): 2098–109.

126   *"Once, every village":* Robert Bateman (@RobertLBateman), "Once, every village had an idiot. It took the internet to bring them all together. —Unknown (well, by me)," Twitter, August 19, 2017, 4:16 p.m., https://twitter.com/RobertLBateman/status/899047467282518017.

A 2016 study: Marc Lynch, Deen Freelon, and Sean Aday, "How Social Media Undermines Transitions to Democracy," Bullets and Blogs, no. 4 (PeaceTech Lab, 2016), https://ipdgc.gwu.edu/sites/g/files/zaxdzs2221/f/downloads/Blogs%20and%20Bullets%20IV.pdf.

"encouraged political society": Ibid., 21.

"echo-chamber qualities": Sean Aday, Deen Freelon, and Marc Lynch, "How Social Media Undermined Egypt's Democratic Transition," *Monkey Cage* (blog), *Washington Post*, October 7, 2016, https://www.washington post.com/news/monkey-cage/wp/2016/10/07/how-social-media-under mined-egypts-democratic-transition/?utm_term=.c6f0a6afc33b.

127　*"There's no such thing"*: Jack Holmes, "A Trump Surrogate Drops the Mic: 'There's No Such Thing as Facts,'" *Esquire*, December 1, 2016, http://www.esquire.com/news-politics/videos/a51152/trump-surrogate-no-such-thing-as-facts/.

shielded their terrified children: Marc Fisher, John Woodrow Cox, and Peter Hermann, "Pizzagate: From Rumor, to Hashtag, to Gunfire in D.C.," *Washington Post*, December 6, 2016, https://www.wash ingtonpost.com/local/pizzagate-from-rumor-to-hashtag-to-gunfire-in-dc/2016/12/06/4c7def50-bbd4-11e6-94ac-3d324840106c_story.html?utm_term=.c84c2847b899.

customers made a run for it: Amanda Robb, "Anatomy of a Fake News Scandal," *Rolling Stone*, November 16, 2017, https://www.rollingstone.com/politics/news/pizzagate-anatomy-of-a-fake-news-scandal-w511904.

an employee holding pizza dough: Fisher, Cox, and Hermann, "Pizzagate."

tiny computer room: Ibid.

128　*"lucid, deadly serious"*: Spencer S. Hsu, "Pizzagate Gunman Says He Was Foolish, Reckless, Mistaken — and Sorry," *Washington Post*, June 14, 2017, https://www.washingtonpost.com/local/public-safety/pizza gate-shooter-apologizes-in-handwritten-letter-for-his-mistakes-ahead-of-sentencing/2017/06/13/f35126b6-5086-11e7-be25-3a519335381c_story.html?utm_term=.63e54b2d390d.

tearful farewell: Grace Hauck, "'Pizzagate' Shooter Sentenced to 4 Years in Prison," CNN, June 22, 2017, https://www.cnn.com/2017/06/22/politics/pizzagate-sentencing/index.html.

sentenced to four years: Ibid.

For James Alefantis: Ibid.

known collectively as #Pizzagate: Fisher, Cox, and Hermann, "Pizzagate."

1.4 million mentions: Robb, "Anatomy of a Fake News Scandal."

"Something's being covered up": Ibid.

Russian sockpuppets working: Ibid.

nearly half of Trump voters: Catherine Rampell, "Americans — Especially but Not Exclusively Trump Voters — Believe Crazy, Wrong Things,"

Washington Post, December 28, 2016, https://www.washingtonpost.com/news/rampage/wp/2016/12/28/americans-especially-but-not-exclusively-trump-voters-believe-crazy-wrong-things/.

129 *"the intel on this":* Adam Goldman, "The Comet Ping Pong Gunman Answers Our Reporter's Questions," *New York Times,* December 7, 2016, https://www.nytimes.com/2016/12/07/us/edgar-welch-comet-pizza-fake-news.html.

Posobiec was relentless: Fisher, Cox, and Hermann, "Pizzagate."

"They want to control": Jack Posobiec (@JackPosobiec), "ANNOUNCING: My next book 4D Warfare: How to Use New Media to Fight and Win the Culture Wars! Published by @VoxDay and Castalia House!," August 3, 2017, 8:03 a.m., https://twitter.com/JackPosobiec/status/893125262958891009.

"False flag": Paul Farhi, "'False Flag' Planted at a Pizza Place? It's Just One More Conspiracy to Digest," *Washington Post,* December 5, 2016, https://www.washingtonpost.com/lifestyle/style/false-flag-planted-at-a-pizza-place-its-just-one-more-conspiracy-to-digest/2016/12/05/fc154b1e-bb09-11e6-94ac-3d324840106c_story.html?utm_term=.7ecbd9f78337.

"Nothing to suggest": Jack Posobiec (@JackPosobiec), "DC Police Chief: 'Nothing to suggest man w/gun at Comet Ping Pong had anything to do with #pizzagate'" (tweet deleted), available at Scoopnest, https://www.scoopnest.com/user/JackPosobiec/805559273426141184-dc-police-chief-nothing-to-suggest-man-w-gun-at-comet-ping-pong-had-anything-to-do-with-pizzagate.

livestreaming from the White House: Jared Holt and Brendan Karet, "Meet Jack Posobiec: The 'Alt-Right' Troll with Press Pass in White House," *Slate,* August 16, 2017, https://www.salon.com/2017/08/16/meet-jack-posobiec-the-alt-right-troll-with-a-press-pass-in-white-house_partner/; Jack Posobiec (@JackPosobiec), "Free our people," Twitter, May 9, 2017, 10:28 a.m., https://twitter.com/jackposobiec/status/861996422920536064.

retweeted multiple times: Colleen Shalby, "Trump Retweets Alt-Right Media Figure Who Published 'Pizzagate' and Seth Rich Conspiracy Theories," *Los Angeles Times,* August 14, 2017, http://www.latimes.com/politics/la-pol-updates-everything-president-trump-retweets-alt-right-blogger-who-1502769297-htmlstory.html; Maya Oppenheim, "Donald Trump Retweets Far-Right Conspiracy Theorist Jack Posobiec Who Took 'Rape Melania' Sign to Rally," *Independent,* January 15, 2018, https://www.independent.co.uk/news/world/americas/donald-trump-jack-posobiec-pizzagate-rape-melania-sign-twitter-conspiracy-theory-far-right-a8159661.html.

"power law": Emma Pierson, "Twitter Data Show That a Few Power-

ful Users Can Control the Conversation," Quartz, May 5, 2015, https://qz.com/396107/twitter-data-show-that-a-few-powerful-users-can-control-the-conversation/.

130   *study of 330 million:* Xu Wei, "Influential Bloggers Set Topics Online," China Daily Asia, December 27, 2013, https://www.chinadailyasia.com/news/2013-12/27/content_15108347.html.

a mere 300 accounts: Ibid.

susceptibility to further falsehoods: Sander van der Linden, "The Conspiracy-Effect: Exposure to Conspiracy Theories (About Global Warming) Decreases Pro-Social Behavior and Science Acceptance," *Personality and Individual Differences* 87 (December 2015): 171–73.

more supportive of "extremism": Sander van der Linden, "The Surprising Power of Conspiracy Theories," *Psychology Today,* August 24, 2015, https://www.psychologytoday.com/blog/socially-relevant/201508/the-surprising-power-of-conspiracy-theories.

spread about six times faster: Brian Dowling, "MIT Scientist Charts Fake News Reach," *Boston Herald,* March 11, 2018, http://www.bostonherald.com/news/local_coverage/2018/03/mit_scientist_charts_fake_news_reach.

131   *"Falsehood diffused":* Soroush Vosoughi, Deb Roy, and Sinan Aral, "The Spread of True and False News Online," *Science* 359, no. 6380 (March 9, 2018): 1146–51.

fake political headlines: Silverman, "This Analysis Shows."

study of 22 million tweets: Philip N. Howard et al., "Social Media, News and Political Information During the US Election: Was Polarizing Content Concentrated in Swing States?" (data memo 2017.8, Computational Propaganda Project, University of Oxford, September 28, 2017), http://comprop.oii.ox.ac.uk/wp-content/uploads/sites/89/2017/09/Polarizing-Content-and-Swing-States.pdf.

"junk news": Ibid.

"Our bodies are programmed": danah boyd, "Streams of Content, Limited Attention: The Flow of Information Through Social Media," quoted in Pariser, *The Filter Bubble,"* 14.

more than 400 times: Brian Stelter, "Trump Averages a 'Fake' Insult Every Day. Really. We Counted," CNNMoney, January 27, 2018, http://money.cnn.com/2018/01/17/media/president-trump-fake-news-count/index.html.

132   *"how social media":* See, but certainly don't buy, Jack Posobiec, *Citizens for Trump: The Inside Story of the People's Movement to Take Back America* (CreateSpace, 2017).

Jestin Coler: Laura Sydell, "We Tracked Down a Fake-News Creator in the Suburbs. Here's What We Learned," *All Things Considered,* NPR, November 23, 2016, https://www.npr.org/sections/alltechconsidered /2016/11/23/503146770/npr-finds-the-head-of-a-covert-fake-news-opera tion-in-the-suburbs.

full-fledged empire: Ibid.

false story of an FBI agent: David Mikkelson, "Fact Check: FBI Agent Suspected in Hillary Email Leaks Found Dead," Snopes, November 5, 2016, http://www.snopes.com/fbi-agent-murder-suicide/.

1.6 million readers: Sydell, "We Tracked Down."

at least 15 million times: Ryan Grenoble, "Here Are Some of Those Fake News Stories That Mark Zuckerberg Isn't Worried About," Huffington Post, November 16, 2016, https://www.huffingtonpost.com/entry/face book-fake-news-stories-zuckerberg_us_5829f34ee4b0c4b63b0da2ea.

"not the safest crowd": Sydell, "We Tracked Down."

133 *1.25 million news stories:* Yochai Benkler et al., "Study: Breitbart-Led Right-Swing Media Ecosystem Altered Broader Media Agenda," *Columbia Journalism Review,* March 3, 2017, https://www.cjr.org/analysis/breitbart-media-trump-harvard-study.php.

founder Andrew Breitbart: Charlie Spiering, "New Andrew Breitbart footage: 'My Goal Is to Destroy the New York Times and CNN,'" *Washington Examiner,* August 6, 2012, http://www.washingtonexaminer.com/new-andrew-breitbart-footage-my-goal-is-to-destroy-the-new-york-times-and-cnn/article/2504131.

"#War": Joseph Bernstein, "Alt-White: How the Breitbart Machine Laundered Racist Hate," BuzzFeed, October 5, 2017, https://www.buzzfeed.com/josephbernstein/heres-how-breitbart-and-milo-smuggled-white-nationalism?utm_term=.eekpAwn4E#.xuoQnyPGK.

Short for "alternative right": "Alt-Right," Southern Poverty Law Center, https://www.splcenter.org/fighting-hate/extremist-files/ideology/alt-right.

the Associated Press put it: John Daniszewski, "How to Describe Extremists Who Rallied in Charlottesville," *The Definitive Source* (blog), Associated Press, August 15, 2017, https://blog.ap.org/behind-the-news/how-to-describe-extremists-who-rallied-in-charlottesville.

"the platform for the alt-right": Sarah Posner, "How Donald Trump's New Campaign Chief Created an Online Haven for White Nationalists," *Mother Jones,* August 22, 2016, https://www.motherjones.com/poli tics/2016/08/stephen-bannon-donald-trump-alt-right-breitbart-news/.

edit their own: Bernstein, "Alt-White."

134 *judged by key measures:* Benkler et al., "Study."

sneaking in via Mexico: Ildefonso Ortiz and Brandon Darby, "Mexico Helping Unvetted African Migrants to U.S. Border, Many from Al-Shabaab Terror Hotbed," Breitbart, September 10, 2016, http://www.breitbart.com/texas/2016/09/10/african-immigrants-working-mexicos-immigration-system-get-free-pass-california/.

Twitter account parodying Trump: John Hayward, "Three Green Berets Killed, Two Wounded in Niger Ambush," Breitbart, October 5, 2017, http://www.breitbart.com/national-security/2017/10/05/three-green-berets-killed-two-wounded-niger-ambush/amp/.

John Herrman had observed: John Herrman, "In the Trenches of the Facebook Election," *The Awl,* November 21, 2014, https://theawl.com/in-the-trenches-of-the-facebook-election-cc0a268cb4f7.

59 percent of all links: Makysm Gabielkov et al., "Social Clicks: What and Who Gets Read on Twitter?" (paper prepared for Sigmetrics '16, Antibes Juan-Les-Pins, France, June 14–18, 2016), https://hal.inria.fr/hal-01281190/document.

135 *one-tenth of professional media coverage:* Thomas E. Patterson, "News Coverage of the 2016 General Election: How the Press Failed the Voters," Shorenstein Center on Media, Politics, and Public Policy, Harvard Kennedy School, December 7, 2016, https://shorensteincenter.org/news-coverage-2016-general-election/.

thirty-two minutes: Andrew Tyndall, "Issues? What Issues?," Tyndall Report, October 25, 2016, http://tyndallreport.com/comment/20/5778/.

"Breitbart of the left": Jonathan Easley, "Top Dem Super PAC Launches Anti-Trump War Room," The Hill, December 6, 2016, http://thehill.com/blogs/ballot-box/308978-top-dem-super-pac-launches-anti-trump-war-room.

secret employ: Casey Michel, "The Bizarre Rise and Dramatic Fall of Louise Mensch and Her 'Blue Detectives,'" ThinkProgress, January 19, 2018, https://thinkprogress.org/blue-detectives-collapse-trump-russia-a42a94537bdf/.

2017 French presidential election: Laura Daniels, "How Russia Hacked the French Election," Politico, April 23, 2017, https://www.politico.eu/article/france-election-2017-russia-hacked-cyberattacks/.

Spain: Vasco Cotovio and Emanuella Grinberg, "Spain: 'Misinformation' on Catalonia Referendum Came from Russia," CNN, November 13, 2017, https://www.cnn.com/2017/11/13/europe/catalonia-russia-connection-referendum/index.html.

Pakistani defense minister: Ben Westcott, "Duped by Fake News Story,

Pakistani Minister Threatens Nuclear War with Israel," CNN, December 26, 2016, https://www.cnn.com/2016/12/26/middleeast/israel-pakistan-fake-news-nuclear/index.html.

false Facebook update: Justin Lynch, "In South Sudan, Fake News Has Deadly Consequences," *Slate,* June 9, 2017, http://www.slate.com/articles/technology/future_tense/2017/06/in_south_sudan_fake_news_has_deadly_consequences.html.

sectarian and ethnic violence: "Social Media and Conflict in South Sudan: A Lexicon of Hate Speech Terms" (report, PeaceTech Lab, n.d.), https://static1.squarespace.com/static/54257189e4b0acod5fca1566/t/5851c214725e25c531901330/1481753114460/PeaceTech+Lab_+SouthSudanLexicon.pdf.

rival Sudanese Facebook groups: Lynch, "In South Sudan."

In India: "Hindutva.Info Runs Fake News of Hindus Thrashing Barkati," ENewsRoom, February 16, 2018, https://enewsroom.in/hindutva-info-runs-fake-story-hindus-rastravadi-muslims-bashing-barkati/.

136 *in Myanmar:* Euan McKirdy, "When Facebook Becomes 'the Beast': Myanmar Activists Say Social Media Aids Genocide," CNN, April 6, 2018, https://www.cnn.com/2018/04/06/asia/myanmar-facebook-social-media-genocide-intl/index.html.

in Sri Lanka: Amanda Taub and Max Fisher, "Where Countries Are Tinderboxes and Facebook Is a Match," *New York Times,* April 21, 2018, https://www.nytimes.com/2018/04/21/world/asia/facebook-sri-lanka-riots.html.

"The germs": Ibid.

trademark look: Michael Lohmuller, "Panic Ensues After MS13 Allegedly Prohibits Blonde Hair in Honduras Markets," InSight Crime, May 27, 2015, https://www.insightcrime.org/news/analysis/ms13-allegedly-prohibits-blonde-hair-in-honduras-markets/.

criminals solemnly denounced: MS-13, press release, trans. Google Translate, InSight Crime, http://www.insightcrime.org/images/PDFs/2015/El-Salvador-Gang-Press-Release.pdf.

force genital mutilation: Ian Black and Fazel Hawramy, "ISIS Denies Ordering That All Girls in Mosul Undergo FGM," *The Guardian,* July 24, 2014, https://www.theguardian.com/world/2014/jul/24/isis-deny-ordering-fgm-girls-mosul.

An ISIS Twitter account: Ibid.

Ninety percent of Americans: Michael Barthel, Amy Mitchell, and Jesse Holcomb, "Many Americans Believe Fake News Is Sowing Confusion," Pew Research Center, December 15, 2016, http://www.journalism.org/2016/12/15/many-americans-believe-fake-news-is-sowing-confu sion/.

Nearly one-quarter of Americans: Ibid.

Washington Post *quietly ended:* Caitlin Dewey, "What Was Fake on the Internet This Week: Why This Is the Final Column," *The Intersect* (blog), *Washington Post,* December 18, 2015, https://www.washingtonpost.com/news/the-intersect/wp/2015/12/18/what-was-fake-on-the-internet-this-week-why-this-is-the-final-column/?utm_term=.05edc76143b3.

mused columnist Caitlin Dewey: Ibid.

137   *"digital wildfires":* "Digital Wildfires in a Hyperconnected World," World Economic Forum, http://reports.weforum.org/global-risks-2013/view/risk-case-1/digital-wildfires-in-a-hyperconnected-world/.

More than half: Sue Shellenbarger, "Most Students Don't Know When News Is Fake, Stanford Study Finds," *Wall Street Journal,* November 21, 2016, https://www.wsj.com/articles/most-students-dont-know-when-news-is-fake-stanford-study-finds-1479752576.

"If it's going viral": Matt McKinney, "'If It's Going Viral, It Must Be True': Hampton Roads Kids Struggle with Fake News, Teachers Say," *Virginian-Pilot,* November 28, 2016, https://pilotonline.com/news/local/education/public-schools/if-it-s-going-viral-it-must-be-true-hampton/article_4a785dfb-3dd3-5229-9578-c4585adfefb4.html.

Angee Dixson was mad: Angee's archived Twitter account provides an excellent snapshot of a Russian botnet in action. See Angee Dixson (@angeelistr), http://archive.today/2017.08.17-214343/https://twitter.com/angeelistr.

some ninety times a day: Isaac Arnsdorf, "Pro-Russian Bots Take Up the Right-Wing Cause After Charlottesville," ProPublica, August 23, 2017, https://www.propublica.org/article/pro-russian-bots-take-up-the-right-wing-cause-after-charlottesville.

#UniteTheRight rally: Christina Caron, "Heather Heyer, Charlottes-ville Victim, Is Recalled as a 'Strong Woman,'" *New York Times,* August 13, 2017, https://www.nytimes.com/2017/08/13/us/heather-heyer-charlottes-ville-victim.html.

138   *claimed "both sides":* Dan Merica, "Trump Says Both Sides to Blame Amid Charlottesville Backlash," CNN, August 16, 2017, https://www.cnn.com/2017/08/15/politics/trump-charlottesville-delay/index.html.

"Dems and Media": Arnsdorf, "Pro-Russian Bots."

"Angee Dixson" was actually: Ibid.

a photograph of Lorena Rae: Casey Michel, "Russia-Linked Propaganda Accounts Banned by Twitter Are Still Active on Facebook," Think-Progress, November 4, 2017, https://thinkprogress.org/russia-linked-propaganda-facebook-1ca727253ccf/.

at least 60,000: Tom Williams, "The Power of Social Media — How Twitter Exposed a Brexiteer, More Influential Than Sky News, as a Russian Troll," *PoliticsMeansPolitics* (blog), Medium, August 31, 2017, https://blog.politicsmeanspolitics.com/the-power-of-social-media-how-twitter-exposed-a-brexiteer-more-influential-than-sky-news-as-a-e0b991129bd9.

139 *Fake followers and "likes":* Nicholas Confessore et al., "The Follower Factory," *New York Times,* January 27, 2018, https://www.nytimes.com/interactive/2018/01/27/technology/social-media-bots.html.

at least $1 billion: Doug Bock Clark, "Inside a Counterfeit Facebook Farm," *The Week,* June 15, 2015, http://theweek.com/articles/560046/inside-counterfeit-facebook-farm.

18 million "likes": Heather Timmons and Josh Horwitz, "China's Propaganda News Outlets Are Absolutely Crushing It on Facebook," Quartz, May 6, 2016, https://qz.com/671211/chinas-propaganda-outlets-have-leaped-the-top-of-facebook-even-though-it-banned-at-home/.

more than a million "fans": Ibid.

4 percent supposedly lived: Jennings Brown and Adi Cohen, "There's Something Odd About Donald Trump's Facebook Page," Vocativ, June 17, 2015, http://www.businessinsider.com/donald-trumps-facebook-followers-2015-6.

Dhaka in Bangladesh: Charles Arthur, "How Low-Paid Workers at 'Click Farms' Create Appearance of Online Popularity," *The Guardian,* August 2, 2013, https://www.theguardian.com/technology/2013/aug/02/click-farms-appearance-online-popularity.

Lapu-Lapu in the Philippines: Clark, "Inside a Counterfeit Facebook Farm."

companies' spam protection: Ibid.

140 *Czech word meaning "slave":* Lydia H. Liu, *The Freudian Robot: Digital Media and the Future of the Unconscious* (University of Chicago Press, 2010), 6.

"Lizynia Zikur": ProPublica (@ProPublica), "The weirdness continues. This Russian account has just 76 followers and tweeted just once: a smear of us, that got . . . 23,400+ retweets," Twitter, August 24, 2017, 6:08 p.m., https://twitter.com/ProPublica/status/900887458400829440.

The "Star Wars" botnet: Bill Brenner, "Twitter's Phantom Menace: A Star Wars Botnet," Naked Security, https://nakedsecurity.sophos.com/2017/01/25/potential-phantom-menace-found-on-twitter-a-star-wars-botnet/.

141 *roughly 15 percent:* Confessore et al., "The Follower Factory."

random soccer statistics: Jillian C. York, "Syria's Twitter spambots," *The Guardian*, April 21, 2011, https://www.theguardian.com/commentis free/2011/apr/21/syria-twitter-spambots-pro-revolution.

beautiful landscape images: Ibid.

hashtags like #FreeTibet: "Twitter Bots Target Tibetan Protests," *Krebs on Security* (blog), March 12, 2012, http://krebsonsecurity.com/2012/03/ twitter-bots-target-tibetan-protests/.

first documented uses: Marion R. Just et al., "'It's Trending on Twitter' — An Analysis of the Twitter Manipulations in the Massachusetts 2010 Special Senate Election" (paper prepared for the Annual Meeting of the American Political Science Association, New Orleans, August 30–September 2, 2012), https://papers.ssrn.com/sol3/papers.cfm?abstract_id=2108272#.

"Swift Boat" negative advertising campaign: Ibid.

142 *"Twitterbomb":* Ibid.

more than a million fake followers: Samuel Woolley and Phil Howard, "Bots Unite to Automate the Presidential Election," *Wired*, May 15, 2016, https://www.wired.com/2016/05/twitterbots-2/.

In Italy: Andrea Vogt, "Hot or Bot? Italian Professor Casts Doubt on Politician's Twitter Popularity," *The Guardian*, July 22, 2012, https://www .theguardian.com/world/2012/jul/22/bot-italian-politician-twitter-grillo.

military cyberwarfare specialists: Choe Sang-Hun, "South Korean Officials Accused of Political Meddling," *New York Times*, December 19, 2013, http://www.nytimes.com/2013/12/20/world/asia/south-korean-cyber warfare-unit-accused-of-political-meddling.html.

nearly 25 million messages: Lee Yoo Eun, "South Korea's Spy Agency, Military Sent 24.2 Million Tweets to Manipulate Election," Global Voices, November 25, 2013, https://globalvoices.org/2013/11/25/south-koreas-spy-agency-military-sent-24-2-million-tweets-to-manipulate-election/.

shifted their attention: Philip N. Howard and Bence Kollanyi, "Bots, #Strongerin, and #Brexit: Computational Propaganda During the UK-EU Referendum," arXiv:1606.06356 [cs.SI], June 20, 2016.

ratio of five to one: Ibid.

linked to Russia: Robert Booth et al., "Russia Used Hundreds of Fake Accounts to Tweet About Brexit, Data Shows," *The Guardian*, November 14, 2017, https://www.theguardian.com/world/2017/nov/14/how-400-russia-run-fake-accounts-posted-bogus-brexit-tweets.

less than 1 percent: Howard and Kollanyi, "Bots, #Strongerin, and #Brexit."

roughly 400,000 bot accounts: Alessandro Bessi and Emilio Ferrara, "Social Bots Distort the 2016 U.S. Presidential Election Online Discussion," *First Monday* 21, no. 11 (November 2016), http://firstmonday.org/ojs/index. php/fm/article/view/7090/5653.

143 *"colonized" pro-Clinton hashtags:* John Markoff, "Automated Pro-Trump Bots Overwhelmed Pro-Clinton Messages, Researchers Say," *New York Times,* November 17, 2016, https://www.nytimes.com/2016/11/18/tech nology/automated-pro-trump-bots-overwhelmed-pro-clinton-messages-researchers-say.html?nytmobile=0&_r=0.

five-to-one ratio: Bence Kollanyi, Philip N. Howard, and Samuel C. Woolley, "Bots and Automation over Twitter During the U.S. Election" (data memo 2016.4, Computational Propaganda Project, University of Oxford, November 2016), http://blogs.oii.ox.ac.uk/politicalbots/wp-con tent/uploads/sites/89/2016/11/Data-Memo-US-Election.pdf.

quote 150 bots: "Chapter 15. Make America Bot Again. Part Three," Sad-BotTrue, http://sadbottrue.com/article/24/.

the online handle "MicroChip": Joseph Bernstein, "Never Mind the Russians, Meet the Bot King Who Helps Trump Win Twitter," BuzzFeed, April 5, 2017, https://www.buzzfeed.com/josephbernstein/from-utah-with-love?utm_term=.dmdwvOGde#.hg6qwrJ28.

more than 30,000 retweets: Ibid.

"how this game works": Ibid.

their "information war": "An Ex St. Petersburg 'Troll' Speaks Out," Meduza, October 15, 2017, https://meduza.io/en/feature/2017/10/15/an-ex-st-petersburg-troll-speaks-out?utm_source=Sailthru&utm_medium=email&utm_campaign=Newpercent20Campaign&utm_term=percent2ASituationpercent20Report.

144 *2.2 million "election-related tweets":* Twitter, Inc., "United States Senate Committee on the Judiciary, Subcommittee on Crime and Terrorism Update on Results of Retrospective Review of Russian-Related Election Activity," January 19, 2018, https://www.judiciary.senate.gov/imo/media/doc/Edgett%20Appendix%20to%20Responses.pdf.

454.7 million times: Ibid.

cited forty-one times: Sheera Frenkel and Katie Bender, "To Stir Discord in 2016, Russians Turned Most Often to Facebook," *New York Times,* February 17, 2018, https://www.nytimes.com/2018/02/17/technology/indict ment-russian-tech-facebook.html.

126 million users: Twitter, Inc., "United States Senate Committee on the Judiciary."

retweeted @realDonaldTrump: Ibid.

collective U.S. intelligence community: "Assessing Russian Activities and

Intentions in Recent US Elections" (Intelligence Community Assessment, Office of the Director of National Intelligence, January 6, 2017), https://www.dni.gov/files/documents/ICA_2017_01.pdf.

five different cybersecurity companies: P. W. Singer, "Cyber-Deterrence and the Goal of Resilience: 30 New Actions That Congress Can Take to Improve U.S. Cybersecurity," testimony before the House Armed Services Committee hearing "Cyber Warfare in the 21st Century: Threats, Challenges, and Opportunities," March 1, 2017, http://docs.house.gov/meetings/AS/AS00/20170301/105607/HHRG-115-AS00-Wstate-Singer P-20170301.pdf.

between 48 percent and 73 percent: Twitter, Inc., "United States Senate Committee on the Judiciary."

"The news media": Jonathon Morgan and Kris Schaffer, "Sockpuppets, Secessionists, and Breitbart," *Data for Democracy* (blog), Medium, March 31, 2017, https://medium.com/data-for-democracy/sockpuppets-secessionists-and-breitbart-7171b1134cd5.

Samuel Woolley: Craig Timberg, "As a Conservative Twitter User Sleeps, His Account Is Hard at Work," *Washington Post,* February 5, 2017, https://www.washingtonpost.com/business/economy/as-a-conservative-twitter-user-sleeps-his-account-is-hard-at-work/2017/02/05/18d5a532-df31-11e6-918c-99ede3c8cafa_story.html?utm_term=.59665eea94ba.

145   *France:* Daniels, "How Russia Hacked the French Election."

Mexico: David Alire Garcia and Noe Torres, "Russia Meddling in Mexican Election: White House Aide McMaster," Reuters, January 7, 2018, https://www.reuters.com/article/us-mexico-russia-usa/russia-meddling-in-mexican-election-white-house-aide-mcmaster-idUSKBN1E WoUD.

one study found: Kirk Semple and Marina Franco, "Bots and Trolls Elbow Into Mexico's Crowded Electoral Field," *New York Times,* May 1, 2018, https://www.nytimes.com/2018/05/01/world/americas/mexico-election-fake-news.html.

Jonathon Morgan and Kris Schaffer: This report numbers among the most fascinating and insightful to come out of the flurry of data-driven, bot-centric journalism that began in 2017. Morgan and Schaffer, "Sockpuppets, Secessionists, and Breitbart."

particular language and culture: Ibid.

"suddenly and simultaneously": Ibid.

146   *a shared playbook:* Ibid.

four times as likely: Ibid.

the word "Jewish": Ibid.

6. WIN THE NET, WIN THE DAY

148 *"Media weapons"*: Charlie Winter, "Media Jihad: The Islamic State's Doctrine for Information Warfare" (report, International Centre for the Study of Radicalisation and Political Violence, 2017), 18, http://icsr.info/wp-content/uploads/2017/02/Media-jihad_web.pdf.

"You can sit": Lorraine Murphy, "The Curious Case of the Jihadist Who Started Out as a Hacktivist," *Vanity Fair,* December 15, 2015, https://www.vanityfair.com/news/2015/12/isis-hacker-junaid-hussain.

"He had hacker cred": Ibid.

easy familiarity with: Del Quentin Wilber, "Here's How the FBI Tracked Down a Tech-Savvy Terrorist Recruiter for the Islamic State," *Los Angeles Times,* April 13, 2017, http://www.latimes.com/politics/la-fg-islamic-state-recruiter-20170406-story.html.

149 *some 30,000 recruits:* Martin Chulov, Jamie Grierson, and Jon Swaine, "Isis Faces Exodus of Foreign Fighters as Its 'Caliphate' Crumbles," *The Guardian,* April 26, 2017, https://www.theguardian.com/world/2017/apr/26/isis-exodus-foreign-fighters-caliphate-crumbles.

"The knives have been": Rukmini Callimachi, "Clues on Twitter Show Ties Between Texas Gunman and ISIS Network," *New York Times,* May 11, 2015, www.nytimes.com/2015/05/12/us/twitter-clues-show-ties-between-isis-and-garland-texas-gunman.html.

He took a wife: Nancy Youssef, "The British Punk Rocker Widow Who Wants to Run ISIS's Hackers," The Daily Beast, September 27, 2015, https://www.thedailybeast.com/the-british-punk-rocker-widow-who-wants-to-run-isiss-hackers.

Pentagon's "kill list": "UK Jihadist Junaid Hussain Killed in Syria Drone Strike, Says US," BBC News, August 27, 2015, http://www.bbc.com/news/uk-34078900.

tricked into clicking: James Cartledge, "Isis Terrorist Junaid Hussain Killed in Drone Attack After Boffins 'Crack Group's Code,'" Birmingham Live, September 16, 2015, http://www.birminghammail.co.uk/news/midlands-news/isis-terrorist-junaid-hussain-killed-10069425.

leave his stepson: Adam Goldman and Eric Schmitt, "One by One, ISIS Social Media Experts Are Killed as Result of F.B.I. Program," *New York Times,* November 24, 2016, https://www.nytimes.com/2016/11/24/world/middleeast/isis-recruiters-social-media.html?mtrref=www.google.com&gwh=D9D7306F189C9D3AD771F097D5C1BD35&gwt=pay.

150 *word of the day:* Jessica Stern and J. M. Berger, *ISIS: The State of Terror* (Ecco, 2015), 120.

apocalyptic interpretation: Quite literally apocalyptic. See William Mc-Cants, *The ISIS Apocalypse: The History, Strategy, and Doomsday Vision of the Islamic State* (St. Martin's, 2015).

"Terrorism is theater": Brian Jenkins, "International Terrorism: A New Kind of Warfare" (RAND Papers Series, no. P-5261, RAND, June 1974), https://www.rand.org/content/dam/rand/pubs/papers/2008/P5261. pdf.

151   *Jihadi Design:* Gilad Shiloach, "How ISIS Supporters Learn to Design Propaganda," The Daily Dot, March 7, 2017, https://www.dailydot.com/layer8/isis-propaganda-graphic-design/.

"With the old methods": Richard Engel, "Sadat's Assassination Plotter Remains Unrepentant," NBC News, July 5, 2011, http://www.nbcnews.com/id/43640995/ns/world_news-mideast_n_africa/t/sadats-assassination-plotter-remains-unrepentant/.

propelled by some 60,000: Steffan Truvé, "ISIS Jumping from Account to Account, Twitter Trying to Keep Up," Recorded Future, September 3, 2014, https://www.recordedfuture.com/isis-twitter-activity/.

dramatic screengrabs: George Brown and CNN Wire, "How Should Media Cover American Beheading," News Channel 3, August 20, 2014, http://wreg.com/2014/08/20/media-outlets-struggle-with-american-beheaded-coverage/.

152   *"Don't share it":* Ibid.

One aspiring politician: Amanda Terkel, "GOP House Candidate Uses James Foley Execution Footage in Campaign Ad," Huffington Post, October 6, 2014, www.huffingtonpost.com/2014/10/06/wendy-rogers-arizona_n_5940346.html.

submerged in a pool: "ISIS Release Brutal Execution Videos of Mosul 'Spies,'" *Newsweek,* June 23, 2015, http://www.newsweek.com/isis-execution-videoisis-iraq-videoislamic-state-execution-videoisis-603331.

"Kill the Jordanian": Gilad Shiloach, "Crowdsourcing Terror: ISIS Asks for Ideas on Killing Jordanian Pilot," Vocativ, December 26, 2014, http://www.vocativ.com/world/isis-2/suggestions-kill-pilot/.

"This is our football": Jay Caspian King, "ISIS's Call of Duty," *The New Yorker,* September 18, 2014, https://www.newyorker.com/tech/elements/isis-video-game?loc=contentwell&lnk=image-of-a-decapitated-head&dom=section-2&irgwc=1&source=affiliate_impactpmx_12f6tote_desktop_VigLink&mbid=affiliate_impactpmx_12f6tote_desktop_VigLink.

"first terrorist group": Jared Cohen, "Digital Counterinsurgency: How to Marginalize the Islamic State Online," *Foreign Affairs,* November/

December 2015, https://www.foreignaffairs.com/articles/middle-east/digital-counterinsurgency.

nearly fifty different: Charlie Winter, "Documenting the Virtual 'Caliphate'" (report, Quilliam, October 2015), 16, http://www.quilliaminternational.com/wp-content/uploads/2015/10/FINAL-documenting-the-virtual-caliphate.pdf.

153 *over a thousand "official":* Ibid., 5.

at least 30,000 civilians: "Iraq Body Count," accessed March 19, 2018, https://www.iraqbodycount.org/database/.

lived with their parents: "Case by Case: ISIS Prosecutions in the United States, March 1, 2014–June 30, 2016" (report, Center on National Security at Fordham Law, July 2016), http://static1.square space.com/static/55dc76f7e4b013c872183fea/t/577c5b43197aea832bd4 86c0/1467767622315/ISIS+Report+-+Case+by+Case+-+July2016.pdf.

pledge his allegiance: Thomas Joscelyn, "Orlando Terrorist Swore Allegiance to Islamic State's Abu Bakr al Baghdadi," *FDD's Long War Journal,* June 20, 2016, http://www.longwarjournal.org/archives/2016/06/orlando-terrorist-swore-allegiance-to-islamic-states-abu-bakr-al-baghdadi.php.

had gone viral: Denver Nicks, "Orlando Shooter Checked Facebook to See If His Attack Went Viral," *Time,* June 16, 2016, http://time.com/4371910/orlando-shooting-omar-mateen-facebook/?xid=emailshare.

virtually every scholar: Jennifer Williams, "How ISIS Uses and Abuses Islam," Vox, November 18, 2015, https://www.vox.com/2015/11/18/9755478/isis-islam.

exaggerated its gains: Daveed Gartenstein-Ross, Nathaniel Barr, and Bridget Moreng, "How the Islamic State's Propaganda Feeds Into Its Global Expansion Efforts," War on the Rocks, April 28, 2016, warontherocks.com/2016/04/how-islamic-states-propaganda-feeds-into-its-global-expansion-efforts/.

claim after the fact: Joshua Keating, "ISIS Is Not Known for Falsely Taking Credit for Attacks — Until Recently," *Slate,* October 2, 2017, http://www.slate.com/blogs/the_slatest/2017/10/02/isis_s_claims_of_respon sibility_are_getting_more_dubious.html.

154 *"If there is":* Mark Mazzetti and Eric Schmitt, "In the Age of ISIS, Who's a Terrorist, and Who's Simply Deranged?," *New York Times,* July 17, 2016, www.nytimes.com/2016/07/18/world/europe/in-the-age-of-isis-whos-a-terrorist-and-whos-simply-deranged.html.

"information jihad": Charlie Winter, "What I Learned from Reading the Islamic State's Propaganda Instruction Manual," *Lawfare* (blog), April

2, 2017, https://lawfareblog.com/what-i-learned-reading-islamic-states-propaganda-instruction-manual.

media "projectiles": Ibid.

"make terrorism sexy": David Francis, "Why Don Draper Would Be Impressed by the Islamic State," *The Cable* (blog), *Foreign Policy,* April 7, 2015, http://foreignpolicy.com/2015/04/07/why-don-draper-would-be-impressed-by-the-islamic-state/.

155 *"became a reality star"*: Spencer Pratt, phone interview with authors, May 23, 2016.

how to make $50,000: Naomi Fry, "The Reality-TV Star Spencer Pratt on America's Addiction to Drama," *The New Yorker,* June 30, 2017, http://www.newyorker.com/culture/persons-of-interest/the-reality-tv-and-snapchat-star-spencer-pratt-on-americas-addiction-to-drama.

"I saw The Osbournes*"*: Pratt interview.

the Kardashians: Suzannah Weiss, "Spencer Pratt Missed His Chance to Make the Kardashians Famous," Refinery29, February 20, 2017, http://www.refinery29.com/2017/02/141810/spencer-pratt-princes-of-malibu-kardashians.

nightclub called Privilege: Michael Sunderland, "Head over Hills: The Undying Love Story of Heidi and Spencer Pratt," *Vice,* February 12, 2016, https://broadly.vice.com/en_us/article/9aepp7/heidi-montag-spencer-pratt-the-hills-profile.

156 *"manipulating the media"*: Pratt interview.

"working with the paparazzi": Ibid.

"Best Villain": "Teen Choice Awards 2009 Nominees," *Los Angeles Times,* June 15, 2009, http://latimesblogs.latimes.com/awards/2009/06/teen-choice-awards-2009-nominees.html.

"shot that scene": Pratt interview.

"just an awful asshole": Ibid.

157 *pioneering 1944 study:* Fritz Heider and Marianne Simmel, "An Experimental Study of Apparent Behavior," *American Journal of Psychology* 57, no. 2 (1944): 243–59.

"psychological political by-products": Matthew Armstrong, "The Past, Present, and Future of the War for Public Opinion," War on the Rocks, January 19, 2017, https://warontherocks.com/2017/01/the-past-present-and-future-of-the-war-for-public-opinion/.

158 *"everyone's an editor"*: Heidi Montag, phone interview with authors, May 23, 2016.

"everyone is a reality star": Pratt interview.

shrunk to eight seconds: "Attention Span Statistics," Statistic Brain, https://www.statisticbrain.com/attention-span-statistics/.

fifth-grade education: "Most Presidential Candidates Speak at Grade 6–8 Level," Cision, March 16, 2016, https://www.prnewswire.com/news-re leases/most-presidential-candidates-speak-at-grade-6-8-level-300237139. html.

159 *the complexity score dipped:* Derek Thompson, "Presidential Speeches Were Once College-Level Rhetoric — Now They're for Sixth-Graders," *The Atlantic,* October 14, 2014, https://www.theatlantic.com/politics/ archive/2014/10/have-presidential-speeches-gotten-less-sophisticated- over-time/381410/.

a shark swimming down: Kaleigh Rogers, "'Shark Swims Down a Flooded Street' Is a Viral Hoax That Won't Die," *Motherboard* (blog), *Vice,* October 5, 2015, https://motherboard.vice.com/en_us/article/jp5ydp/ shark-swims-down-a-flooded-street-is-a-viral-hoax-that-wont-die.

160 *more unyieldingly hyperpartisan:* Adam Hughes and Onyi Lam, "Highly Ideological Members of Congress Have More Facebook Followers Than Moderates Do," Pew Research Center, August 21, 2017, http://www. pewresearch.org/fact-tank/2017/08/21/highly-ideological-members-of- congress-have-more-facebook-followers-than-moderates-do/.

why conspiracy theories: Marco Guerini and Carlo Strapparava, "Why Do Urban Legends Go Viral?," *Information Processing and Management* 52, no. 1 (January 2016): 163–72.

a jar of Nutella: Scott Campbell, "ISIS Using Kittens and NUTELLA to Lure Jihadi Wannabes into Evil Death Cult," *Mirror,* May 27, 2016, http://www.mirror.co.uk/news/world-news/isis-using-kittens-nutella- lure-8061303.

from far-right political leaders: Andrew Marantz, "Trolls for Trump," *The New Yorker,* October 31, 2016, https://www.newyorker.com/maga zine/2016/10/31/trolls-for-trump.

women's rights activists: Laura Durnell, "All Women Must Control the Narrative in and After the Trumpian Age," Huffington Post, February 28, 2017, https://www.huffingtonpost.com/entry/women-must-control-the- narrative-in-the-age-of-trump_us_58b609fae4b02f3f81e44dc7.

the Kardashian clan: Nathan Heller, "The Multitasking Celebrity Takes Center Stage," *The New Yorker,* June 23, 2016, http://www.newyorker.com/ culture/cultural-comment/the-organizational-celebrity.

Omar Hammami: Simon Cottee, "Why It's So Hard to Stop ISIS Propa- ganda," *The Atlantic,* March 2, 2015, https://www.theatlantic.com/interna tional/archive/2015/03/why-its-so-hard-to-stop-isis-propaganda/386216/.

161   *"we'll lose the battle":* Meeting (not for attribution), Department of State, Washington, DC, October 7, 2015.

Gunner Stone: Korey Lane, "What Does Gunner Stone Mean? Heidi Montag and Spencer Pratt's First Child Is Already Pretty Interesting," Romper, October 4, 2017, https://www.romper.com/p/what-does-gun ner-stone-mean-heidi-montag-spencer-pratts-first-child-is-already-pretty-interesting-2790177.

"When we do not know": T. S. Eliot, "The Perfect Critics," in *The Sacred Wood: Essays on Poetry and Criticism* (Knopf, 1921), 9.

"vast accumulations of knowledge": Ibid.

162   *the stronger the emotions:* Jonah Berger and Katherine Milkman, "What Makes Online Content Viral?," *Journal of Marketing Research* 49, no. 2 (2012): 192–205; Marco Guerini and Jacopo Staiano, "Deep Feelings: A Massive Cross-Lingual Study on the Relation Between Emotions and Virality," arXiv:1503.04723 [cs.SI], March 2015.

70 million messages: Rui Fan et al., "Anger Is More Influential Than Joy: Sentiment Correlation in Weibo," *PLoS ONE* 9, no. 10 (2014), e:110184, http://journals.plos.org/plosone/article/file?id=10.1371/journal. pone.0110184&type=printable.

"Anger is more influential": Ibid.

ramp up their language: Ibid.

nearly 700,000 users: Adam D. L. Kramer, Jamie E. Guillory, and Jeffrey T. Hancock, "Experimental Evidence of Massive-Scale Emotional Contagion Through Social Networks," *PNAS* 111, no. 24 (2014): 8788–90.

"emotional contagion": Ibid.

163   *"Our lives matter":* Elazar Sontag, "To This Black Lives Matter Co-founder, Activism Begins in the Kitchen," *Washington Post,* March 26, 2018, https://www.washingtonpost.com/lifestyle/food/to-this-black-lives-matter-co-founder-activism-begins-in-the-kitchen/2018/03/26/964ec51a-2df1-11e8-b0b0-f706877db618_story.html?utm_term=.150be2273ebc.

adding the hashtag #BlackLivesMatter: Jon Schuppe and Safia Samee Ali, "Cities Don't Want Justice Department to Back Off Police Reforms," NBC News, April 5, 2017, http://www.nbcnews.com/news/us-news/cities-dont-want-justice-department-back-police-reforms-n742661.

"trolling for MiGs": Andy Bodle, "Trolls: Where Do They Come From?," *The Guardian,* April 19, 2012, https://www.theguardian.com/media/mind-your-language/2012/apr/19/trolls-where-come-from.

164   *"If you don't fall":* Steven Supmante, "Trolling the Web: A Guide," *Urban75 Magazine,* n.d., http://www.urban75.com/Mag/troll.html#one.

emotional manipulation: Natalie Sest and Evita March, "Constructing the Cyber-Troll: Psychopathy, Sadism, and Empathy," *Personality and Individual Differences* 119 (December 2017): 69–72.

philosopher Jean-Paul Sartre: Anti-Semite and Jew (Schocken, 1948), 13.

"Ironghazi": Ben Collins and Joseph Cox, "Jenna Abrams, Russia's Clown Troll Princess, Duped the Mainstream Media and the World," The Daily Beast, November 2, 2017, https://www.thedailybeast.com/jenna-abrams-russias-clown-troll-princess-duped-the-mainstream-media-and-the-world.

psychopathy and sadism: Sest and March, "Constructing the Cyber-Troll."

165 *twice as likely to engage:* Justin Cheng et al., "Anyone Can Become a Troll: Causes of Trolling Behavior in Online Discussions," arXiv:1702.01119 [cs. SI], February 2017, 8.

"can be contagious": Ibid., 2.

166 *"You have the prettiest":* Jonathan Vankin, "Taylor Swift the New Dear Abby? Gives Lovelorn Fan Wise-Beyond-Her-Years Advice," Inquisitr, July 25, 2014, http://www.inquisitr.com/1373553/taylor-swift-the-new-dear-abby-gives-lovelorn-fan-wise-beyond-her-years-advice/.

another, 16-year-old fan: Rebecca Borison, "Taylor Swift Is Incredibly Good at Being a Celebrity," Business Insider, September 10, 2014, http://www.businessinsider.com/taylor-swift-is-a-business-genius-2014-9.

#Taylurking: Lindsey Weber, "Taylor Swift Is the Reigning Queen of Celebrity Social-Media," Vulture, October 29, 2014, http://www.vulture.com/2014/10/taylor-swift-queen-of-celebrity-social-media.html.

"this new site": Taylor Swift, "For Taylor Swift, the Future of Music Is a Love Story," *Wall Street Journal,* July 7, 2014, https://www.wsj.com/articles/for-taylor-swift-the-future-of-music-is-a-love-story-1404763219.

"In the future": Ibid.

she strategically copyrighted: Dan Rys, "Taylor Swift Files Nine Trademarks for the Word 'Swifties,' but Why?," *Billboard,* March 15, 2017, https://www.billboard.com/articles/business/7727743/taylor-swift-trademarks-swifties-but-why.

40 million albums: "Taylor Swift Named IFPI Global Recording Artist of 2014," IFPI, February 23, 2015, http://www.ifpi.org/news/Taylor-Swift-named-IFPI-global-recording-artists-of-2014.

digital streaming records: Lisa Marie Segarra, "Taylor Swift's Spotify Songs Made an Insane Amount of Money in a Week," *Fortune,* June 23, 2017, http://fortune.com/2017/06/23/taylor-swift-spotify-songs-money/.

youngest of Forbes: Zack O'Malley Greenburg, "Taylor Swift Is the Youngest of America's Richest Self-Made Women," *Forbes,* June 1, 2016,

https://www.forbes.com/sites/zackomalleygreenburg/2016/06/01/taylor-swift-is-the-youngest-of-americas-richest-self-made-women/#3c2593c07c1a.

weren't very candid: Tim Teeman, "Why Taylor Swift's Parties Look Like Utter Hell," The Daily Beast, July 6, 2016, http://www.thedailybeast.com/why-taylor-swifts-parties-look-like-utter-hell.

"There's no simple answer": Amy Zimmerman, "How Kim Kardashian Beat Taylor Swift at Her Own Game," The Daily Beast, July 18, 2016, http://www.thedailybeast.com/how-kim-kardashian-beat-taylor-swift-at-her-own-game.

167 *an impromptu concert:* Anna Silman, "Taylor Swift Gives the Gift of Taylor Swift for Christmas," The Cut, December 27, 2016, https://www.thecut.com/2016/12/taylor-swift-gives-the-gift-of-taylor-swift-for-christmas.html.

random Christmas gifts: Kristin Harris, "Taylor Swift Surprised Her Fans with Christmas Presents and Their Reactions Are Hysterical," BuzzFeed, November 13, 2014, https://www.buzzfeed.com/kristinharris/taylor-swift-sent-fans-surprise-christmas-presents?utm_term=.aqzdZLPZB2#.xxNbKEBK2k.

"the element of surprise": Swift, "For Taylor Swift."

birthday parties: Christiaan Triebert (@trbrtc), "IS fighters having dinner and cake. Photo found on mobile phone killed fighter. (h/t@AfarinMamosta/@theOSINTblog)," Twitter, May 18, 2016, 12:35 p.m., https://twitter.com/trbrtc/status/733018236481089537.

their cats: "Cat got your gun? Iraq, Syria, jihadist pictures go viral," Al Arabiya, June 22, 2014, https://english.alarabiya.net/en/variety/2014/06/22/ISIS-fighters-big-on-cats.html.

168 Jumanji: John Hall, "'We Are Humans Like You . . . Why Shouldn't We See Jumanji?,'" *Daily Mail,* August 12, 2014, http://www.dailymail.co.uk/news/article-2722878/Bizarre-Twitter-outburst-ISIS-fighters-reveal-love-late-Robin-Williams-blockbuster-hit-Jumanji.html.

169 *"People feel like":* "Longform Podcast #254: Maggie Haberman," *Longform,* July 26, 2017, https://longform.org/posts/longform-podcast-254-maggie-haberman.

team of eleven staffers: Kyle Cheney, "The Staff Army Behind a Clinton Tweet," *Politico Live Blog,* Politico, October 15, 2016, https://www.politico.com/live-blog-updates/2016/10/john-podesta-hillary-clinton-emails-wikileaks-000011.

"MODERN DAY PRESIDENTIAL": Donald Trump (@realDonaldTrump), "My use of social media is not Presidential — it's MODERN DAY PRESI-

DENTIAL. Make America Great Again!," Twitter, July 1, 2017, 3:41 p.m., https://twitter.com/realDonaldTrump/status/881281755017355264.

La Meute: Julia Carrie Wong, "How Facebook Groups Bring People Closer Together — Neo-Nazis Included," *The Guardian,* July 31, 2017, https://www.theguardian.com/technology/2017/jul/31/extremists-neo-nazis-facebook-groups-social-media-islam.

"bring the world": Josh Constine, "Facebook Changes Mission State-ment to 'Bring the World Closer Together,'" TechCrunch, June 22, 2017, https://techcrunch.com/2017/06/22/bring-the-world-closer-together/.

170 *ballooned 600 percent:* J. M. Berger, "Nazis vs. ISIS on Twitter: A Com-parative Study of White Nationalist and ISIS Online Social Media Net-works" (paper, Program on Extremism, George Washington University, September 2016), https://cchs.gwu.edu/sites/g/files/zaxdzs2371/f/down loads/Nazis%20v.%20ISIS%20Final_0.pdf.

some 1,600: "Hate and Extremism," Southern Poverty Law Center, ac-cessed March 20, 2018, https://www.splcenter.org/issues/hate-and-ex tremism.

350 far-right extremist terror attacks: Weiyi Cai and Simone Landon, "Attacks by White Extremists Are Growing. So Are Their Connections," *New York Times,* April 3, 2019, https://www.nytimes.com/interactive/2019/04/03/world/white-extremist-terrorism-christchurch.html.

fifty Americans: "Right-Wing Extremism Linked to Every 2018 Extremist Murder in the U.S., ADL Finds," Anti-Defamation League, January 23, 2019, https://www.adl.org/news/press-releases/right-wing-extremism-linked-to-every-2018-extremist-murder-in-the-us-adl-finds.

1.5 million times: Mike Snider, "No One Reported New Zealand Mosque Shooting Livestream As It Happened, Facebook Says," *USA Today,* March 19, 2019, https://www.usatoday.com/story/tech/news/2019/03/19/facebook-new-zealand-shooter-livestream-not-reported-during-massa cre/3209751002/.

Instagram celebrity: Jack Stubbs, "17 Minutes of Carnage: How New Zea-land Gunman Broadcast His Killings on Facebook," *Reuters,* March 15, 2019, https://www.reuters.com/article/us-newzealand-shootout-live streaming/17-minutes-of-carnage-how-new-zealand-gunman-broadcast-his-killings-on-facebook-idUSKCN1QW294.

Sunday school teacher: Rukmini Callimachi, "ISIS and the Lonely Young American," *New York Times,* June 27, 2015, https://www.nytimes.com/2015/06/28/world/americas/isis-online-recruiting-american.html.

"It's a closed community": Scott Shane, Matt Apuzzo, and Eric Schmitt, "Americans Attracted to ISIS Find an 'Echo Chamber' on

Social Media," *New York Times,* December 8, 2015, http://www.nytimes.com/2015/12/09/us/americans-attracted-to-isis-find-an-echo-chamber-on-social-media.html.

171   *"Isolation may be":* Hannah Arendt, *Totalitarianism: Part Three of the Origins of Totalitarianism* (Harcourt, 1968), 172.

Farah Pandith is a pioneer: Farah Pandith, interview with authors, Washington, DC, November 20, 2015.

"Only peer-to-peer relations": Ibid.

172   *"fully focusing on millennials":* Ibid.

"no government is credible": Ibid.

"Dumbledore's Army": Ibid.

Online Civil Courage Initiative: "Online Civil Courage Initiative," ISD, accessed March 19, 2018, https://www.isdglobal.org/programmes/communications-technology/online-civil-courage-initiative-2/.

Gen Next: Adam Popescu, "This Online Group of Former Islamic Extremists Deradicalizes Jihadists," Bloomberg, October 23, 2017, https://www.bloomberg.com/news/articles/2017-10-23/this-online-group-of-former-islamic-extremists-deradicalizes-jihadists.

Creative Minds for Social Good: "Hedayah Launches 'Creative Minds for Social Good' Initiative to Counter Terrorist Propaganda," Hedayah, October 25, 2016, http://www.hedayahcenter.org/media-details/49/news/73/press-releases/688/hedayah-launches-creative-minds-for-social-good-initiative-to-counter-terrorist-propaganda.

"swarm the content": Pandith interview.

173   *"digital world war":* Haroon K. Ullah, *Digital World War: Islamists, Extremists, and the Fight for Cyber Supremacy* (Yale University Press, 2017).

nearly 8 million times: Paul Stollery, "Mentions of 'Fuck' Online," @PaulStollery (blog), Medium, https://medium.com/@PaulStollery/all-of-the-fucks-given-online-in-2016-58c60edd6e44.

"Trump ran against": David Wong [Jason Pargin], "Don't Panic," *Cracked,* November 9, 2016, http://www.cracked.com/blog/dont-panic/.

174   *$5 billion worth:* Mary Harris, "A Media Post-Mortem on the 2016 Presidential Election," Media Quant, November 14, 2016, http:///www.mediaquant.net/2016/11/a-media-post-mortem-on-the-2016-presidential-election/.

"Trump understands": Eli Stokols and Josh Dawsey, "Trump's Twitter Feed Traumatizes Washington," Politico, January 5, 2017, http://www.politico.com/story/2017/01/trump-twitter-feed-fears-233242.

"I think that social media": Rebecca Morin, "Trump Says Social Media Was

Key to Victory," Politico, November 12, 2016, https://www.politico.com/story/2016/11/donald-trump-social-media-231285.

"Poe's Law": For a good definition, see "Poe's Law," RationalWiki, accessed March 20, 2018, http://rationalwiki.org/wiki/Poe%27s_Law.

175 *saw someone like them:* Dale Beran, "4chan: The Skeleton Key to the Rise of Trump," @*DaleBeran* (blog), Medium, February 14, 2017, https://medium.com/@DaleBeran/4chan-the-skeleton-key-to-the-rise-of-trump-624e7cb798cb.

The discussion board /r/The_Donald: "/r/The_Donald metrics," Reddit Metrics, accessed March 20, 2018, http://redditmetrics.com/r/The_Donald#disqus_thread.

Trump aides who: Ben Schreckinger, "World War Meme," *Politico Magazine,* March/April 2017, https://www.politico.com/magazine/story/2017/03/memes-4chan-trump-supporters-trolls-internet-214856.

pattern that continued: Charlie Warzel, "From Reddit to Trump's Twitter — in Less Than 24 Hours," BuzzFeed, March 3, 2017, https://www.buzzfeed.com/charliewarzel/from-reddit-to-trumps-twitter-in-less-than-24-hours?utm_term=.nq3Op34p7m#.odynXpYXQ1.

176 *"They are endlessly available":* Charlie Warzel, "'Journalistic Integrity Is Dead.' Is the Mainstream Media Next?," *Infowarzel,* TinyLetter, July 23, 2017, tinyletter.com/Infowarzel/letters/journalistic-integrity-is-dead-is-the-mainstream-media-next.

the firm Brandwatch: Adam Tiouririne, "Hillary Clinton's Twitter Chart of Doom," Bloomberg, November 16, 2016, www.bloomberg.com/politics/articles/2016-11-16/hillary-clinton-s-twitter-chart-of-doom.

"We have an army": Michael Flynn, speech (Young Americans Forum, Washington, DC, November 12, 2016). Video available at "Lieutenant General Michael T. Flynn," Young America's Foundation, November 12, 2016, accessed March 20, 2018, https://www.yaf.org/videos/lieutenant-general-michael-t-flynn/.

its strategic efforts: Steven Bertoni, "Exclusive Interview: How Jared Kushner Won Trump the White House," *Forbes,* November 22, 2016, https://www.forbes.com/sites/stevenbertoni/2016/11/22/exclusive-interview-how-jared-kushner-won-trump-the-white-house/#633d9c9a3af6.

177 *his first $2 million:* Sue Halpern, "How He Used Facebook to Win," *New York Review of Books,* June 8, 2017, http://www.nybooks.com/articles/2017/06/08/how-trump-used-facebook-to-win/.

"Who controls Facebook": Tudor Mihailescu, "Trump Campaign Chief Outlines $1 Billion Strategy for 2020 on Trip to Romania," *Forbes,* March 24, 2019, https://www.forbes.com/sites/tudormihailescu/2019/03/24/

trump-campaign-chief-outlines-1-billion-strategy-for-2020-on-trip-to-romania/#6888c5831647.

every last bit: Halpern, "How He Used Facebook to Win."

8 trillion pieces: Ibid.

5,000 data points: Ibid.

"sex compass": Alyssa Newcomb, "'Sex Compass' App Harvested User Data, Former Cambridge Analytica Employee Says," NBC News, April 17, 2018, https://www.nbcnews.com/tech/tech-news/sex-compass-app-harvested-user-data-former-cambridge-analytica-employee-n866666.

not just 87 million users: Sean Burch, "Facebook Now Says 87 Million Users Hit by Cambridge Analytica Leak," The Wrap, April 4, 2018, https://www.thewrap.com/facebook-87-million-cambridge-analytica-leak/.

direct messages: Alex Hern and Carole Cadwalladr, "Revealed: Aleksander Kogan Collected Facebook Users' Direct Messages," *The Guardian*, April 13, 2018, https://www.theguardian.com/uk-news/2018/apr/13/revealed-aleksandr-kogan-collected-facebook-users-direct-messages.

"gold mine": Natasha Bertrand, "A Long-Overlooked Player Is Emerging as a Key Figure in the Trump-Russia Investigation," Business Insider, June 23, 2017, http://www.businessinsider.com/brad-parscale-trump-russia-investigation-2017-6.

178　*patterns of Facebook "likes":* Hannes Grassegger and Mikael Krogerus, "The Data That Turned the World Upside Down," *Motherboard* (blog), *Vice*, January 28, 2017, https://motherboard.vice.com/en_us/article/mg9vvn/how-our-likes-helped-trump-win.

only ten "likes": Ibid.

"We exploited Facebook": Carole Cadwalladr and Emma Graham-Harrison, "Revealed: 50 Million Facebook Profiles Harvested for Cambridge Analytica in Major Data Breach," *The Guardian*, March 17, 2018, https://www.theguardian.com/news/2018/mar/17/cambridge-analytica-facebook-influence-us-election.

Lookalike Audiences: Halpern, "How He Used Facebook to Win."

"fifteen people": Brad Parscale, interview, "How Facebook Ads Helped Elect Trump," *60 Minutes*, CBS, October 6, 2017, https://www.cbsnews.com/news/how-facebook-ads-helped-elect-trump/.

"perfect" messages: Issie Laprowsky, "Here's How Facebook *Actually* Won Trump the Presidency," *Wired*, November 15, 2016, https://www.wired.com/2016/11/facebook-won-trump-election-not-just-fake-news/.

approached 200,000: Ibid.

179　*almost 6 million different versions:* Sarah Frier, "Trump's Campaign Said It Was Better at Facebook. Facebook Agrees," Bloomberg, April

3, 2018, https://www.bloomberg.com/news/articles/2018-04-03/trump-s-campaign-said-it-was-better-at-facebook-facebook-agrees.

just 66,000: Tudor Mihailescu, "Trump Campaign Chief Outlines $1 Billion Strategy for 2020 on Trip to Romania," *Forbes,* March 24, 2019, https://www.forbes.com/sites/tudormihailescu/2019/03/24/trump-campaign-chief-outlines-1-billion-strategy-for-2020-on-trip-to-romania/#6888c5831647.

even the topics: Bertoni, "Exclusive Interview."

"all being orchestrated": Ibid.

"15 Hedgehogs": "15 Hedgehogs with Things That Look Like Hedgehogs," BuzzFeed, February 18, 2013, https://www.buzzfeed.com/babymantis/15-hedgehogs-with-things-that-look-like-hedgehogs-10pu?utm_term=.sdy7MQdMKY#.tg3jLoeLm4.

"Which Ousted": Miriam Berger, "Which Ousted Arab Spring Ruler Are You?," BuzzFeed, March 6, 2014, https://www.buzzfeed.com/miriamberger/which-ousted-arab-spring-ruler-are-you?utm_term=.dfDjeoRyZJ#.camRoZnQ8B.

more than 200 articles: Robinson Meyer, "How Many Stories Do Newspapers Publish Per Day?" *The Atlantic,* May 26, 2016, https://www.theatlantic.com/technology/archive/2016/05/how-many-stories-do-newspapers-publish-per-day/483845/.

monitored the performance: David Rowan, "How BuzzFeed Mastered Social Sharing to Become a Media Giant for a New Era," *Wired,* January 2, 2014, http://www.wired.co.uk/article/buzzfeed.

"12 Extremely Disappointing": Dave Sopera, "12 Extremely Disappointing Facts About Popular Music," BuzzFeed, October 11, 2011, https://www.buzzfeed.com/daves4/12-extremely-disappointing-facts-about-popular-mus#.qr9a4wzrG.

"Leonardo DiCaprio Might Be": Lauren Yapalater, "Leonardo DiCaprio Might Be a Human Puppy," BuzzFeed, April 26, 2013, https://www.buzzfeed.com/lyapalater/leonardo-dicaprio-might-be-a-human-puppy?utm_term=.ge7NqKQqan#.rpPL9oD9kZ.

180 *"firehose of falsehood":* Christopher Paul and Miriam Matthews, "The Russian 'Firehose of Falsehood' Propaganda Model," Perspective (report, RAND, 2016), https://www.rand.org/content/dam/rand/pubs/perspectives/PE100/PE198/RAND_PE198.pdf.

7. LIKEWAR

181 *"the first non-linear war":* Surkov is famed less for his writing than as a principal architect of the modern Russian state. Natan Dubovitsky

[Vladislav Surkov], "Without Sky," quoted in Peter Pomerantsev, *Nothing Is True and Everything Is Possible: The Surreal Heart of the New Russia* (PublicAffairs, 2014), 2.

crew of the CS Alert: "British Cable Steamer Cuts Germany's Trans-Atlantic Submarine Cables," Palestine: Information with Provenance Database, accessed March 20, 2018, http://cosmos.ucc.ie/cs1064/jabowen/IPSC/php/event.php?eid=4289.

all seven of Germany's: The exact order by which the lines were cut remains a matter of contention. See Chad R. Fuldwider, *German Propaganda and U.S. Neutrality in World War I* (University of Missouri Press, 2016), 196; "British Cable Steamer."

German word kadaver: Garth S. Jowett and Victoria O'Donnell, *Propaganda and Persuasion,* 5th ed. (Sage, 2012), 167.

182   *When the United States:* Ibid., 217–25.

Pentagon immediately classified: David Ronfeldt, phone interview with author, December 4, 2014.

"Cyberwar Is Coming!": John Arquilla and David Ronfeldt, "Cyberwar Is Coming!," *Comparative Strategy* 12, no. 2 (1993): 141–65, https://www.rand.org/content/dam/rand/pubs/reprints/2007/RAND_RP223.pdf.

"information is becoming": Ibid.

183   *"It means trying":* Ibid.

essentially a dead topic: Ronfeldt interview.

"Our hope was": John Arquilla, phone interview with author, November 3, 2014.

184   *"global information warfare":* Jolanta Darczewska, *The Anatomy of Russian Information Warfare: The Crimean Operation, A Case Study,* Point of View, no. 42 (Centre for Eastern Studies, May 2014).

release of an atomic bomb: Ulrik Franke, "War by Non-military Means: Understanding Russian Information Warfare" (report, Swedish Ministry of Defense, March 2015), 27, http://johnhelmer.net/wp-content/uploads/2015/09/Sweden-FOI-Mar-2015-War-by-non-military-means.pdf.

"blur the traditional": Ministry of Foreign Affairs of the Russian Federation, "Doctrine of Information Security of the Russian Federation," December 5, 2016, http://www.mid.ru/en/foreign_policy/official_documents/-/asset_publisher/CptICkB6BZ29/content/id/2563163.

"a system of spiritual": Franke, "War by Non-military Means," 12.

"measures aiming to pre-empt": Ibid., 11.

"three warfares": *Information at War: From China's Three Warfares to NATO's Narratives,* Beyond Propaganda (Transitions Forum, Legatum Institute, September 2015), https://stratcomcoe.org/legatum-institute-information-war-chinas-three-warfares-natos-narratives.

185 *"War is accelerating"*: State Council Information Office of the People's Republic of China, "China's Military Strategy (2015)" (report, May 2015), https://jamestown.org/wp-content/uploads/2016/07/China%E2%80%99s-Military-Strategy-2015.pdf.

Operation Earnest Voice: Nick Fielding and Ian Cobain, "Revealed: US Spy Operation That Manipulates Social Media," *The Guardian*, March 17, 2011, https://www.theguardian.com/technology/2011/mar/17/us-spy-operation-social-networks.

"allow one U.S. serviceman": Ibid.

In 2015, Britain formed: Ewen MacAskill, "British Army Creates Team of Facebook Warriors," *The Guardian*, January 31, 2015, https://www.theguardian.com/uk-news/2015/jan/31/british-army-facebook-warriors-77th-brigade.

"agent of change": "77th Brigade," British Army, accessed October 5, 2017, http://www.army.mod.uk/structure/39492.aspx?t=/77thBrigade (page deleted). See also David Hume Footsoldier, "Introducing 77 Brigade and a New Way of Business," Think Defence, February 29, 2015, https://www.thinkdefence.co.uk/2015/02/introducing-77-brigade-new-way-business/.

"weaponization of social media": "NATO Installs Information Warfare Center in Latvia," Telesur, March 29, 2015, https://www.telesurtv.net/english/news/NATO-Installs-Information-Warfare-Center-in-Latvia-20150329-0017.html.

186 *"chilled-out [dude]"*: Jessica Chia, "Pepe Croaks: Cartoonist Kills Off Frog Character That Spawned Thousands of Memes Before It Was Hijacked by the Far Right and Branded a Hate Symbol," *Daily Mail*, May 9, 2017, http://www.dailymail.co.uk/news/article-4487364/Pepe-croaks-Cartoonist-kills-frog-turned-hate-symbol.html.

187 *declared a hate symbol*: "Pepe the Frog," Anti-Defamation League, accessed March 20, 2018, https://www.adl.org/education/references/hate-symbols/pepe-the-frog.

Trump tweeted a picture: "Pepe the Frog," Know Your Meme, accessed March 20, 2018, http://knowyourmeme.com/memes/pepe-the-frog#fn2.

"somewhere in the American": "Wanted: Put Up a Pepe Billboard," WeSearchr, updated October 4, 2016, https://web.archive.org/web/20180122193031/http://wesearchr.com/bounties/put-up-a-pepe-billboard.

Russia's UK embassy: James Vincent, "Russian Embassy Trolls UK Prime Minister with Pepe Cartoon," The Verge, January 10, 2017, https://www.theverge.com/2017/1/10/14222780/russian-embassy-trolls-uk-prime-minister-with-pepe.

"drinkin', stinkin'": Matt Furie, *Boy's Club #1* (Buenaventura, 2006).

"Feels good man": "Feels Good Man," Know Your Meme, accessed March 20, 2018, http://knowyourmeme.com/memes/feels-good-man.

began to make their own: Katie Notopoulos, "1,272 Rare Pepes," BuzzFeed, May 11, 2015, https://www.buzzfeed.com/katienotopoulos/1272-rare-pepes?utm_term=.cbOagq2gN8#.qu9X8EA8ob.

the unofficial mascot: "Pepe the Frog," Know Your Meme.

188 *thirty "shitposters"*: Olivia Nuzzi, "How Pepe the Frog Became a Nazi Trump Supporter and Alt-Right Symbol," The Daily Beast, May 26, 2016, http://www.thedailybeast.com/how-pepe-the-frog-became-a-nazi-trump-supporter-and-alt-right-symbol.

a literal Nazi: Ibid.

punched him in the face: Paul P. Murphy, "White Nationalist Richard Spencer Punched During Interview," CNN, January 21, 2017, https://www.cnn.com/2017/01/20/politics/white-nationalist-richard-spencer-punched/.

189 *"Meme War Veteran"*: "Meme War Veteran Hat," Snake Hound Machine, accessed March 20, 2018, http://www.snakehoundmachine.com/product/meme-war-veteran-hat/.

far-right militias: Jack Smith IV, "America's Militia Movement Found Its Next Generation of Soldiers: Teenage 4chan Trolls," Mic, May 23, 2017, https://mic.com/articles/177106/americas-militia-movement-alt-right-teenage-4chan-trolls-boston-free-speech-rally.

peppered with Pepe memes: Claudia Koerner and Cora Lewis, "Here's What We Know About the Man Accused of Killing a Woman at a White Supremacist Rally," BuzzFeed, August 12, 2017, https://www.buzzfeed.com/claudiakoerner/what-we-know-about-james-alex-fields-charlottesville-crash?utm_term=.otXDA8VA9l#.ik6Dmlbmd9.

Ku Klux Klan mask: "From Charlottesville. Pepe did nothing wrong!," /r/The_Donald, Reddit, accessed November 10, 2017, https://www.reddit.com/r/The_Donald/comments/6t9ryg/from_charlottesville_pepe_did_nothing_wrong/.

evolutionary biologist Richard Dawkins: James Gleick, "What Defines a Meme?," *Smithsonian*, May 2011, http://www.smithsonianmag.com/arts-culture/what-defines-a-meme-1904778/?all.

"The computers": Richard Dawkins, *The Selfish Gene*, 40th anniv. ed. (Oxford University Press, 2016), 255.

190 *Russian secret police*: "Anti-Semitism in the United States: Henry Ford Invents a Jewish Conspiracy," Jewish Virtual Library, http://www.jewishvirtuallibrary.org/henry-ford-invents-a-jewish-conspiracy.

"It is a perfect milieu": "Memes: The New Replicators — Chapter 11 from

Richard Dawkins, 'The Selfish Gene,'" *Evolution etc* (blog), November 23, 2015, http://evolutionetc.blogspot.com/2015/11/memes-new-replicators-chapter-11-from.html.

a first class ecology: "Richard Dawkins," Know Your Meme, accessed March 20, 2018, http://knowyourmeme.com/memes/people/richard-dawkins.

a toxic Twitter troll: David Freeman and Eliza Sankar-Gorton, "15 of Richard Dawkins' Most Controversial Tweets," Huffington Post, September 22, 2015, https://www.huffingtonpost.com/entry/15-of-richard-dawkins-most-controversial-tweets_us_56004360e4b00310edf7eaf6.

191 *"Digital content can travel":* Whitney Phillips and Ryan Milner, "The Complex Ethics of Online Memes," The Ethics Centre, October 26, 2016, http://www.ethics.org.au/On-Ethics/blog/October-2016/the-complex-ethics-of-online-memes.

went so far as to sue: Matthew Gault, "Pepe the Frog's Creator Goes Legally Nuclear Against the Alt-Right," *Motherboard* (blog), *Vice,* September 18, 2017, https://motherboard.vice.com/en_us/article/8x8gaa/pepe-the-frogs-creator-lawsuits-dmca-matt-furie-alt-right.

"nonlinear battlefield": Michael B. Prosser, "Memetics — a Growth Industry in US Military Operations" (master's thesis, Marine Corps School of Advanced Warfighting, 2006), http://www.dtic.mil/dtic/tr/fulltext/u2/a507172.pdf.

"military memetics": Jacob Siegel, "Is America Prepared for Meme Warfare?," *Motherboard* (blog), *Vice,* January 31, 2017, https://motherboard.vice.com/en_us/article/xyvwdk/meme-warfare.

"Exploring the Utility": Vera Zakem, Megan K. McBride, and Kate Hammerberg, "Exploring the Utility of Memes for U.S. Government Influence Campaigns" (report, Center for Naval Analyses, April 2018), https://www.cna.org/CNA_files/PDF/DRM-2018-U-017433-Final.pdf.

192 *Bureau of Memetic Warfare:* /bmw/ — The Bureau of Memetic Warfare (website), 8chan, https://8ch.net/bmw/index.html.

One user grandly summarized: "Control the Memes, Control the World," /bmw/ — The Bureau of Memetic Warfare, 8chan, https://8ch.net/bmw/res/1.html#1.

form of Jeff Giesea: Joseph Bernstein, "This Man Helped Build the Trump Meme Army — Now He Wants to Reform It," BuzzFeed, January 18, 2017, https://www.buzzfeed.com/josephbernstein/this-man-helped-build-the-trump-meme-army-and-now-he-wants-t?utm_term=.usYQYkNYy6#.ogmlxRJxMO.

put his thoughts to paper: Jeff Giesea, "It's Time to Embrace Memetic

Warfare," *Defence Strategic Communications* 1, no. 1 (Winter 2015): 68, https://www.stratcomcoe.org/jeff-giesea-its-time-embrace-memetic -warfare.

"It's time to adopt": Ibid., 76.

193   *"Own the moment"*: August Cole (@august_cole), "Own the moment, own the hour. Own the hour, own the country. @selected wisdom @peter wsinger," Twitter, January 23, 2018, 12:27 p.m., https://twitter.com/august _cole/status/955899876700717056.

bombs on packed school buses: Aluf Benn, "Israel Killed Its Subcontractor in Gaza," *Haaretz*, November 14, 2012, https://www.haaretz.com/.pre mium-death-of-israel-s-subcontractor-1.5198285.

five assassination attempts: "The Maestro of Difficult Missions," *Majalla*, October 30, 2009, http://eng.majalla.com/2009/10/article559851/ the-maestro-of-difficult-missions.

Israeli Heron drone: Nick Meo, "How Israel Killed Ahmed Jabari, Its Toughest Enemy in Gaza," November 17, 2012, *The Telegraph*, http://www. telegraph.co.uk/news/worldnews/middleeast/israel/9685598/How-Is rael-killed-Ahmed-Jabari-its-toughest-enemy-in-Gaza.html.

"The IDF has begun": IDF (@IDFSpokesperson), "The IDF has begun a widespread campaign on terror sites & operatives in the #Gaza Strip, chief among them #Hamas & Islamic Jihad targets," Twitter, November 14, 2012, 6:29 a.m., https://twitter.com/IDFSpokesperson/sta tus/268722403989925888.

"In case you missed": IDF (@IDFSpokesperson), "In case you missed it — VIDEO — IDF Pinpoint Strike on Ahmed Jabari, Head of #Hamas Military Wing http://youtu.be/P6U2ZQoEhN4#PillarOfDefense," Twitter, November 14, 2012, 11:12 a.m., https://twitter.com/idfspokesperson/ status/268793527943708673.

nearly 5 million times: "IDF Pinpoint Strike on Ahmed Jabari, Head of Hamas Military Wing," YouTube video, 0:09, uploaded by Israel Defense Forces, November 14, 2012, https://www.youtube.com/ watch?v=P6U2ZQoEhN4.

share it with all: Yaakov Lappin, "IAF Strike Kills Hamas Military Chief Jabari," *Jerusalem Post*, November 14 2012, http://www.jpost.com/ Defense/IAF-strike-kills-Hamas-military-chief-Jabari.

194   *dozens of missile caches:* "Ahmed Jabari Funeral Draws Large Crowds, but No Hamas Leaders," *Israel Hayom*, November 15, 2012, http://www .israelhayom.com/site/newsletter_article.php?id=6425.

"We recommend": Brian Fung, "Military Strikes Go Viral: Israel Is Live-Tweeting Its Own Offensive into Gaza," *The Atlantic*, November 14, 2012, http://www.theatlantic.com/international/archive/2012/11/mil

itary-strikes-go-viral-israel-is-live-tweeting-its-own-offensive-into-gaza/265227/.

"Our blessed hands": Ibid.

Two IDF soldiers: "Israel Under Fire — November 2012," Israel Ministry of Foreign Affairs, November 22, 2012, http://www.mfa.gov.il/mfa/foreignpolicy/terrorism/pages/israel_under_fire-november_2012.aspx.

"the world of social networks": "War Fighting in the Information Age — the Israeli Way," IMSL, January 16, 2013, http://intelmsl.com/insights/in-the-news/war-fighting-in-the-information-age-the-israeli-way/.

more than 10 million heated messages: Chris Moody, "Gaza Goes Viral: An Analysis of Influence — Google Ideas Summit" (speech, Google Ideas Conflict in a Connected World Conference, New York, October 21, 2013), https://blog.gdeltproject.org/gaza-goes-viral-an-analysis-of-influence-google-ideas-summit/.

even a Pinterest page: "War Fighting in the Information Age."

a "Consistent" badge: Jon Mitchell, "Unbelievable! The IDF Has Gamified Its War Blog," Readwrite, November 15, 2012, http://readwrite.com/2012/11/15/unbelievable-the-idf-has-gamified-its-war-blog/.

195   *"What Would You Do?"*: Luke Justin Heemsbergen and Simon Lindgren, "The Power of Precision Air Strikes and Social Media Feeds in the 2012 Israel-Hamas Conflict: 'Targeting Transparency,'" *Australian Journal of International Affairs* 68, no. 5 (2014): 581.

"Playing war games": Peter Schorsch, "Israeli Defense Forces Live-Tweets War," *SaintPetersBlog*, http://saintpetersblog.com/israeli-defense-forces-live-tweets-war/.

"Israeli hunger for pizza": Ran Boker, "Hamas Hacks Channel 10: 'Prepare for Drawn-Out Stay in Bomb Shelters,'" Ynetnews, July 15, 2014, http://www.ynetnews.com/articles/0,7340,L-4543596,00.html.

narrate the conflict: Thomas Zeitzoff, "Does Social Media Influence Conflict? Evidence from the 2012 Gaza Conflict," *Journal of Conflict Resolution* 62, no. 1 (2016): 29–63.

the Hamas spokesperson: Lahav Harkov, "IDF and Hamas Wage War on Twitter," *Jerusalem Post*, November 21, 2012, http://www.jpost.com/Features/In-Thespotlight/IDF-and-Hamas-wage-war-on-Twitter.

"short bursts": David Sarno, "Twitter Creator Jack Dorsey Illuminates the Site's Founding Document. Part I," *Technology* (blog), *Los Angeles Times*, February 18, 2009, http://latimesblogs.latimes.com/technology/2009/02/twitter-creator.html.

196   *more than halved*: Zeitzoff, "Does Social Media Influence Conflict?," 35.

Hamas actively solicited: Stuart Winer, "Netanyahu: Hamas Wants 'Telegenically Dead Palestinians,'" *Times of Israel*, July 20, 2014, http://

www.timesofisrael.com/netanyahu-hamas-wants-telegenically-dead-pal
estinians/.

"There is nothing wrong": Sarah Fowler, "Hamas and Israel Step Up Cyber Battle for Hearts and Minds," BBC News, July 15, 2014, http://www
.bbc.com/news/world-middle-east-28292908.

"People don't need": Ian Burrell, "Israel-Gaza Conflict: Social Media Becomes the Latest Battleground in Middle East Aggression — but
Beware of Propaganda and Misinformation," *Independent,* July 14, 2014,
http://www.independent.co.uk/news/world/middle-east/israel-gaza-
conflict-social-media-becomes-the-latest-battleground-in-middle-east-
aggression-but-9605952.html.

197   *more than 4 million times:* Paul Mason, "Why Israel Is Losing the Social Media War over Gaza," Huffington Post, July 23, 2014, http://www
.huffingtonpost.co.uk/paul-mason/israel-gaza-social-media_b_5612510.
html?utm_hp_ref=uk.

many Israelis were furious: Mirren Gidda, "Poll: 92% of Israeli Jews Say
Operation Protective Edge Was Justified," *Time,* August 19, 2014, http://
time.com/3144232/israeli-jews-poll-gaza-protective-edge/.

nine out of ten: Ibid.

one-third of Palestinians: Harriet Salem, "Facebook Is Being Sued by
20,000 Israelis for Inciting Palestinian Terror," *Vice,* October 27, 2015,
https://news.vice.com/article/facebook-is-being-sued-by-20000-israelis-
for-inciting-palestinian-terror.

just 7 years old: Janna Jihad, "10-Year-Old Palestinian Journalist Covers
Violence in the West Bank," Women in the World, June 1, 2016, https://
womenintheworld.com/2016/06/01/10-year-old-palestinian-journalist-
covers-violence-in-the-west-bank/?refresh.

driven not by opportunity: "Ten-Year Old Is Youngest Palestinian Amateur Journalist," The Palestine Chronicle, June 1, 2016, http://www.pal
estinechronicle.com/ten-year-old-is-youngest-palestinian-amateur-jour
nalist/.

death of a cousin: Jihad, "10-Year-Old Palestinian Journalist."

her mom's iPhone: Ibid.

"They're killing us!": Janna Jihad, "Spirit of My Land," Facebook, December 10, 2017, https://www.facebook.com/Janna.Jihad/?hc_ref=ARSF
_77Y2ygZMxH_IB7KL9aJA6VMdCXGCoIXHsF87wYBVsRlyKmhs2I
AKD5PNYWXeo4&fref=nf.

"My camera is my gun": Tarunika Rajesh, "Meet 13-Year-Old Journalist
Janna Jihad Who Records Her Messages from the World's Most Dangerous
War Zone," Meaww, January 16, 2018, https://meaww.com/read/women/

meet-13-year-old-journalist-janna-jihad-who-records-her-messages-from-the-worlds-most-dangerous-war-zone.

198 *"upload the pictures"*: "Digital Intifada," *Vice*, March 24, 2016, https://news.vice.com/video/digital-intifada-full-length.

"I will attack you": "Hebrew Music Video Glorifying the Killing of Israeli Jews Circulates on Palestinian Social Media," Middle East Media Research Institute, July 28, 2017, https://www.memri.org/tv/hebrew-music-video-glorifying-killing-israeli-jews-on-palestinian-social-media.

most vulnerable arteries: "Social Media as a Platform for Palestinian Incitement — Part II: Video Tutorials, Tips for Achieving More 'Effective' Attacks," Middle East Media Research Institute, October 14, 2015, https://www.memri.org/reports/social-media-platform-palestinian-incitement-%E2%80%93-part-ii-video-tutorials-tips-achieving-more.

posed their toddlers: "Social Media as a Platform for Incitement — Part III: Posting Pictures of Small Children Wielding Knives as Praise, Encouragement for Terrorism," Middle East Media Research Institute, October 22, 2015, https://www.memri.org/reports/social-media-platform-incitement-%E2%80%93-part-iii-posting-pictures-small-children-wielding-knives.

#killajew: "Incitement to Violence Against Jews Spreads Online," Anti-Defamation League, October 9, 2015, https://www.adl.org/blog/incitement-to-violence-against-jews-spreads-online.

"job to save my son": "Digital Intifada."

Jewish Internet Defense Force: Sharon Udasin, "Internet Activist No Friend of Facebook," *Jewish Week*, September 29, 2009, http://jewishweek.timesofisrael.com/internet-activist-no-friend-of-facebook/.

hasbara fellowships: "Wikipedia, Israel and the Internet: Hasbara Fellowships," *Deconditioning Our Minds* (blog), http://decondition.blogspot.com/2007/08/wikipedia-israel-and-internet.html.

"information directorate": "Cabinet Communique," Israel Ministry of Foreign Affairs, July 8, 2007, http://mfa.gov.il/MFA/PressRoom/2007/Pages/Cabinet%20Communique%208-Jul-2007.aspx.

fund an "internet warfare squad": Rona Kuperboim, "Thought-Police Is Here," Ynetnews, July 10, 2009, https://www.ynetnews.com/articles/0,7340,L-3744516,00.html.

199 *Freedom Flotilla*: Isabel Kershner, "Israel Intercepts Gaza Flotilla; Violence Reported," *New York Times*, May 30, 2010, http://www.nytimes.com/2010/05/31/world/middleeast/31flotilla.html.

a bloody boarding: Noah Schactman, "Israel Turns to YouTube, Twitter After Flotilla Fiasco," *Wired*, June 1, 2010, https://www.

wired.com/2010/06/israel-turns-to-youtube-twitter-to-rescue-info-war/.

"They tried to block": Gilad Lotan, phone interview with author, November 7, 2014.

"we are going to lose": Joshua Mitnick, "Gaza Aid Flotilla: Why Israel Expects to Lose the PR War," *Christian Science Monitor,* May 28, 2010, https://www.csmonitor.com/World/Middle-East/2010/0528/Gaza-aid-flotilla-Why-Israel-expects-to-lose-the-PR-war.

a fun side project: Noah Schactman, "Israel's Accidental YouTube War," *Wired,* January 21, 2009, https://www.wired.com/2009/01/israels-acciden/; Allison Hoffman, "The 'Kids' Behind IDF's Media," *Tablet,* November 20, 2012, http://www.tabletmag.com/jewish-news-and-poli tics/117235/the-kids-behind-idf-media.

one energetic 26-year-old: Hoffman, "The 'Kids' Behind IDF's Media."

"I say it's magic": Ibid.

"join the international": "Join the International Social Media Desk," *IDF Blog,* Israeli Defense Forces, https://web.archive.org/web/20170129003601/https://www.idfblog.com/join/.

"Create. Engage. Influence": Ibid.

400 volunteers: Lidar Gravé-Lazi, "IDC Fights War on Another Front," *Jerusalem Post,* July 15, 2014, http://www.jpost.com/printarticle.aspx?id=362804.

200 *"This war doesn't end":* Reuven Weiss, "A Lesson in Hasbara," Ynetnews, June 27, 2017, https://www.ynetnews.com/articles/0,7340,L-4981081,00.html.

"Iron Dome of Truth": Allison Kaplan Sommer, "Sexy Women, 'Missions' and Bad Satire: Israeli Government App Recruits Online Soldiers in Anti-BDS Fight," *Haaretz,* June 13, 2017, https://www.haaretz.com/israel-news/premium-how-israel-recruits-online-foot-soldiers-to-fight-bds-1.5483038.

"You are going to tell": "The Right Thing . . . The Easy Way!," YouTube video, 02:29, uploaded by 4IL, June 7, 2017, https://www.youtube.com/watch?v=HxKrn8Aqa0A.

monitored social networks: Tia Goldenberg, "Israel Takes On Facebook in Battle Against Incitement," AP News, July 21, 2016, https://apnews.com/e08f5c12f80143f986c02df2e45c1dec/israel-takes-facebook-battle-against-incitement.

Humvees prowled: "Digital Intifada."

more than 400 Arabs: Goldenberg, "Israel Takes On Facebook."

conviction rate of 99 percent: "Digital Intifada."

201 *weird moments of impropriety:* Paul Szoldra, "The Taliban Is Waging an Online Flame War Against the United States," Task & Purpose, March 15, 2018, https://taskandpurpose.com/taliban-online-trolls-flame-war/.

a NATO commander's mistress: That commander was General David Petraeus, who was called out in an infamous but now deleted tweet: Abdulqahar Balkhi (@balkhi_a), "Dear Petraeus, Afghanistan is no Paula on whom you can unleash 'precision strike' assets & wish problems away," Twitter, January 14, 2016, 9:43 p.m., https://twitter.com/balkhi_a/status/687872651801395201.

"Both are named Lana": Imed Lamloum, "Libya's Facebook Militias," *Correspondent* (blog), Agence France-Presse, January 11, 2017, https://correspondent.afp.com/libyas-facebook-militias.

"90 percent of information": Ibid.

throw at each other: David Stern, "The Twitter War: Social Media's Role in Ukraine Unrest," *National Geographic,* May 11, 2014, https://news.nationalgeographic.com/news/2014/05/140510-ukraine-odessa-russia-kiev-twitter-world/.

Russian-backed separatist commander: "Funeral of DPR Battalion Commander 'Givi' Took Place in Donetsk City — Approximately 55 Thousand Attend," DONI Press, February 10, 2017, https://dnipress.com/en/posts/funeral-of-dpr-battalion-commander-givi-took-place-in-donetsk-city-approximately-55-thousand-attend/.

202 *laughing and crying emojis:* Christopher Miller (@ChristopherJM), "Mix of crying & laughing emojis from @lifenews_ru Facebook live viewers of Givi's Donetsk funeral, https://www.facebook.com/lifenews.ru/?fref=ts . . . ," Twitter, February 10, 2017, 12:09 a.m., https://twitter.com/christopherjm/status/829965418295070720.

television series South Park: @Russia, "Whoever comes to us with #sanctions, from sanctions will perish. We dedicate this video to those who try to hurt us with new sanctions!," Twitter, August 2, 2017, 12:38 p.m., https://twitter.com/ukraine/status/892864248799428609.

killed a Ukrainian soldier: ATO [Anti-Terrorist Operation] Press Center, trans. Facebook built-in translator, Facebook, August 1, 2017, https://www.facebook.com/ato.news/posts/1650531408290992 (post deleted).

killed 40 Indian soldiers: Maria Abi-Habib, Sameer Yasir, and Hari Kumar, "India Blames Pakistan for Attack in Kashmir, Promising a Response," *New York Times,* February 15, 2019, https://www.nytimes.com/2019/02/15/world/asia/kashmir-attack-pulwama.html.

armies of digital activists: Pranav Dixit and Nishita Jha, "While Two Nuclear Powers Were on the Brink of War, a Full-Blown Online Misinformation Battle Was Underway," Buzzfeed, February 28, 2019, https://www.buzzfeednews.com/article/pranavdixit/india-and-pakistan-were-on-the-brink-of-war-but-a-full?wpisrc=nl_cybersecurity202&wpmm=1.

#IndiaWantsRevenge: Nikhil Rampal, "Abhinandan returns: How India and Pakistan Fought a War on Social Media," *India Today,* March 1, 2019, https://www.indiatoday.in/india/story/abhinandan-returns-india-versus-pakistan-twitter-facebook-war-iaf-airstrike-pulwama-terror-attack-1468524-2019-03-01.

one leader tweeting threats: "North Korea Calls Trump Tweet 'a Declaration of War,'" CBS News, September 25, 2017, https://www.cbsnews.com/news/north-korea-trump-statement-declaration-of-war-live-updates/.

203   *"a 'pro-Russian drift'":* "Novaya Gazeta's 'Kremlin Papers' Article: Full Text in English," UNIAN, February 25, 2015, https://www.unian.info/politics/1048525-novaya-gazetas-kremlin-papers-article-full-text-in-english.html.

"centrifugal aspirations": Ibid.

name of this Ukrainian revolt: Jim Heintz, "Ukraine's Euromaidan: What's in a Name?," Yahoo!, December 2, 2013, https://www.yahoo.com/news/ukraines-euromaidan-whats-name-090717845.html.

explained by Dmitry Peskov: Jim Rutenberg, "RT, Sputnik and Russia's New Theory of War," *New York Times,* September 13, 2017, https://www.nytimes.com/2017/09/13/magazine/rt-sputnik-and-russias-new-theory-of-war.html.

204   *"an informational disaster":* Ibid.

Negative Russian-language news: Szabolcs Panyi, "Orbán Is a Tool in Putin's Information War Against the West," Rovatok, February 4, 2017, http://index.hu/english/2017/02/04/orban_is_a_tool_for_putin_in_his_information_war_against_the_west/.

At least thirty-one people: Howard Amos and Harriet Salem, "Ukraine Clashes: Dozens Dead After Odessa Building Fire," *The Guardian,* May 2, 2014, https://www.theguardian.com/world/2014/may/02/ukraine-dead-odessa-building-fire.

"17-year-old hooligans": "Odessa Tragedy Survivor: 'Many People Strangled After Escaping the Fire,'" RT, May 7, 2014, https://www.rt.com/news/157256-odessa-witness-massacre-ukraine/.

"Mass Murder in Odessa": Daniel McAdams, "US Media Covers Up Mass Murder in Odessa," Infowars, May 5, 2014, https://www.infowars.com/us-media-covers-up-mass-murder-in-odessa/.

"not to allow fascism": "Odessa Tragedy 'Fascism in Action' — Lavrov," RT, May 7, 2014, https://www.rt.com/news/157292-lavrov-odessa-ukraine-fascism/.

205 *strapping the mother*: Anna Nemtsova, "There's No Evidence the Ukrainian Army Crucified a Child in Slovyansk," The Daily Beast, July 15, 2014, https://www.thedailybeast.com/theres-no-evidence-the-ukrainian-army-crucified-a-child-in-slovyansk.

"Sweetest guys": Jessica Misener, "People in Crimea Are Taking Pictures and Selfies with Soldiers," BuzzFeed, March 2, 2014, https://www.buzzfeed.com/jessicamisener/people-in-crimea-are-taking-selfies-with-soldiers?utm_term=.gkqZE1XE0O#.imVmqpXqMd.

"little green men": Vitaly Shevchenko, "'Little Green Men' or 'Russian Invaders?,'" BBC News, March 11, 2014, http://www.bbc.com/news/world-europe-26532154.

"We pounded Ukraine": Max Seddon, "Does This Soldier's Instagram Account Prove Russia Is Covertly Operating in Ukraine?," BuzzFeed, July 30, 2014, https://www.buzzfeed.com/maxseddon/does-this-soldiers-instagram-account-prove-russia-is-covertl?utm_term=.tiEGwg5wz7#.vfN1z VLzQa.

more than 10,000 medals: "Russia's War in Ukraine: The Medals and Treacherous Numbers," Bellingcat, August 31, 2016, https://www.bellingcat.com/news/uk-and-europe/2016/08/31/russias-war-ukraine-medals-treacherous-numbers/.

"it was hybrid warfare": Ivo H. Daalder, "Responding to Russia's Resurgence," Foreign Affairs, November/December 2017, https://www.foreignaffairs.com/articles/russia-fsu/2017-10-16/responding-russias-resurgence?cid=int-fls&pgtype=hpg.

"amazing information blitzkrieg": John Vandiver, "SACEUR: Allies Must Prepare for Russia 'Hybrid War,'" Stars and Stripes, September 4, 2014, https://www.stripes.com/news/saceur-allies-must-prepare-for-russia-hybrid-war-1.301464.

206 *"I began to understand"*: David Patrikarakos, *War in 140 Characters* (Basic Books, 2017), loc. 122, Kindle.

Baltic states of Latvia: "Internet Trolling as a Hybrid Warfare Tool: The Case of Latvia" (report, Strategic Communications Centre of Excellence, n.d.), https://www.stratcomcoe.org/internet-trolling-hybrid-warfare-tool-case-latvia-0; Teri Schultz, "Latvia Faces Hybrid Threat as EU, NATO Boost Defenses," DW, April 12, 2016, http://www.dw.com/en/latvia-faces-hybrid-threat-as-eu-nato-boost-defenses/a-36616678.

raped a 15-year-old: "Lithuania Looking for Source of False Accusation of Rape by German Troops," Reuters, February 17, 2017, https://www.reuters.com/article/us-lithuania-nato-idUSKBN15W1JO.

swarmed Latvian news portals: "Internet Trolling as a Hybrid Warfare Tool," 48.

confiscate Estonian cars: Inga Springe et al., "Sputnik's Unknown Brother," Re:Baltica, April 6, 2017, https://en.rebaltica.lv/2017/04/sputniks-unknown-brother/.

ethnic-Russian enclaves: Emma Graham-Harrison and Daniel Boffey, "Lith-uania Fears Russian Propaganda Is Prelude to Eventual Invasion," *The Guardian,* April 3, 2017, https://www.theguardian.com/world/2017/apr/03/lithuania-fears-russian-propaganda-is-prelude-to-eventual-invasion.

"If we lose": Ibid.

207 *no limit on the number:* Alberto Nardelli, "Angela Merkel's Stance on Refugees Means She Stands Alone Against Catastrophe," *The Guardian,* November 8, 2015, https://www.theguardian.com/commentisfree/2015/nov/08/angela-merkel-refugee-crisis-europe.

covered only by RT: Rutenberg, "RT, Sputnik and Russia's New Theory of War."

story was a hoax: Nadine Schmidt and Tim Hume, "Berlin Teen Admits Fabricating Migrant Gang-Rape Story, Official Says," CNN, February 1, 2016, https://www.cnn.com/2016/02/01/europe/germany-teen-migrant-rape-false/index.html.

"It is clear": Tim Hume and Carolin Schmid, "Russia Cries Cover-Up in Alleged Migrant Rape of 13-Year-Old in Germany," CNN, January 27, 2016, https://www.cnn.com/2016/01/27/europe/russia-germany-berlin-rape/index.html.

nearly sixty years: Kate Connolly, "German Election: Merkel Wins Fourth Term but Far-Right AfD Surges to Third," *The Guardian,* September 24, 2017, https://www.theguardian.com/world/2017/sep/24/angela-merkel-fourth-term-far-right-afd-third-german-election.

Scottish independence referendum: "Russia Meddled in Scottish Independence Referendum to Aid SNP, Claims Security Expert," Express, January 5, 2017, https://www.express.co.uk/news/uk/754332/russia-spies-scotland-independence-referendum-aid-snp-claims-security-expert.

the UK's Brexit proposal: Ben Nimmo, "Lobbying for Brexit: How the Kremlin's Media Are Distorting the UK's Debate," Institute for Statecraft, February 13, 2016, http://www.statecraft.org.uk/research/lobbying-brexit-how-kremlins-media-are-distorting-uks-debate.

Catalonia teetered: David Alandete, "Russian Meddling Machine Sets Sights on Catalonia," *El País,* September 28, 2017, https://elpais.com/el pais/2017/09/26/inenglish/1506413477_994601.html.

208 *assassinate the prime minister:* Christo Gorzev, "Balkan Gambit: Part 2. The Montenegro Zugzwang," Bellingcat, March 25, 2017, https://www.bellingcat.com/news/uk-and-europe/2017/03/25/balkan-gambit-part-2-montenegro-zugzwang/.

local police discovered: Gordana Andric, "Serbian PM 'Moved to Safety' After Weapons Find," Balkan Insight, October 29, 2016, https://www.balkaninsight.com/en/article/police-finds-weapons-near-serbian-pm-house-10-29-2016.

European Parliament declared: "European Parliament Urges Action Against Russian 'Hostile Propaganda,'" DW, November 23, 2016, http://www.dw.com/en/european-parliament-urges-action-against-russian-hostile-propaganda/a-36498112.

Putin answered with disdain: Ibid.

"every disinformation message": Ladislav Bittman, *The KGB and Soviet Disinformation,* quoted in Stanley B. Cunningham, *The Idea of Propaganda* (Praeger, 2002), 110.

209 *"make the sonofabitch":* Hunter S. Thompson, *Fear and Loathing on the Campaign Trail '72* (Simon & Schuster, 1973), 227.

210 *"Rape Melania":* Joseph Bernstein, "Inside the Alt-Right's Campaign to Smear Trump Protesters as Anarchists," BuzzFeed, January 11, 2017, https://www.buzzfeed.com/josephbernstein/inside-the-alt-rights-campaign-to-smear-trump-protesters-as?utm_term=.qvpx9MR9Pa#.xd4XkonkNa.

top-trending hashtags: Ibid.

Breitbart's smug headline: Katie McHugh, "Twitter Allows 'Rape Melania' to Trend After Site Explodes with Trump Assassination Threats," Breitbart, November 13, 2016, http://www.breitbart.com/big-government/2016/11/13/twitter-allows-rape-melania-to-trend-after-site-explodes-with-trump-assassination-threats/.

Posobiec also admitted: Bernstein, "Inside the Alt-Right's Campaign."

"character assassination": "Talk:Jack Posobiec," Wikipedia, accessed March 20, 2018, https://en.wikipedia.org/wiki/Talk:Jack_Posobiec#Character_assassination.

211 *volunteer "Internet Army":* "Yuriy Stets, Minister of Information Policy Wants to Create 'Internet Army,'" *Euromaidan Press,* January 28, 2015, http://euromaidanpress.com/2015/01/28/yuriy-stets-minister-of-information-policy-wants-to-create-internet-army/.

turned it into a joke: "Ukrainian Internet Army Report Card: 'A' for Effort 'F' for Achievement," Sputnik, June 3, 2015, https://sputniknews.com/science/201503061019154259/.

Center of Defense Against Disinformation: Rachel Stern, "Germany's Plan to Fight Fake News," *Christian Science Monitor,* January 9, 2017, https://www.csmonitor.com/World/Passcode/2017/0109/Germany-s-plan-to-fight-fake-news.

"ministry of truth": "Ministry of Truth? Berlin Reportedly Plans 'Center of Defense Against Fake News,'" YouTube video, 7:21, uploaded by RT, December 24, 2016, https://www.youtube.com/watch?v=8YaN_kaov50.

plot by the "deep state": "Hannity: Deep State's Massive Effort to Destroy Trump," Fox News, June 17, 2017, http://www.foxnews.com/politics/2017/06/17/hannity-deep-states-massive-effort-to-destroy-trump.html.

212   *The shrouded figure gazes:* "Anonymous: Operation Ice ISIS (#OpIce ISIS)," YouTube video, 4:29, uploaded by *AnonJournal,* June 21, 2014, https://www.youtube.com/watch?v=_kJtvFUMELM.

We are Anonymous: Emerson Brooking, "Anonymous vs. the Islamic State," *Foreign Policy,* November 13, 2015, http://foreignpolicy.com/2015/11/13/anonymous-hackers-islamic-state-isis-chan-online-war/.

213   *"cyber terrain vigilance":* Ghost Security Group, home page, https://ghostsecuritygroup.com/.

214   *think tank wonks:* Laura Rosenberger and J. M. Berger, "Hamilton 68: A New Tool to Track Russian Disinformation on Twitter," German Marshall Fund, August 2, 2017, http://securingdemocracy.gmfus.org/blog/2017/08/02/hamilton-68-new-tool-track-russian-disinformation-twitter.

215   *"I need peace":* Bana Alabed (@AlabedBana), "I need peace," Twitter, September 24, 2016, 5:07 a.m., https://twitter.com/AlabedBana/status/779653424145113088.

"capturing us now": Bana Alabed (@AlabedBana), "We are sure the army is capturing us now. We will see each other another day dear world. Bye. — Fatemah #Aleppo," Twitter, December 4, 2016, 10:38 a.m., https://twitter.com/AlabedBana/status/805481415458623489.

"I miss school": Bana Alabed (@AlabedBana), "I miss school so much — Bana #Aleppo," Twitter, October 6, 2016, 11:54 a.m., https://twitter.com/AlabedBana/status/784104553440501761.

200,000 Twitter followers: Caitlin Gibson, "How a 7-Year-Old Aleppo Girl on Twitter Became Our Era's Anne Frank," *Washington Post,* December 6, 2016, https://www.washingtonpost.com/lifestyle/style/how-a-7-year-old-aleppo-girl-on-twitter-became-our-eras-anne-frank/2016/12/06/

b474af5c-bb09-11e6-91ee-1adddfe36cbe_story.html?utm_term=.
e2356256891a.

author J. K. Rowling: Ibid.

Critics alleged that Bana: Nick Waters, "Finding Bana — Proving the Existence of a 7-Year-Old Girl in Eastern Aleppo," Bellingcat, December 14, 2016, https://www.bellingcat.com/news/mena/2016/12/14/bana-al abed-verification-using-open-source-information/.

"This bombs": Bana Alabed (@AlabedBana), "I am very afraid I will die tonight. This bombs will kill me now. — Bana #Aleppo," Twitter, October 2, 2016, 10:00 a.m., https://twitter.com/AlabedBana/sta tus/782626282291036160.

"game of propaganda": "President al-Assad Interview: 'The Moderate Opposition Is a Myth . . . We Won't Accept That Terrorists Take Control of Any Part of Syria,'" Global Research, October 7, 2016, https://www. globalresearch.ca/president-al-assad-interview-the-moderate-opposition-is-a-myth-we-wont-accept-that-terrorists-take-control-of-any-part-of-syria/5549743.

a Bellingcat investigation: Waters, "Finding Bana."

her English-literate mother: Ibid.

216 *"use force to unify Taiwan":* Ben Blanchard and Faith Hung, "South China Sea? For Beijing, Taiwan Is the No. 1 Security Issue," Reuters, January 17, 2016, https://www.reuters.com/article/us-taiwan-election-security/south-china-sea-for-beijing-taiwan-is-the-no-1-security-issue-idUSKCN0UV064.

"Even if no grass": Peter Navarro, "Senkaku Suicide Scenarios: China vs. Ameripan," HuffPost, December 6, 2017, https://www. huffingtonpost.com/peter-navarro-and-greg-autry/senkaku-suicide-scenarios_b_9583586.html.

censors and state media: Bethany Allen-Ebrahimian, "After South China Sea Ruling, China Censors Online Calls for War," *Tea Leaf Nation* (blog), *Foreign Policy,* July 12, 2016, https://foreignpolicy.com/2016/07/12/after-south-china-sea-ruling-china-censors-online-calls-for-war-unclos-tribu-nal/.

"Stop boasting": Ibid.

217 *"Gone are the days":* Thomas J. Christensen, "The Advantages of an Assertive China," *Foreign Affairs,* March/April 2011, https://www.foreignaf fairs.com/articles/east-asia/2011-02-21/advantages-assertive-china.

Confiding in their diaries: Jan Ian Chong and Todd H. Hall, "The Lessons of 1914 for East Asia Today: Missing the Trees for the Forest," *International Security* 39, no. 1 (2014): 7–43.

8. MASTERS OF THE UNIVERSE

218   *Janet Jackson's nipple:* Jim Hopkins, "Surprise! There's a Third YouTube Co-founder," *USA Today,* October 11, 2006, http://usatoday30.usatoday.com/tech/news/2006-10-11-youtube-karim_x.htm.

nine-sixteenths of a second: Daniel Kreps, "Nipple Ripples: 10 Years of Fallout from Janet Jackson's Halftime Show," *Rolling Stone,* January 30, 2014, https://www.rollingstone.com/culture/news/nipple-ripples-10-years-of-fallout-from-janet-jacksons-halftime-show-20140130.

140 million viewers: Hugh McIntyre, "How Janet Jackson's Super Bowl 'Wardrobe Malfunction' Helped Start YouTube," *Forbes,* February 1, 2015, https://www.forbes.com/sites/hughmcintyre/2015/02/01/how-janet-jacksons-super-bowl-wardrobe-malfunction-helped-start-youtube/#7299c00019ca.

540,000 complaints: Anahad O'Connor, "Court Throws Out Super Bowl Fine," July 22, 2008, http://www.nytimes.com/2008/07/22/business/media/22FCC.html.

America Online: Kreps, "Nipple Ripples."

219   *YouTube was born:* Hopkins, "Surprise!"

San Francisco rave scene: Nick Bilton, *Hatching Twitter: A True Story of Money, Power, Friendship, and Betrayal* (Penguin, 2013), 84.

Even Google started: John Battelle, "The Birth of Google," *Wired,* August 1, 2005, https://www.wired.com/2005/08/battelle/.

"original sin of Silicon Valley": Mike Monteiro, "One Person's History of Twitter, from Beginning to End," *@monteiro* (blog), Medium, October 15, 2017, https://medium.com/@monteiro/one-persons-history-of-twitter-from-beginning-to-end-5b41abed6c20.

220   *color of agitation:* David Robinson, "How the Colour Red Warps the Mind," BBC News, September 1, 2014, http://www.bbc.com/future/story/20140827-how-the-colour-red-warps-the-mind.

like opening a present: Paul Lewis, "'Our Minds Can Be Hijacked': The Tech Insiders Who Fear a Smartphone Dystopia," *The Guardian,* October 6, 2017, https://www.theguardian.com/technology/2017/oct/05/smartphone-addiction-silicon-valley-dystopia.

2,617 times each day: Michael Winnick, "Putting a Finger on Our Phone Obsession," dscout, June 16, 2016, https://blog.dscout.com/mobile-touches.

"We connect people": Ryan Mac, Charlie Warzel, and Alex Kantrowitz, "Growth at Any Cost: Top Facebook Executive Defended Data Collection in 2016 Memo — and Warned That Facebook Could Get People

Killed," BuzzFeed, March 29, 2018, https://www.buzzfeed.com/ryanmac/growth-at-any-cost-top-facebook-executive-defended-data?utm_term=.xlReZ2OZoo#.ag5ZJRzJ10.

221 *"One Policy"*: Adam D. Thierer, "Unnatural Monopoly: Critical Moments in the Development of the Bell System Monopoly," *Cato Journal* 14, no. 2 (1992): 267–85.

replacing comment boxes: Heather Kelly, "Facebook's Adding Text Bubbles and Round Profiles," CNN, August 15, 2017, http://money.cnn.com/2017/08/15/technology/facebook-newsfeed-updates/index.html.

tiny shifts: Alexis C. Madrigal, "What Facebook Did to American Democracy," *The Atlantic,* October 12, 2017, https://www.theatlantic.com/technology/archive/2017/10/what-facebook-did/542502/.

"process of building": John Herrman, "How Hate Groups Forced Online Platforms to Reveal Their True Nature," *New York Times Magazine,* August 21, 2017, https://www.nytimes.com/2017/08/21/magazine/how-hate-groups-forced-online-platforms-to-reveal-their-true-nature.html?nytmobile=0&_r=0.

222 *"You're so focused":* Deepa Seetharaman, Robert McMillan, and Georgia Wells, "Tone-Deaf: How Facebook Misread America's Mood on Russia," *Wall Street Journal,* March 2, 2018, https://www.wsj.com/articles/tone-deaf-how-facebook-misread-americas-mood-on-russia-1520006034.

"Political stories": "Tell HN: Political Detox Week," Hacker News, Y Combinator, accessed March 20, 2018, https://news.ycombinator.com/item?id=13108404. This is a fascinating discussion thread that digs deep into the Silicon Valley zeitgeist.

"If we could use code": Authors' interview with senior social media company official, Washington, DC, July 14, 2016.

223 *pleas from Ukrainian activists:* Volodymyr Scherbachenko, "We Support Ukraine on Facebook!," trans. Facebook built-in translator, Facebook, August 28, 2014, https://www.facebook.com/uspikh/posts/355353931293866?fref=nf.

"integrity of the German elections": "Read Mark Zuckerberg's Full Remarks on Russian Ads That Impacted the 2016 Elections," CNBC, September 21, 2017, https://www.cnbc.com/2017/09/21/zuckerbergs-full-remarks-on-russian-ads-that-impacted-2016-election.html?view=story&%24DEVICE%24=native-android-tablet.

"spreading prosperity": Mark Zuckerberg, "Building Global Community," Facebook, February 16, 2017, https://www.facebook.com/notes/mark-zuckerberg/building-global-community/10154544292806634/.

224 *"CYBERPORN":* Philip Elmer-Dewitt, "Finding Marty Rimm," *For-*

tune, July 1, 2015, http://fortune.com/2015/07/01/cyberporn-time-marty-rimm/.

Rimm vanished: Ibid.

"a red-light district": Robert Cannon, "The Legislative History of Senator Exon's Communications Decency Act: Regulating Barbarians on the Information Superhighway," *Field Communications Law Journal* 49, no. 1 (1996): 53.

Communications Decency Act: Ibid., 58.

fine of $100,000: Ibid.

one crucial tweak: Christopher Zara, "The Most Important Law in Tech Has a Problem," *Wired,* January 3, 2017, https://www.wired.com/2017/01/the-most-important-law-in-tech-has-a-problem/.

"the most important law": Ibid.

225 *"protection for 'Good Samaritan'":* 47 U.S.C. § 230 (1996).

no list of rules: "The Big FAQ," Blogger, (2000), accessed March 20, 2018, https://web.archive.org/web/20010904030704/http://ex.blogger.com:80/howto/faq.pyra#30366.

Digital Millennium Copyright Act: 17 U.S.C. § 1204 (1998).

226 *"safe harbor" provision:* David Kravets, "10 Years Later, Misunderstood DMCA Is the Law That Saved the Web," *Wired,* October 27, 2008, https://www.wired.com/2008/10/ten-years-later/.

ten-minute limit: Ken Fisher, "YouTube Caps Video Lengths to Reduce Infringement," *Ars Technica,* March 29, 2006, https://arstechnica.com/uncategorized/2006/03/6481-2/.

$1.7 billion: Matt Marshall, "They Did It! YouTube Bought by Google for $1.65B in Less Than Two Years," VentureBeat, October 9, 2006, https://venturebeat.com/2006/10/09/they-did-it-youtube-gets-bought-by-gooogle-for-165b-in-less-than-two-years/.

"Content ID" system: Kevin J. Delaney, "YouTube to Test Software to Ease Licensing Fights," *Wall Street Journal,* June 12, 2007, https://www.wsj.com/articles/SB118161295626932114.

John McCain complained: Sarah Lai Stirland, "YouTube to McCain: You Made Your DMCA Bed, Lie in It," *Wired,* October 15, 2008, https://www.wired.com/2008/10/youtube-to-mcca/.

Digital rights activists: Ibid.

pushing a toy stroller: "'Let's Go Crazy' #1," YouTube video, 0:29, uploaded by Stephanie Lenz, February 7, 2007, https://www.youtube.com/watch?v=N1KfJHFWlhQ.

plead "fair use": "Lenz v. Universal Music Corp.," *Harvard Law Review* 129, no. 2289 (June 2016), https://harvardlawreview.org/2016/06/lenz-v-uni versal-music-corp/.

227 *PhotoDNA:* "New Technology Fights Child Porn by Tracking Its 'PhotoDNA,'" Microsoft, December 15, 2009, https://news.microsoft.com/ 2009/12/15/new-technology-fights-child-porn-by-tracking-its-photodna/ #sm.0001mpmupctevct7pjn11vtwrw6xj.

more than a million instances: Tracy Ith, "Microsoft's PhotoDNA: Protecting Children and Businesses in the Cloud," Microsoft, July 15, 2015, https://news.microsoft.com/features/microsofts-photodna-protecting-children-and-businesses-in-the-cloud/.

half of all American teenagers: Amanda Lenhart et al., "Social Media and Young Adults," Pew Research Center, February 3, 2010, http://www .pewinternet.org/2010/02/03/social-media-and-young-adults/.

16-year-old Josh Evans: Lauren Collins, "Friend Game," *The New Yorker,* January 21, 2008, https://www.newyorker.com/magazine/2008/01/21/ friend-game.

"meet a great girl": Ibid.

"an online Frankenstein's monster": Ibid.

joined the Drew family: Ibid.

228 *"You're a shitty person":* Ibid.

convicted, but then acquitted: Kim Zetter, "Judge Acquits Lori Drew in Cyberbullying Case, Overrules Jury," *Wired,* July 2, 2009, https://www .wired.com/2009/07/drew-court/.

Myspace was technically: "Woman Indicted in Cyber-Bully Suicide," CBS News, May 15, 2008, https://www.cbsnews.com/news/woman-indicted-in-cyber-bully-suicide/.

229 *"will not censor":* Sarah Jeong, "The History of Twitter's Rules," *Motherboard* (blog), *Vice,* January 14, 2016, motherboard.vice.com/read/the-history-of-twitters-rules.

"the free speech wing": Josh Halliday, "Twitter's Tony Wang: We Are the Free Speech Wing of the Free Speech Party," *The Guardian,* March 22, 2012, https://www.theguardian.com/media/2012/mar/22/twitter-tony-wang-free-speech.

"honeypot for assholes": Charlie Warzel, "'A Honeypot for Assholes': Inside Twitter's 10-Year Failure to Stop Harassment," BuzzFeed, August 11, 2016, https://www.buzzfeed.com/charliewarzel/a-honeypot-for-ass holes-inside-twitters-10-year-failure-to-s?utm_term=.yb3RlEBl8O#.wb wxNORNzy.

sustained harassment: Ibid.

"not a mediator of content": Ibid.

report abusive tweets: Jeong, "The History of Twitter's Rules."

"Gamergate": Aja Romano, "The Data Behind Gamergate Reveals Its Ugly Truth," The Daily Dot, December 11, 2015, https://www.dailydot.com/parsec/72-hours-of-gamergate-twitter-analysis/.

inquest by the United Nations: Allegra Frank, "Anita Sarkeesian, Zoe Quinn, and More Take Aim at Cyber Harassment Against Women," Polygon, September 25, 2015, https://www.polygon.com/2015/9/25/9399169/united-nations-women-cyber-violence-anita-sarkeesian-zoe-quinn.

230 *"Freedom of expression":* Vijaya Gadde, "Twitter Executive: Here's How We're Trying to Stop Abuse While Preserving Free Speech," *PostEverything* (blog), *Washington Post,* April 4, 2016, https://www.washingtonpost.com/posteverything/wp/2015/04/16/twitter-executive-heres-how-were-trying-to-stop-abuse-while-preserving-free-speech/?utm_term=.5350fac72e36.

vanished from its mission statement: Jeong, "The History of Twitter's Rules."

banned "unlawful, obscene": "Terms of Use," YouTube, 2005, accessed March 20, 2018, https://web.archive.org/web/20050428210756/http://www.youtube.com:80/terms.php.

Mexican drug cartels: Manuel Roig-Franzia, "Mexican Drug Cartels Leave a Bloody Trail on YouTube," *Washington Post,* April 9, 2007, http://www.washingtonpost.com/wp-dyn/content/article/2007/04/08/AR2007040801005.html.

Egyptian anti-torture activist: "YouTube Shuts Down Egyptian Anti-torture Activist's Account," CNN, November 29, 2007, http://www.cnn.com/2007/WORLD/meast/11/29/youtube.activist/.

loss of its "exclusive footage": "Israel Brings Battle with Hamas to YouTube," Fox News, December 31, 2008, http://www.foxnews.com/story/2008/12/31/israel-brings-battle-with-hamas-to-youtube.amp.html.

wanted to avoid: Julia Angwin and Hannes Grassegger, "Facebook's Secret Censorship Rules Protect White Men from Hate Speech but Not Black Children," ProPublica, June 28, 2017, https://www.propublica.org/article/facebook-hate-speech-censorship-internal-documents-algorithms.

15,000 words: Catherine Buni and Soraya Chemaly, "The Secret Rules of the Internet," The Verge, April 13, 2016, https://www.theverge.com/2016/4/13/11387934/internet-moderator-history-youtube-facebook-reddit-censorship-free-speech.

"incitement of violence": Nick Hopkins, "Revealed: Facebook's Internal

Rulebook on Sex, Terrorism and Violence," *The Guardian,* May 21, 2017, https://www.theguardian.com/news/2017/may/21/revealed-facebook-internal-rulebook-sex-terrorism-violence.

"kick a person": Ibid.

"cut your tongue out": Ibid.

231 *protests of historians:* Ibid.

images of breastfeeding: Maya Rhodan, "Facebook Lifts Ban on Exposed Nipples in Breastfeeding Pictures," *Time,* June 13, 2014, http://time.com/2869849/facebook-breastfeeding-nipples/.

#freethenipple: Alex Bruce-Smith, "Instagram Blocks the #Curvy Hashtag for Nudity Reasons," Pedestrian.TV, July 17, 2015, https://www.pedestrian.tv/news/instagram-blocks-the-curvy-hashtag-for-nudity-reasons/.

portrayals of breastfeeding: Soraya Chemaly, "#FreeTheNipple: Facebook Changes Breastfeeding Mothers Photo Policy," Huffington Post, June 9, 2014, https://www.huffingtonpost.com/soraya-chemaly/freethenipple-facebook-changes_b_5473467.html.

not the principal *focus:* Mythili Sampathkumar, "Facebook Bans Woman Who Shared Article on Breastfeeding," *Independent,* October 6, 2017, https://www.independent.co.uk/news/world/australasia/facebook-breastfeeding-ban-woman-shared-article-a7985111.html.

subject to U.S. laws: Xeni Jardin, "More on Orkut and Law Enforcement: Brazil," *Boing Boing,* March 13, 2007, https://boingboing.net/2007/03/13/more-on-Orkut-and-la.html.

dozens of national jurisdictions: Glyn Moody, "Facebook Hit with Fines and Investigations in Six EU Countries over Privacy Law Breaches," Privacy News Online, May 18, 2017, https://www.privateinternetaccess.com/blog/2017/05/facebook-hit-fines-investigations-six-eu-countries-privacy-law-breaches/.

"Push-Button Publishing": Bilton, *Hatching Twitter,* 16.

submit censorship requests: Jocelyn Richard, "Google Will Censor Blogger Blogs on 'Per Country' Basis," Huffington Post, February 1, 2012, https://www.huffingtonpost.com/2012/02/01/google-blogger-censorship_n_1247380.html.

232 *"an American company":* Austin Carr, "Can Alphabet's Jigsaw Solve Google's Most Vexing Problems?," Fast Company, October 22, 2017, https://www.fastcompany.com/40474738/can-alphabets-jigsaw-solve-the-internets-most-dangerous-puzzles.

"I woke up": Kate Conger, "Cloudflare CEO on Terminating Service

to Neo-Nazi Site: 'The Daily Stormer Are Assholes,'" Gizmodo, August 16, 2017, https://gizmodo.com/cloudflare-ceo-on-terminating-service-to-neo-nazi-site-1797915295.

celebrating the murder: Judd Legum, "White Supremacists Cheer Trump's Response to Charlottesville Violence," ThinkProgress, August 12, 2017, https://thinkprogress.org/white-supremacists-cheer-trumps-re sponse-to-charlottesville-violence-3d0d50196c52/.

"censoring the internet": Jon Brodkin, "Cloudflare Changes Abuse Policy but Refuses to 'Censor the Internet,'" *Ars Technica*, May 8, 2017, https://arstechnica.com/tech-policy/2017/05/cloudflare-changes-abuse-policy-but-refuses-to-censor-the-internet/.

233   *"This was my decision":* Conger, "Cloudflare CEO."

some fifty countries passed laws: Paul Mozur, Mark Scott, and Mike Isaac, "Facebook Faces a New World as Officials Rein In a Wild Web," *New York Times*, September 17, 2017, https://www.nytimes.com/2017/09/17/technol ogy/facebook-government-regulations.html.

234   *snipers killing U.S. soldiers:* "Islamic Terrorists Using YouTube to Spread Propaganda," Fox News, February 13, 2007, http://www.foxnews. com/story/2007/02/13/islamic-terrorists-using-youtube-to-spread-propa ganda.html.

slow to remove the clips: Ibid.

700-video library: Brian Bennett, "YouTube Is Letting Users Decide on Terrorism-Related Videos," *Los Angeles Times*, December 12, 2010, http:// articles.latimes.com/2010/dec/12/nation/la-na-youtube-terror-20101213.

Fort Hood, Texas: Scott Shane, "The Lessons of Anwar al-Awlaki," *New York Times Magazine*, August 27, 2015, https://www.nytimes. com/2015/08/30/magazine/the-lessons-of-anwar-al-awlaki.html?_r=0.

YouTube algorithm: Bennett, "YouTube Is Letting Users Decide."

his online propaganda: Eric Holder to Patrick Leahy, letter (unclassi-fied), May 22, 2013, https://www.justice.gov/slideshow/AG-letter-5-22-13. pdf.

another six years: Alex Hern, "'YouTube Islamist' Anwar al-Awlaki Vid-eos Removed in Extremism Clampdown," *The Guardian*, November 13, 2017, https://www.theguardian.com/technology/2017/nov/13/youtube-is lamist-anwar-al-awlaki-videos-removed-google-extremism-clampdown.

235   *Twitter brushed off:* Ben Farmer, "Congress Calls on Twitter to Block Taliban," *The Telegraph*, December 25, 2011, https://www.telegraph.co.uk/ technology/twitter/8972884/Congress-calls-on-Twitter-to-block-Tali ban.html.

why not the Taliban: Jon Boone, "Taliban Join the Twitter Revolution,"

The Guardian, May 12, 2011, https://www.theguardian.com/world/2011/may/12/taliban-join-twitter-revolution.

obsessive Twitter adopters: Jessica Stern and J. M. Berger, *ISIS: The State of Terror* (Ecco, 2015), 63–64.

snapped by the gunmen: "Kenya Attack Unfolded in Up and Down Twitter Feeds," NDTV, September 25, 2013, www.ndtv.com/world-news/kenya-attack-unfolded-in-up-and-down-twitter-feeds-535648.

"#Westgate": Harriet Alexander, "Tweeting Terrorism: How Al Shabaab Live Blogged the Nairobi Attacks," *The Telegraph,* September 22, 2013, https://www.telegraph.co.uk/news/worldnews/africaandindianocean/kenya/10326863/Tweeting-terrorism-How-al-Shabaab-live-blogged-the-Nairobi-attacks.html.

spread misinformation: Journalist Josh Kron, email to authors, January 3, 2017.

Twitter intervened: Alexander, "Tweeting Terrorism."

registered new ones: J. M. Berger, "Twitter's Week of Reckoning," *Foreign Policy,* October 1, 2013, http://foreignpolicy.com/2013/10/01/twitters-week-of-reckoning/.

at least 70,000 Twitter accounts: J. M. Berger and Jonathon Morgan, "The ISIS Twitter Census: Defining and Describing the Population of ISIS Supporters on Twitter" (analysis paper no. 20, Brookings Project on U.S. Relations with the Islamic World, Brookings Institution, March 2015), 4.

236 *definition of "terrorist activity":* David Fidler, "The War on Terrorists' Tweets," Defense One, July 17, 2015, http://www.defenseone.com/technology/2015/07/war-terrorists-tweets/118087/.

"closing that internet up": Andrew Griffin, "Donald Trump Wants to Ban the Internet, Plans to Ask Bill Gates to 'Close It Up,'" *Independent,* December 8, 2015, http://www.independent.co.uk/news/people/donald-trump-wants-to-ban-the-internet-will-ask-bill-gates-to-close-it-up-a6764396.html.

Twitter blocklists: David Auerbach, "Beware the Blocklists," *Slate,* August 11, 2015, http://www.slate.com/articles/technology/bitwise/2015/08/twitter_blocklists_they_can_stop_harassment_and_they_can_create_entirely.html.

hundreds of times: Nichole Perlroth and Mike Isaac, "Terrorists Mock Bids to End Use of Social Media," *New York Times,* December 7, 2015, www.nytimes.com/2015/12/08/technology/terrorists-mock-bids-to-end-use-of-social-media.html.

a birthday cake: Ibid.

detecting 95 percent: Rachel Kaser, "Twitter Claims It's Removed 95%

of Extremist Content with No One Noticing," The Next Web, September 19, 2017, https://thenextweb.com/socialmedia/2017/09/19/twitter-claims-removed-extremist-content-no-one-noticing/.

used the advertising space: "Google's Clever Plan to Stop Aspiring ISIS Recruits," *Wired,* September 9, 2016, https://www.wired.com/2016/09/googles-clever-plan-stop-aspiring-isis-recruits/.

237 *150-person counterterrorism force:* Seth Fiegerman, "Facebook Grows Its Counterterrorism Team," CNNMoney, June 15, 2017, http://money.cnn.com/2017/06/15/technology/business/facebook-terrorism-content/index.html.

"knowingly provided material support": Gwen Ackerman, "Facebook Accused in $1 Billion Suit of Being Hamas Tool," Bloomberg, July 11, 2016, http://www.bloomberg.com/news/articles/2016-07-11/facebook-sued-for-1b-for-alleged-hamas-use-of-medium-for-terror.

feared they might *suffer:* Jonathan Stempel, "Facebook Wins Dismissal of U.S. Lawsuits Linked to Terrorism," Reuters, May 18, 2017, https://www.reuters.com/article/us-facebook-lawsuit/facebook-wins-dismissal-of-u-s-lawsuits-linked-to-terrorism-idUSKCN18E2GF.

"Facebook and Twitter": Harriet Salem, "Facebook Is Being Sued by 20,000 Israelis for Inciting Palestinian Terror," https://news.vice.com/article/facebook-is-being-sued-by-20000-israelis-for-inciting-palestinian-terror.

prompted further lawsuits: Nina Iacono Brown, "Should Social Networks Be Held Liable for Terrorism?," *Slate,* June 16, 2017, http://www.slate.com/articles/technology/future_tense/2017/06/a_new_legal_theory_for_holding_social_networks_liable_for_terrorism.html.

ultranationalists, white supremacists: J. M. Berger, "Nazis vs. ISIS on Twitter: A Comparative Study of White Nationalist and ISIS Online Social Media Networks" (paper, Program on Extremism, George Washington University, September 2016), https://cchs.gwu.edu/sites/g/files/zaxdzs2371/f/downloads/Nazis%20v.%20ISIS%20Final_0.pdf.

238 *"Smith" became "(((Smith)))":* Cooper Fleishman and Anthony Smith, "(((Echoes))), Exposed: The Secret Symbol Neo-Nazis Use to Target Jews Online," Mic, June 1, 2016, https://mic.com/articles/144228/echoes-exposed-the-secret-symbol-neo-nazis-use-to-target-jews-online.

use of the same language: Herrman, "How Hate Groups Forced."

Milo Yiannopoulos: Charlie Warzel, "Twitter Permanently Suspends Conservative Writer Milo Yiannopoulos," BuzzFeed, July 20, 2016, https://www.buzzfeed.com/charliewarzel/twitter-just-permanently-suspended-conservative-writer-milo?utm_term=.eneooAGoO4#.gwQ4zmjzkn.

Ghostbusters *remake:* Ibid.

239 *"LongKnives1290":* Joseph Bernstein, "Alt-White: How the Breitbart Machine Laundered Racist Hate," BuzzFeed, October 5, 2017, https://www.buzzfeed.com/josephbernstein/heres-how-breitbart-and-milo-smuggled-white-nationalism?utm_term=.eekpAwn4E#.xu0QnyPGK.

more than 700 hate crimes: Carter Evans, "Hate, Harassment Incidents Spike Since Trump Election," CBS News, November 19, 2016, https://www.cbsnews.com/news/hate-harassment-incidents-spike-since-donald-trump-election/.

a dramatic rebuttal: "The Knight of Long Knives," YouTube video, 03:13, uploaded by NPI/Radix, November 15, 2016, https://www.youtube.com/watch?v=qiADHzBOqZo.

"a great purge going on": David Scharfenberg, "Should Twitter Ban the Alt-Right? The Case for Online Censorship," *Boston Globe,* December 11, 2016, https://www.bostonglobe.com/ideas/2016/12/11/should-twitter-ban-alt-right-the-case-for-online-censorship/aTt7la9oS2krWhQEYHKM7J/story.html.

"free speech" events: Terrence McCoy, "The Road to Hate: For Six Young Men, Charlottesville Is Only the Beginning," *Washington Post,* August 19, 2017, https://www.washingtonpost.com/local/social-issues/the-road-to-hate-for-six-young-men-of-the-alt-right-charlottesville-is-only-the-beginning/2017/08/19/cd1a3624-8392-11e7-b359-15a3617c767b_story.html.

"We have been spreading": Herrman, "How Hate Groups Forced."

Facebook removed pages: Associated Press, "Facebook Bans White Nationalist's Accounts over Hate Speech," AP News, August 16, 2017, https://apnews.com/3e725b8c8f62460cb71d576edc6ca61c?utm_campaign=SocialFlow&utm_source=Twitter&utm_medium=AP.

Reddit rewrote: Brianna Sacks, "Reddit Is Removing Nazi and Alt-Right Groups as Part of a New Policy and Some Users Are Confused," BuzzFeed, October 25, 2017, https://www.buzzfeed.com/briannasacks/reddit-is-banning-nazi-and-alt-right-groups-as-part-of-a?utm_term=.iqY20Eo0dK#.aydLp5Vp2X.

room-sharing service: Matt Stevens, "After Charlottesville, Even Dating Apps Are Cracking Down on Hate," *New York Times,* August 24, 2017, https://www.nytimes.com/2017/08/24/technology/okcupid-christopher-cantwell.html?smid=tw-nytimes&smtyp=cur%20Most%20deeply%20impacted:%20; https://www.wsj.com/articles/facebook-employees-pushed-to-remove-trump-posts-as-hate-speech-1477075392.

same standard: Joseph Cox and Jason Koebler, "Facebook Bans White Nationalism and White Separatism," *Motherboard* (blog), *Vice,* March 27, 2019, https://motherboard.vice.com/en_us/article/nexpbx/facebook-bans-white-nationalism-and-white-separatism?utm_campaign=sharebutton.

240 *Most deeply impacted:* Deepa Seetharaman, "Facebook Employees Pushed to Remove Trump's Posts as Hate Speech," October 21, 2016, https://www.wsj.com/articles/facebook-employees-pushed-to-remove-trump-posts-as-hate-speech-1477075392.

"a pretty crazy idea": Mark Zuckerberg, "Live from the Techonomy Conference," Facebook, November 17, 2016, https://www.facebook.com/zuck/videos/10103248351713921/.

a private scolding: Adam Entous, Elizabeth Dwoskin, and Craig Timberg, "Obama Tried to Give Zuckerberg a Wake-Up Call over Fake News on Facebook," *Washington Post,* September 24, 2017, https://www.washingtonpost.com/business/economy/obama-tried-to-give-zuckerberg-a-wake-up-call-over-fake-news-on-facebook/2017/09/24/15d19b12-ddac-4ad5-ac6e-ef909e1c1284_story.html.

241 *tried to reassure users:* Mark Zuckerberg, https://www.facebook.com/zuck/posts/10103253901916271.

crowdsource solutions: Sheera Frenkel, "Renegade Facebook Employees Form Task Force to Battle Fake News," BuzzFeed, November 14, 2016, https://www.buzzfeed.com/sheerafrenkel/renegade-facebook-employees-form-task-force-to-battle-fake-n?utm_term=.pqkjwVXwK0#.lpooRvnRBr.

fear of violating: Mike Isaac, "Facebook, in Cross Hairs After Election, Is Said to Question Its Influence," *New York Times,* November 12, 2016, https://www.nytimes.com/2016/11/14/technology/facebook-is-said-to-question-its-influence-in-election.html?_r=0.

"Information Operations and Facebook": Jen Weedon, William Nuland, and Alex Stamos, "Information Operations and Facebook" (report, Facebook Security, April 27, 2017), https://fbnewsroomus.files.wordpress.com/2017/04/facebook-and-information-operations-v1.pdf.

named its adversary: Alex Stamos, "An Update on Information Operations on Facebook," Facebook Newsroom, September 6, 2017, https://newsroom.fb.com/news/2017/09/information-operations-update/.

a crucial nine months: Seetharaman, McMillan, and Wells, "Tone-Deaf."

French and German governments: Dustin Volz and Jonathan Landay, "Twitter to Brief Congress on Possible Russia-Backed Ads: U.S. Senator," Reuters, September 7, 2017, https://www.reuters.com/article/us-twitter-propoganda/twitter-to-brief-congress-on-possible-russia-backed-ads-u-s-senator-idUSKCN1BI22R.

"war room": Sheera Frenkel and Mike Isaac, "Inside Facebook's Election

'War Room,'" *New York Times*, September 19, 2018, https://www.nytimes.com/2018/09/19/technology/facebook-election-war-room.html.

Zuckerberg apologized: Sam Levin, "Mark Zuckerberg: I Regret Ridiculing Fears over Facebook's Effect on Election," *The Guardian*, September 27, 2017, https://www.theguardian.com/technology/2017/sep/27/mark-zuckerberg-facebook-2016-election-fake-news.

"I don't want anyone": Kurt Wager, "Read Mark Zuckerberg's Full Speech on How Facebook Is Fighting Back Against Russia's Election Interference," Recode, September 21, 2017, https://www.recode.net/2017/9/21/16347036/mark-zuckerberg-facebook-russia-election-interference-full-speech.

"the biggest risk we face": Steve Huffman, "In Response to Recent Reports About the Integrity of Reddit, I'd Like to Share Our Thinking," Reddit, March 5, 2018, https://www.reddit.com/r/announcements/comments/827zqc/in_response_to_recent_reports_about_the_integrity/.

242 *$57 million:* Melissa Eddy and Mark Scott, "Delete Hate Speech or Pay Up, Germany Tells Social Media Companies," *New York Times*, https://www.nytimes.com/2017/06/30/business/germany-facebook-google-twitter.html.

forty-four nations: Daniel Funke, "A Guide to Anti-Misinformation Actions Around the World," *Poynter*, accessed April 30, 2019, https://www.poynter.org/ifcn/anti-misinformation-actions/.

three-year prison sentences: Damien Cave, "Australia Passes Law to Punish Social Media Companies for Violent Posts," *New York Times*, April 3, 2019, https://www.nytimes.com/2019/04/03/world/australia/social-media-law.html.

Federal Election Commission disclosure rules: Heather Timmons, "The US Want to Regulate Political Advertising on Social Media," World Economic Forum, October 19, 2017, https://www.weforum.org/agenda/2017/10/the-us-want-to-regulate-political-advertising-on-social-media.

same exemptions as skywriting: Donie O'Sullivan, "Facebook Sought Exception from Political Ad Disclaimer Rules in 2011," CNNMoney, September 27, 2017, *CNN Money*, http://money.cnn.com/2017/09/27/technology/business/facebook-political-ad-rules/index.html.

As Zuckerberg confessed: Kevin Roose and Sheera Frenkel, "Mark Zuckerberg's Reckoning: 'This Is a Major Trust Issue,'" *New York Times*, March 21, 2018, https://www.nytimes.com/2018/03/21/technology/mark-zuckerberg-q-and-a.html?mtrref=www.theringer.com.

243 *"violence to resist occupation":* Angwin and Grassegger, "Facebook's Secret Censorship Rules."

A Chinese billionaire: Alexandra Stevenson, "Facebook Blocks Chinese

Billionaire Who Tells Tales of Corruption," *New York Times,* October 1, 2017, https://www.nytimes.com/2017/10/01/business/facebook-china-guo-wengui.html.

Rohingya Muslim minority: Betsy Woodruff, "Exclusive: Facebook Silences Rohingya Reports of Ethnic Cleansing," The Daily Beast, September 18, 2017, https://www.thedailybeast.com/exclusive-rohingya-activists-say-facebook-silences-them.

every bot made Twitter: Selina Wang, "Twitter Is Crawling with Bots and Lacks Incentive to Expel Them," Bloomberg, October 13, 2017, https://www.bloomberg.com/news/articles/2017-10-13/twitter-is-crawling-with-bots-and-lacks-incentive-to-expel-them?cmpid=socialflow-twitter-business&utm_content=business&utm_campaign=socialflow-organic&utm_source=twitter&utm_medium=social.

"more concerned with growth numbers": Selina Wang, "Twitter Sidestepped Russian Account Warnings, Former Worker Says," Bloomberg, November 3, 2017, https://www.bloomberg.com/news/articles/2017-11-03/former-twitter-employee-says-fake-russian-accounts-were-not-taken-seriously.

conservative users: Seetharaman, "Facebook Employees Pushed."

"It is difficult": Upton Sinclair, *I, Candidate for Governor, and How I Got Licked* (Farrar & Rinehart, 1935), 105.

244 *26 million subscribers:* Keith Collins and David Ingold, "Through Years of Tumult, AOL Sticks Around," Bloomberg, May 12, 2015, https://www.bloomberg.com/graphics/2015-verizon-aol-deal/.

245 *"500 Hours Free!":* Today, there is a thriving collectors' market for AOL free-trial CDs. See Arielle Pardes, "Inside the Intense, Insular World of AOL Disc Collecting," *Vice,* October 7, 2015, https://www.vice.com/en_us/article/kwxngw/inside-the-weird-world-of-aol-disc-collecting-511; "AOL," Yard Sale, The Beanie News, http://thebeanienews.com/Yardsale/AOL.html.

half of all the CDs: Dan Lewis, "Remember All Those AOL CDs? There Were More Than You Think," *Now I Know,* December 10, 2012, http://nowiknow.com/remember-all-those-aol-cds-there-were-more-than-you-think/.

special screen names: Lisa Margonelli, "Inside AOL's 'Cyber-Sweatshop,'" *Wired,* October 1, 1999, https://www.wired.com/1999/10/volunteers/.

three-month training process: Jim Hu, "Former AOL Volunteers File Labor Suit," CNET, January 2, 2002, https://www.cnet.com/news/former-aol-volunteers-file-labor-suit/.

minimum of four hours: Ibid.

14,000 volunteers: Margonelli, "Inside AOL's 'Cyber-Sweatshop.'"

"cyber-sweatshop": Ibid.

$15 million: Lauren Kirchner, "AOL Settled with Unpaid 'Volunteers' for $15 Million," *Columbia Journalism Review,* February 20, 2011, http://ar chives.cjr.org/the_news_frontier/aol_settled_with_unpaid_volunt. php.

246 *a thousand graphic images:* Olivia Solon, "Underpaid and Overburdened: The Life of a Facebook Moderator," *The Guardian,* May 25, 2017, https://www.theguardian.com/news/2017/may/25/facebook-moderator-underpaid-overburdened-extreme-content.

a million pieces of content: Buni and Chemaly, "The Secret Rules of the Internet."

a 74-year-old grandfather: Olivia Solon, "Facebook Killing Video Puts Moderation Policies Under the Microscope, Again," *The Guardian,* April 17, 2017, https://www.theguardian.com/technology/2017/apr/17/face book-live-murder-crime-policy.

an estimated 150,000 workers: Benjamin Powers, "The Human Cost of Monitoring the Internet," *Rolling Stone,* September 9, 2017, https://www .rollingstone.com/culture/features/the-human-cost-of-monitoring-the-internet-w496279.

247 *India and the Philippines:* Adrian Chen, "The Laborers Who Keep Dick Pics and Beheadings out of Your Facebook Feed," *Wired,* October 23, 2014, https://www.wired.com/2014/10/content-moderation/.

bright young college graduates: Sarah T. Roberts, "Behind the Screen: The People and Politics of Commercial Content Moderation" (presentation at re:publica 2016, Berlin, May 2, 2016), transcript available at Open Transcripts, http://opentranscripts.org/transcript/politics-of-commer cial-content-moderation/.

reduced libido: Brad Stone, "Policing the Web's Lurid Precincts," *New York Times,* July 28, 2010, http://www.nytimes.com/2010/07/19/ technology/19screen.html.

regular psychological counseling: Abby Ohlheiser, "The Work of Monitoring Violence Online Can Cause Real Trauma. And Facebook Is Hiring," *Washington Post,* May 4, 2017, https://www.washingtonpost. com/news/the-intersect/wp/2017/05/04/the-work-of-monitoring-vi olence-online-can-cause-real-trauma-and-facebook-is-hiring/?utm_ term=.3fb95a5143da.

"compassion fatigue": Greg Hadley, "Forced to Watch Child Porn for Their Job, Microsoft Employees Developed PTSD, They Say," McClatchy,

January 11, 2017, http://www.mcclatchydc.com/news/nation-world/na
tional/article125953194.html.

"internal video screen": Ibid.

just 55 employees: Cade Metz, "Why WhatsApp Only Needs 50 Engineers for Its 900m Users," *Wired,* September 15, 2015, https://www.wired.
com/2015/09/whatsapp-serves-900-million-users-50-engineers/.

248 *none of whom spoke Arabic:* Heather Timmons, "Why It Remains Difficult to Shut Down Jihadist Propaganda Online," Defense One, January
12, 2015, http://www.defenseone.com/threats/2015/01/why-it-remains-
difficult-shut-down-jihadist-propaganda-online/102684/.

500,000 new comments: "The Top 20 Valuable Facebook Statistics —
Updated March 2018," Zephoria Digital Marketing, accessed March 18,
2018, https://zephoria.com/top-15-valuable-facebook-statistics/.

500 hours of video: Alexandria Walden, "Hate Crimes and the Rise of White
Nationalism," testimony before the House Judiciary Committee, April 9,
2019, https://judiciary.house.gov/sites/democrats.judiciary.house.gov/
files/documents/Testimony%20of%20Alexandria%20Walden.pdf.

350,000 tweets: "Twitter Usage Statistics," Internet Live Stats, accessed
April 23, 2019, https://www.internetlivestats.com/twitter-statistics/.

regenerating Twitter network: Rita Katz, "Want to Know How Isis Cheats
Death on Twitter? Read This," *International Business Times,* April 10, 2015,
http://www.ibtimes.co.uk/want-know-how-isis-cheats-death-twitter-
read-this-1495750.

"pissing in the ocean": Christopher Mims, "Facebook Is Still in Denial
About Its Biggest Problem," *Wall Street Journal,* October 1, 2017, https://
www.wsj.com/articles/facebook-is-still-in-denial-about-its-biggest-prob
lem-1506855607.

"YOU LOOK LIKE A THING": Janelle Shane, "The Neural Network Generated Pickup Lines That Are Actually Kind of Adorable," *AI Weirdness*
(blog), http://aiweirdness.com/post/159302925452/the-neural-network-
generated-pickup-lines-that-are.

249 *the 1940s:* Warren S. McCulloch and Walter H. Pitts, "A Logical Calculus of the Ideas Immanent in Nervous Activity," *Bulletin of Mathematical
Biophysics* 5 (1943): 115–13.

multiple "layers": Gideon Lewis-Kraus, "The Great A.I. Awakening,"
New York Times Magazine, December 14, 2016, https://www.nytimes.
com/2016/12/14/magazine/the-great-ai-awakening.html.

"giant machine democracy": Ibid.

Google Brain project: Quoc V. Le et al., "Building High-Level Features Using Large Scale Unsupervised Learning," in *Proceedings of the 29th
International Conference on Machine Learning* (Edinburgh, 2012).

250 *pictures of cats:* John Markoff, "How Many Computers to Identity a Cat? 16,000," *New York Times,* June 25, 2012, http://www.nytimes.com/2012/06/26/technology/in-a-big-network-of-computers-evidence-of-machine-learning.html.

"We never told it": Ibid.

where to put the traffic lights: Erik Brynjolfsson and Andrew McAfee, "The Business of Artificial Intelligence," *Harvard Business Review,* July 2017, https://hbr.org/2017/07/the-business-of-artificial-intelligence.

more than a million: Joaquin Quiñonero Candela, "Building Scalable Systems to Understand Content," Facebook Code, February 2, 2017, https://code.facebook.com/posts/1259786714075766/building-scalable-systems-to-understand-content/.

wearing a black shirt: Ibid.

251 *80 percent:* "An Update on Our Commitment to Fight Violent Extremist Content Online," *Official Blog,* YouTube, October 17, 2017, https://youtube.googleblog.com/2017/10/an-update-on-our-commitment-to-fight.html.

"attack scale": Andy Greenberg, "Inside Google's Internet Justice League and Its AI-Powered War on Trolls," *Wired,* September 19, 2016, https://www.wired.com/2016/09/inside-googles-internet-justice-league-ai-powered-war-trolls/.

about 90 percent: Ibid.

thoughts of suicide: Vanessa Callison-Burch, Jennifer Guadagno, and Antigone Davis, "Building a Safer Community with New Suicide Prevention Tools," Facebook Newsroom, March 1, 2017, https://newsroom.fb.com/news/2017/03/building-a-safer-community-with-new-suicide-prevention-tools/%20https://newsroom.fb.com/news/2017/03/building-a-safer-community-with-new-suicide-prevention-tools/.

database of facts: Jonathan Stray, "The Age of the Cyborg," *Columbia Journalism Review,* Fall/Winter 2016, https://www.cjr.org/analysis/cyborg_virtual_reality_reuters_tracer.php.

managing the "trade-offs": Kurt Wagner, "Facebook's AI Boss: Facebook Could Fix Its Filter Bubble If It Wanted To," Recode, December 1, 2016, https://www.recode.net/2016/12/1/13800270/facebook-filter-bubble-fix-technology-yann-lecun.

252 *Their more advanced version:* For a good overview, see "Cleverbot Data for Machine Learning," Existor, accessed March 20, 2018, https://www.existor.com/products/cleverbot-data-for-machine-learning/.

machine-driven communications tools: Matt Chessen, "Understanding the Psychology Behind Computational Propaganda," in *Can Public Diplo-*

macy Survive the Internet?, ed. Shawn Powers and Markos Kounalakis (U.S. Advisory Commission on Public Diplomacy, May 2017), 41.

"RACE WAR NOW": Sophie Kleeman, "Here Are the Microsoft Twitter Bot's Craziest Racist Rants," Gizmodo, March 24, 2016, https://gizmodo.com/here-are-the-microsoft-twitter-bot-s-craziest-racist-ra-1766820160.

put to sleep: James Vincent, "Twitter Taught Microsoft's AI Chatbot to Be a Racist Asshole in Less Than a Day," The Verge, March 24, 2016, https://www.theverge.com/2016/3/24/11297050/tay-microsoft-chatbot-racist.

253 *take its prognostication:* Will Knight, "The Dark Secret at the Heart of AI," *MIT Technology Review,* April 11, 2017, https://www.technologyreview.com/s/604087/the-dark-secret-at-the-heart-of-ai/.

steal a page from: Ibid.

10,000 different words: Alexander G. Huth et al., "Natural Speech Reveals the Semantic Maps That Tile Human Cerebral Cortex," *Nature* 532 (April 2016): 453–58.

a "smart" censorship system: Mike Isaac, "Facebook Said to Create Censorship Tool to Get Back into China," *New York Times,* November 22, 2016, https://www.nytimes.com/2016/11/22/technology/facebook-censorship-tool-china.html.

Sun Microsystems and Cisco: Jack Linchuan Qiu, "Virtual Censorship in China: Keeping the Gate Between the Cyberspaces," *International Journal of Communications Law and Policy* 4 (Winter 1999/2000): 1–23.

free, open-source tools: See, for example, TensorFlow, https://www.tensorflow.org/.

mimic a speaker's voice: Bahar Gholipour, "New AI Tech Can Mimic Any Voice," *Scientific American,* May 2, 2017, https://www.scientificamerican.com/article/new-ai-tech-can-mimic-any-voice/.

254 *essentially perfect:* Ibid.

Lyrebird: Natasha Lomas, "Lyrebird Is a Voice Mimic for the Fake News Era," TechCrunch, April 25, 2017, https://techcrunch.com/2017/04/25/lyrebird-is-a-voice-mimic-for-the-fake-news-era/.

"Photoshop for audio": Craig Stewart, "Adobe Prototypes 'Photoshop for Audio,'" Creative Bloq, November 3, 2016, https://www.creativebloq.com/news/adobe-prototypes-photoshop-for-audio.

a two-dimensional photograph: Shunsuke Saito et al., "Photorealistic Facial Texture Inference Using Deep Neural Networks," arXiv:1612.00523 [cs.CV], December 2016.

captured the "facial identity": Justus Thies et al., "Face2Face: Real-Time Face Capture and Reenactment of RGB Videos" (unpublished paper, January 2016), http://niessnerlab.org/papers/2016/1facetoface/thies2016face.pdf.

"deformation transfer": Ibid.

"hard to distinguish": "Face2Face: Real-Time Face Capture and Reenactment of RGB Videos (CVPR 2016 Oral)," YouTube video, uploaded by Matthias Niessner, March 17, 2016, https://www.youtube.com/watch?v=ohmajJTcpNk.

255 *"generative networks"*: Anh Nguyen et al., "Plug and Play Generative Networks: Conditional Iterative Generation of Images in Latent Space," arXiv:1612.00005 [cs.CV], November 2016.

no earthly counterparts: Ibid.

same thing with video: Carl Vondrick, Hamed Pirsiavash, and Antonio Torralba, "Generating Videos with Scene Dynamics" (paper presented at the 29th Conference on Neural Information Processing, Barcelona, 2016), http://carlvondrick.com/tinyvideo/paper.pdf.

a predictive future: Ibid.

256 *the first to respond:* Chessen, "Understanding the Psychology," 42.

"determine the fate": Ibid., 45.

"start programming us": Matt Chessen, "Machines Will Soon Program People," @mattlesnake (blog), Medium, May 16, 2017, https://medium.com/@mattlesnake/machines-will-soon-program-people-73929e84c4c4.

"We are so screwed": Charlie Warzel, "Infocalypse Now," BuzzFeed, February 11, 2018, https://www.buzzfeed.com/charliewarzel/the-terrifying-future-of-fake-news?utm_term=.viEmNOlN3o#.xtPNkBWkwD.

257 *"generative adversarial networks"*: Cade Metz, "Google's Dueling Neural Networks Spar to Get Smarter, No Humans Required," *Wired,* April 11, 2017, https://www.wired.com/2017/04/googles-dueling-neural-networks-spar-get-smarter-no-humans-required/.

9. CONCLUSION

258 *"We are as gods"*: Stewart Brand, "We Are as Gods," *Whole Earth Catalog,* Fall 1968, http://www.wholeearth.com/issue/1010/article/195/we.are.as.gods. Although the words are now indelibly his, Brand actually lifted the line from the British anthropologist Edmund Leach.

Long before the military: Authors' phone interview with representatives of the Joint Readiness Training Center, November 14, 2014.

260 *"built to accomplish"*: "Zuckerberg's Letter to Investors," Reuters, Feb-

ruary 1, 2012, https://www.reuters.com/article/us-facebook-letter-idUS
TRE8102MT20120201.

261 *youth is no defense:* Brooke Donald, "Stanford Researchers Find Students Have Trouble Judging the Credibility of Information Online,"
Stanford Graduate School of Education News Center, November 22, 2016,
https://ed.stanford.edu/news/stanford-researchers-find-students-have-
trouble-judging-credibility-information-online.

263 *other nations now look:* Michael Birnbaum, "Sweden Is Taking On Russian Interference Ahead of Fall Elections. The White House Might Take
Note," *Washington Post,* February 22, 2018, https://www.washingtonpost.
com/world/europe/sweden-looks-at-russias-electoral-interference-in-
the-us-and-takes-steps-not-to-be-another-victim/2018/02/21/9e58ee48-0
768-11e8-aa61-f3391373867e_story.html?utm_term=.3b666b5148d2.

clearest "losers": Ibid.

over fifteen years: "President Trump Unveils America's First Cybersecurity
Strategy in 15 Years," The White House, https://www.whitehouse.gov/
articles/president-trump-unveils-americas-first-cybersecurity-strategy-
15-years/.

Finland, Estonia: Reid Standish, "Russia's Neighbors Respond to Putin's 'Hybrid War,'" *Foreign Policy,* October 12, 2017, http://foreignpolicy.
com/2017/10/12/russias-neighbors-respond-to-putins-hybrid-warlatvia-
estonia-lithuania-finland/.

264 *first cabinet-level meeting:* Ellen Nakashima, "Trump Chairs Election Security Meeting but Gives No New Orders to Repel Russian Interference,"
Washington Post, July 28, 2018, https://www.washingtonpost.com/world/
national-security/trump-chairs-election-security-meeting-but-gives-no-
new-orders-to-repel-russian-interference/2018/07/27/ebf85f50-91b8-11e8-
8322-b5482bf5e0f5_story.html?utm_term=.22eb748aaba1.

"like pulling teeth": Eric Schmitt, David Sanger, and Maggie Haberman, "In Push for 2020 Election Security, Top Official Was Warned:
Don't Tell Trump," *New York Times,* April 24, 2019, https://www.ny-
times.com/2019/04/24/us/politics/russia-2020-election-trump.
html?smid=nytcore-ios-share.

"a goddamn hoax": John Dawsey, Ellen Nakashima, and Shane Harris,
"As Security Officials Prepare for Russian Attack on 2020 Presidential
Race, Trump and Aides Play Down Threat," *Washington Post,* April 30,
2019,   https://www.washingtonpost.com/world/national-security/as-
security-officials-prepare-for-russian-attack-on-2020-presidential-race-
trump-and-aides-play-down-threat/2019/04/29/275af37c-6766-11e9-82b
a-fcfeff232e8f_story.html?utm_term=.149cc01cebfc#click=https://t.co/
XX2Ok6Vooj.

avoid discussing the issue: Jake Tapper and Jim Acosta, "'Like Pulling Teeth' to Get White House to Focus on Russian Election Interference, Official Says," CNN, April 24, 2019, https://www.cnn.com/2019/04/24/politics/kirstjen-nielsen-trump-election-security/index.html.

cannot start early enough: Lisa Guernsey, "It's Never Too Early to Start Teaching Kids Media Literacy," *Slate,* November 8, 2017, http://www.slate.com/articles/technology/future_tense/2017/11/in_the_age_of_fake_news_it_s_never_too_early_to_teach_kids_media_literacy.

265 *at least a dozen:* Michael Rosenwald, "Making Media Literacy Great Again," *Columbia Journalism Review,* Fall 2017, https://www.cjr.org/special_report/media-literacy-trump-fake-news.php.

Calling Bullshit: "Calling Bullshit: Data Reasoning in a Digital World" (course syllabus, University of Washington, Autumn 2017), http://calling-bullshit.org/syllabus.html.

"conscientious objector": Thuy Ong, "Sean Parker on Facebook: 'God Only Knows What It's Doing to Our Children's Brains,'" The Verge, November 9, 2017, https://www.theverge.com/2017/11/9/16627724/sean-parker-facebook-childrens-brains-feedback-loop.

"our children's brains": Ibid.

266 *Such "technocracy" views:* Parag Khanna, "To Beat Populism, Blend Democracy and Technocracy, S'pore Style," *Straits Times,* January 21, 2017, http://www.straitstimes.com/opinion/to-beat-populism-blend-democracy-and-technocracy-spore-style.

the Flux movement: Mark Kaye and Nathan Spataro, "Redefining Democracy: On a Democratic System Designed for the 21st Century, and Disrupting Democracy for Good" (unpublished paper, January 2017), https://voteflux.org/pdf/Redefining%20Democracy%20-%20Kaye%20&%20Spataro%201.0.2.pdf.

267 *"dangerous speech":* "Understanding Dangerous Speech," Dangerous Speech Project, accessed March 12, 2018, https://dangerousspeech.org/faq/.

268 *"The more we connect":* Chris Matyszcyk, "Facebook's New Ads Aren't as Friendly as They Seem," CNET, February 16, 2015, https://www.cnet.com/news/facebooks-new-ads-arent-as-friendly-as-they-seem/.

like Mark Zuckerberg: Mark Zuckerberg, "I Wanted to Share Some Thoughts on Facebook and the Election," Facebook, November 12, 2016, https://www.facebook.com/zuck/posts/10103253901916271; Callum Borchers, "Twitter Executive on Fake News: 'We Are Not the Arbiters of Truth,'" *The Fix* (blog), *Washington Post,* February 8, 2018, https://www.washingtonpost.com/news/the-fix/wp/2018/02/08/twitter-ex-

ecutive-on-fake-news-we-are-not-the-arbiters-of-truth/?utm_
term=.084a121450e4.

269 *"answers to simple questions":* Lorenzo Franceschi-Bicchierai, "If Facebook Actually Wants to Be Transparent, It Should Talk to Journalists,"
Motherboard (blog), *Vice,* November 10, 2017, https://motherboard.vice.
com/en_us/article/ywbe3g/facebook-should-talk-to-journalists?utm_
campaign=sharebutton%3Futm_campaign%3Dsharebutton.

the word "transparency": Charlie Warzel, "Twitter Would Like You to
Know It Is Committed to Being More Transparent," BuzzFeed, October 12, 2017, https://www.buzzfeed.com/charliewarzel/twitter-would-
like-you-to-know-it-is-committed-to-being?utm_term=.hy109e063E#.
fjV316OXpa.

Reddit is the only one: Caroline O., "Russian Propaganda on Reddit,"
Arc (blog), Medium, April 17, 2018, https://arcdigital.media/russian-pro
paganda-on-reddit-7945dc04eb7b.

effective information literacy education: John Cook, Stephan Lewan-
dowsky, and Ullrich K. H. Ecker, "Neutralizing Misinformation Through
Inoculation: Exposing Misleading Argumentation Techniques Reduces
Their Influence," *PLoS ONE* 12, no. 5 (May 2017): e0175799, http://jour
nals.plos.org/plosone/article?id=10.1371/journal.pone.0175799.

270 *not a single social media firm:* Alexis Madrigal, "15 Things We
Learned from the Internet Giants," *The Atlantic,* November 2, 2017, https://
www.theatlantic.com/technology/archive/2017/11/a-list-of-what-we-re
ally-learned-during-techs-congressional-hearings/544730/.

outside researchers raised concerns: Charlie Warzel, "Researchers Are
Upset That Twitter Is Dismissing Their Work on Election Interference,"
BuzzFeed, October 3, 2017, https://www.buzzfeed.com/charliewarzel/
researchers-are-upset-that-twitter-is-dismissing-their-work?utm_term=.
ut379NjnD#.mioWe4mZL.

"Facebook is only": Zeynep Tufekci, "It's the (Democracy Poisoning)
Age of Free Speech," *Wired,* January 16, 2018, https://www.wired.com/
story/free-speech-issue-tech-turmoil-new-censorship/.

271 *the least informed:* Christoph Aymanns, Jakob Foerster, and Co-Pierre
Georg, "Fake News in Social Networks," arXiv:1708.06233 [cs.AI], August
2017; Mark Buchanan, "Why Fake News Spreads So Quickly on Facebook,"
Sydney Morning Herald, September 1, 2017, https://www.smh.com.au/
world/why-fake-news-spreads-so-quickly-on-facebook-20170901-gy8je4.
html.

"easily manipulated": Sam Wineburg and Sarah McGre, "Lateral Reading: Reading Less and Learning More When Evaluating Digital Information" (working paper no. 2017-A1, Stanford History Education Group,

Stanford University, October 2017), https://papers.ssrn.com/sol3/papers.cfm?abstract_id=3048994.

approached the task "laterally": Ibid.

"a maze": Carrie Spector, "Stanford Scholars Observe 'Experts' to See How They Evaluate the Credibility of Information Online," Stanford News Service, Stanford University, October 24, 2017, https://news.stanford.edu/press-releases/2017/10/24/fact-checkers-ouline-information/.

"seeking context and perspective": Ibid.

272 *Reality is one:* Paul J. Griffiths, *An Apology for Apologetics: A Study in the Logic of Interreligious Dialogue* (Wipf and Stock, 2007), 46.

most important insights: Plato, "The Allegory of the Cave," *Republic,* 7.514a2–517a7, trans. Thomas Sheehan, https://web.stanford.edu/class/ihum40/cave.pdf.

273 *"believe whatever you want":* The Matrix, directed by the Wachowski Brothers (Warner Bros., 1999).

Index